DISCRIMINANT ANALYSIS AND APPLICATIONS

ACADEMIC PRESS RAPID MANUSCRIPT REPRODUCTION

Proceedings of a NATO Advanced Study Institute
on Discriminant Analysis and Applications
Athens, Greece
June 8-20, 1972

DISCRIMINANT ANALYSIS AND APPLICATIONS

Edited by

T. CACOULLOS

UNIVERSITY OF ATHENS
ATHENS, GREECE

Academic Press New York and London **1973**

A Subsidiary of Harcourt Brace Jovanovich, Publishers

ACADEMIC PRESS, INC.
111 Fifth Avenue, New York, New York 10003

United Kingdom Edition published by
ACADEMIC PRESS, INC. (LONDON) LTD.
24/28 Oval Road, London NW1

Library of Congress Cataloging in Publication Data

NATO Advanced Study Institute on Discriminant Analysis and Applications, Athens, 1972.
Discriminant analysis and applications.

Proceedings of the conference held June 8-20, 1972.
Bibliography: p.
1. Discriminant analysis–Congresses. I. Cacoullos, Theophilos, ed. II. Title.
QA278.65.N37 1972 519.5′3 73-806
ISBN 0–12–154050–2

PRINTED IN THE UNITED STATES OF AMERICA

In memory of

PRASANTA CHANDRA MAHALANOBIS

(1893–1972)

CONTENTS

CONTRIBUTORS

J.A. Anderson, Department of Biomathematics, University of Oxford, Oxford, England

T.W. Anderson, Department of Statistics, Stanford University, Stanford, California

D.F. Andrews, Department of Mathematics, University of Toronto, Toronto, Ontario, Canada

T. Cacoullos, Statistical Unit, University of Athens, Athens, Greece

Somesh Das Gupta, School of Statistics, University of Minnesota, Minneapolis, Minnesota

M.M. Desu, Department of Statistics, State University of New York, Buffalo, New York

W.J. Dixon, Department of Biomathematics, University of California, Los Angeles, California

Daniel Dugué, Institut de Statistique, Université de Paris, Paris, France

S. Geisser, School of Statistics, University of Minnesota, Minneapolis, Minnesota

R.I. Jennrich, Department of Biomathematics, University of California, Los Angeles, California

M.G. Kendall, Scientific Control Systems Ltd., London, England

Peter A. Lachenbruch, Department of Biostatistics, University of North Carolina, Chapel Hill, North Carolina

Kameo Matusita, Institute of Statistical Mathematics, Tokyo, Japan

J. Michaelis, Institut für Medizinische Statistik und Dokumentation der Universität Mainz, Mainz, West Germany

Elliott Nebenzahl, Department of Statistics, California State University, Hayward, California

R.A. Reyment, Paleontologiska Institut, Uppsala Universitet, Uppsala, Sweden

Willem Schaafsma, Mathematisch Instituut, Risj Universiteit, Groningen, The Netherlands

Rosedith Sitgreaves, Department of Statistics, California State University, Hayward, California

Milton Sobel, School of Statistics, University of Minnesota, Minneapolis, Minnesota

George P. H. Styan, Department of Mathematics, McGill University, Montreal, Quebec, Canada

J. Tiago de Oliveira, Faculdade de Ciências Matemática, Lisbon, Portugal

PREFACE

This book comprises the proceedings of the NATO Advanced Study Institute on Discriminant Analysis and Applications held in Kifissia, Athens, Greece, June 8-20, 1972. Participants in the Institute came from almost all NATO countries; there were 19 lecturers and 60 additional participants. It was unfortunate that Professor C. R. Rao was unable to participate due to the serious illness of P. C. Mahalanobis during those days; Mahalanobis died on June 28, 1972. In memoriam and appreciation of his pioneering contributions in the field of discriminant analysis, this volume is dedicated to Prasanta Chandra Mahalanobis (1893-1972), founder of the Indian Statistical Institute and of *Sankhyā*.

The term *discriminant analysis* in the title of these Proceedings is an alternative to the term *classification* which is used as a synonym for *discrimination* by many statisticians. This choice is to avoid ambiguity, since classification is often also used in the sense of *taxonomy* or construction of classes. For a discussion of these differences the reader is referred to the paper by Professor M. G. Kendall in this volume.

Both theory and applications are discussed. A comprehensive survey and a historical development of discriminant analysis methods are given in the review paper by Professor Somesh Das Gupta. Here the reader will also find a fairly extensive up-to-date bibliography arranged by subject matter and then chronologically within subject matter. This bibliography contains articles and books published up to 1972 and some unpublished papers as well. At the end of this book, there is a bibliography, presumably exhaustive, with authors in alphabetical order, prepared by Professor G. P. H. Styan (McGill University) and the Editor. It includes papers published up to 1971 and books up to 1972.

Several papers deal with the exact, asymptotic, and empirical (Monte Carlo) evaluation of the probabilities of misclassification, especially under the assumption of normal alternative populations with the same or different dispersion matrices. Logistic and quasi-linear discrimination are also covered. Distance functions are considered especially as regards discrimination in the Gaussian case. Applications, particularly in medicine and biology, are also discussed in some of these papers. Moreover, computer graphical analysis and graphical techniques for multidimensional data are discussed and illustrated by examples in two papers. It is regretted that final versions of the papers by Professors Ingram Olkin (Stanford University) and M. S. Srivastava (University of Toronto) did not arrive in time to be included in these Proceedings.

I wish to thank the Scientific Directors of the Institute, Professors T. W. Anderson and M. G. Kendall, for their encouragement and suggestions throughout the organization of the Symposium. Thanks are also due to my Assistants at the University of Athens, Ch. Charalambides, Urania Chrysaphinou, A. Dallas, and Amalia Veini, for their assistance with the local arrangements and for carrying out the correspondence for the Institute.

The editing of this book was completed during my visit at McGill University. Special thanks go to my colleague G. P. H. Styan for valuable suggestions; to Leslie Ann Hathaway for helping with proofreading and the compilation of the bibliography at the end of this volume, I express my sincere appreciation. I am particularly grateful to Maria Lam for patiently preparing the difficult typescript; she did a superb job.

Thanks are also due to the contributors for their cooperation and to Academic Press whose interest in publishing these Proceedings was expressed in the early stages of the Institute.

It goes without saying that the Institute would not have materialized without the financial support of the NATO Science Committee. I also acknowledge with gratitute the support of the Greek Ministry of Culture and Science for sponsoring the social activities of the Institute, by no means less contributory to its success.

T. Cacoullos

INTRODUCTORY ADDRESS*

T. Cacoullos
Director of the NATO Advanced Study Institute
on Discriminant Analysis and Applications

In a NATO report on nonmilitary cooperation issued in 1956, discussing science and technology, the three foreign ministers of Canada, Italy, and Norway recommended the creation of an ad hoc working group to study topics of special importance to NATO and specifically the way in which the Organization might support or supplement the scientific activities of other organizations. A Task Force was formed in June 1957 to consider these recommendations. The psychological shock of the launching of the first Soviet Sputnik in October 1957 resulted in the immediate presentation of the Task Force's Report to the Heads of Governments of the Alliance at their meeting in December 1957. The recommendations of the former were accepted and both a Science Adviser and a Science Committee were appointed to follow up the proposals of the Task Force.

The Science Committee met first in March 1958 and ever since its activities have grown to significant proportions. Its programs of Fellowships, Research Grants, and Advanced Study Institutes have contributed a great deal to the promotion of common understanding and cooperation between the scientific communities of NATO members.

The Advanced Study Institutes Program has been sponsoring about 50 Advanced Study Institutes annually since 1959. The Study Institutes, or Summer Schools as they are called, range in subject from mathematics and physics through biological and medical topics to such areas as psychological measurement theory and programming languages for computers. The principal feature of such meetings is the bringing together of research workers, often in secluded surroundings, for about two weeks to study a specific topic. It is exactly within this framework that the present Institute on Discriminant Analysis and Applications was organized. It is therefore with pleasure that we acknowledge the support of the NATO Science Committee which made this meeting possible.

*Read at the Opening Ceremony, June 8, 1972.

On behalf of the participants I should also express our appreciation to the Greek Ministry of Culture and Science for its cosponsoring of the Institute; in particular, most of its social activities which include tonight's cocktails and buffet dinner, the Delphi excursion this coming Sunday, and the Hydra-Aegina cruise next weekend. We are honored by the presence of his Exellency, the Minister of Culture and Science, Mr. C. A. Panayiotakis, who has accepted my invitation to deliver the opening address of the Institute.

Before I introduce his Excellency, the Minister, allow me to spend a few minutes introducing the Institute's topic. Discriminant analysis, otherwise known as classification theory, is one of the most important areas of multivariate statistical analysis.

A statistical discrimination or classification problem consists in assigning or classifying an individual or group of individuals to one of several known or unknown alternative populations on the basis of several measurements on the individual and samples from the unknown populations. In the simplest case of two normal populations, for example, a linear combination of the measurements, called linear discriminator or discriminant function, is constructed, and on the basis of its value the individual is assigned to one or the other of the two populations. This decision process involves two kinds of error, namely, that the individual is assigned to population No. 2 while it really belongs to population No. 1 and vice versa. The researcher wishes to minimize in some sense the chances of a wrong decision. R. A. Fisher was the first to suggest the use of a linear discriminant function in 1936. Its early applications led to some important anthropometric discoveries such as sex differences in mandibles, the extraction from a dated series of the particular compound of cranial measurements showing secular trends, and solutions of taxonomic problems in general.

At about the same time that Fisher introduced the discriminant function, Mahalanobis (in India) proposed his generalized distance which was applied with success to craniometric and other anthropological studies.

Let us mention a few of the early applications of discriminant analysis. Professor C. R. Rao, Director of Research and Training at the Indian Statistical Institute, considered the assignment of an individual to one of the three Indian castes, Brahmin, Artisan, or Korwa, on the basis of four measurements, namely height, sitting height, nasal depth, and nasal height. More details about this and other applications can be found in the well-known multivariate analysis textbooks of two distinguished participants of the present Institute, Professors T. W. Anderson of Stanford University and M. G. Kendall, formerly of the University of London, now Chairman of Scientific Control Systems, whom we are happy to have with us in the role of Scientific Directors and Lecturers for the Institute.

A further case history of multiple discrimination is given by another distinguished member of the audience, Professor Richard Reyment of the Palaeontological Institute of Uppsala. Blocks of limestone containing the ostracod Stensloffina were found in mid-eastern Sweden and Gotland as a result of transportational activities of the ice sheets during the Ice Age. These blocks were associated with three infraspecific groups based on in situ occurrences. The characteristics measured on the ostracod carapaces were the following: length, distance of the lateral spine from the anterior margin, and its distance from the dorsal margin. By using the discriminators it turned out that, of five samples from the Baltic area, two were identified with the first group and three with the third one, a palaeontologically sound result. More case histories from anthropometric and palaeontological studies can be found in Professor Reyment's book *Multivariate Morphometrics,* coauthored with R. E. Blackith.

We are meeting in Attica which we all appreciate as the cradle of some of the most fertile ideas of the western mind. Yet it is curious and perhaps worthwhile to note that the scientific discipline known to us as probability and statistics was almost outside the sphere of classical Greek scientific speculation. I would like to dwell briefly on this fact.

It is a little surprising that whereas statistics means primarily numbers and according to the Pythagoreans the very essence of things resided in numbers, yet the geometrical way of thinking of the Greeks and the prevailing state of knowledge did not lead very far as regards the quantitative aspect of phenomena. The Pythagoreans, however, may be regarded as the fathers of modern numerical taxonomy since they characterized animals and plants by numbers, counting how many junctions there are between the lines of an outline sketch of the corresponding object. Thus the number for man was 250 and for a plant 360.

The Pythagorean school was opposed by Aristotle who was inclined to think in terms more qualitative than quantitative; and although his contributions to what we should now describe as taxonomic theory are worthwhile, he did destroy the concern for quantitative aspects underlying external form, a concern so characteristic of the Pythagoreans. Though Aristotle significantly attenuated the Platonic Theory of Forms (he was its first important critic), he was in the final analysis interested in the essence or form of an object which he considered one of its four causes, the others being the matter out of which an object is made, the purpose for which it is made, and the maker. The fact that Aristotle spoke of several causes of an object or an event indicates that he diverged significantly from the doctrines of his great teacher, Plato. Philosophers refer to Aristotle as a Naturalist. Yet it is still true to say that it was the Platonic Aristotle who influenced the thinking of later generations. The Greeks and the Medievals remained interested in the essence or form of things and taxonomic characters were selected on a priori principles, namely, according to their hypothetical weight in determining a plant or animal. This continued to be the case until the eighteenth century when Adanson proposed allotting equal a priori weighting to all characters.

Aristotle's speculations about the nature of process and change, especially in Book II of *The Physics,* provide us with a fundamental concept in the Aristotelian system, namely, the notion of chance. I should like to review briefly his analysis of chance (*τύχη*), for I think it is here that we get important evidence for the claim I want to make: that the Ancient Greeks for all their firm adherence to the rational, the determined, and the causally explicable were nonetheless profoundly aware of the random and sometimes inexplicable element in nature. Apart from Aristotle, the sayings of Heraclitus and Empedocles bear witness to this awareness. Thus, although probability and statistics were not systematically developed by the Greeks to the extent that other sciences had been, still the Greeks had a rudimentary understanding of ideas underlying modern probability and statistical procedures.

Aristotle reckons chance and spontaneity among the causes of events, that is to say, many things are said both to be and to come to be as a result of chance and spontaneity. He refers to his predecessors and specifically the father of modern atomic theory, Democritos, who questions whether chance is real or not. He notes that these early thinkers maintain that nothing happens by chance, that everything which we ascribe to chance or spontaneity has some definite cause. Thus, coming by "chance" to the market and finding a man whom one wanted to see but did not expect to meet is due to one's wish to shop in the market. Similarly, it is possible for all so-called chance events to find the cause which explains why and how the event occurred.

It is true that the early Ionian natural philosophers found no place for chance among the causes they recognized, love, strife, mind, fire, and the like. Aristotle finds this strange; he wonders whether they supposed that there is no such thing as chance or whether they thought that there is but simply chose to ignore it even when they sometimes employ the idea as when Empedocles writes that air is not always separated into the highest region but "as it may chance." The latter further says in his cosmology that "it happened to run that way at that time, but it often ran otherwise." Others, Aristotle notes and he apparently has Democritos in mind again, believe that chance is a cause but that it is inscrutable to human intelligence; it is rather a divine thing and full of mystery.

In discussing the nature of chance and spontaneity, Aristotle remarks that "some things always come to pass in the same way, and others for the most part. It is clearly of neither of these that chance is said to be the cause, nor can the 'effect of chance' be identified with any of the things that come to pass by necessity and always or for the most part." Clearly this defines what we call sure or almost sure events in modern probability language. And Aristotle continues: "But as there is a third class of events besides these two events which all say are 'by chance,' it is plain that there is such a thing as chance and spontaneity." Furthermore, even among the things which are outside of the necessary and the normal, there are some for which the phrase "for the sake of something" is applicable. Things of this kind then, when they happen incidentally, are said to occur "by chance." That which is per se cause of the effect is determinate, but the incidental cause is indeterminable, for the possible attributes of an individual are innumerable. In his teleological approach Aristotle could not avoid associating chance with some kind of *τέλος*, i.e., purpose. Chance, then, "is an incidental cause in the sphere of those actions for the sake of something which involve purposes."

Aristotle gives the following example: A man is engaged in collecting subscriptions for a feast. He would have gone to such and such a place for the purpose of getting the money if he had known. He goes to this place but for another purpose and incidentally gets his money by going there. It is important to note that he does not go to this place as a rule or necessarily nor is his getting the money a reason (cause) for his going there nor is it the case that he is always engaged in collecting subscriptions in the fashion of a robot. When these conditions are satisfied, the man is said to have gone "by chance." If he had some definite purpose in mind, namely, that of collecting subscriptions and if he always or normally went to this place whenever he was collecting payments, then he could not be said to have gone by chance. At this point Aristotle does not distinguish between "incidental" and "by chance."

Proceeding further he says, "no doubt, it is necessary that the causes of what comes to pass by chance be indefinite and that is why chance is supposed to be inscrutable to man, and why it might be thought, in a way, nothing occurs by chance." Returning to his example he explains that "the causes of the man's coming and getting the money (when he did not come for the sake of that) are innumerable. He may have wished to see somebody or been following somebody or avoiding somebody or may have gone to see a spectacle. Thus to say that chance is a thing contrary to rule is correct." It appears that Aristotle's notion of chance fluctuates between the haphazard, the incidental, and the unintentionally occurring on the one hand and the indefinite or unpredictable, as we would say today, on the other hand. The latter, of course, is essentially in agreement with the definition of chance events as generally accepted nowadays. Attempts at numerical computations of probabilities are not known to have been made by Aristotle or any of the Greek philosophers and mathematicians. We know that these did not start until the Renaissance and especially with the work of Pascal and Fermat.

It is also worth noting that the historian Thucydides reveals an elementary grasp of statistical reasoning and procedures. In the *History of the Peloponnesian War*, he discusses ways of evaluating averages, modal values, and estimating, for example, the perimeter of an irregular island, The Athenians, according to Thucydides, were "capable of taking risks and of estimating them beforehand . . . " and they criticized their enemy for basing their judgment on wishful thinking rather than on a sound calculation of probabilities.

I would like to end my short discussion of the probabilistic and statistical aspects of ancient Greek thought by quoting Thucydides. He refers to a Spartan spokesman when the latter considers a strategy in the face of uncertainty (see also Rubin, 1971):

> We are taught that there is not a great deal of difference between the way we think and the way others think and that it is impossible to calculate accurately events that are determined by chance. The practical measures that we take are always based on the assumption that our enemies are not unintelligent. And it is right and proper for us to put our hopes in the reliability of our own precautions rather than in the possibility of our opponent making mistakes.

A close examination of these comments indicates some of the basic ideas underlying the recently developed theory of games and decisions. One can replace the enemy strategy of the quotation by the "state of nature" in the terminology of modern decision theory.

Allow me to welcome once again the foreign participants to the homeland of all those men whose thinking helped shape the course of western civilization.

REFERENCE

Rubin, Ernest (1971), Quantitative commentary in Thucydides, *The American Statistician*, 25, No. 5, 52-54.

DISCRIMINANT ANALYSIS AND APPLICATIONS

LOGISTIC DISCRIMINATION WITH MEDICAL APPLICATIONS

J. A. Anderson
University of Oxford

Summary

Consideration is given to the problem of discriminating between populations ($H_1, H_2, \ldots, H_k$) when some or all of the observations ($\underset{\sim}{x}$) are qualitative. For two populations, logistic discrimination was suggested by Cox (1966) and Day and Kerridge (1967) with the restriction that estimation of the discriminator was based on samples from the mixture of the populations. This method was recently extended (Anderson, 1972) to more than two populations and to the more usual plan of sampling from each distribution separately, using Aitchison and Silvey's (1958) method of constrained maximum likelihood estimation.

Logistic discriminators can be used in a simple linear form and when the likelihood ratios of the populations are linear in the observations, they are optimal irrespective of the actual likelihoods. They are thus optimal for a much wider class of distributions than standard linear discriminators. The method of logistic discrimination can be extended to the case where the likelihood ratios are quadratic in the observations.

The maximum likelihood equations for the coefficients of the logistic discriminators cannot be solved explicitly, so the Newton-Raphson procedure is used. In practice, this has given very good convergence using zeros as starting values for all the coefficients. However, there is a problem if the sample points from the different populations can be separated by hyperplanes (complete separation). In this case, there are sets of values for the coefficients each of which maximizes the likelihood and can be used to give the discriminator. Fortunately, the Newton-Raphson procedure converges to one or other of these solutions, if they exist.

Two examples of the use of the method are in the differential diagnosis of kerato-conjunctivitis sicca and in the estimation of relative risks in epidemiology.

1. *Introduction*

There are many situations where discrimination is required between two or more populations on the basis of discrete variables, possibly with some continuous variables as well. An example in medicine is differential diagnosis based on symptoms and signs, which are present or absent, and perhaps on some clinical tests whose values vary on a continuous scale (Anderson and Boyle, 1968). In taxonomy, present or absent attributes are used to discriminate between species. In a commercial context, the decision whether or not to issue a credit card is based on continuous and discrete data. Thus, discrete variables crop up quite naturally in discriminatory work. However, the bulk of statistical activity has been concerned directly with multivariate normal distributions or with techniques like Fisher's linear discriminant function which are optimal for normal distributions.

There have been some attempts to establish discriminatory methods for discrete variables (Linhart, 1959; Johns, 1961; Cochran and Hopkins, 1961) but these have tended to be restrictive in the number of variables permitted. More recently Martin and Bradley (1972) introduced a method which does not seem to have this restriction but it is still limited to discrete variables. Perhaps the most useful and general approach is that of logistic discrimination introduced by Cox (1966) and Day and Kerridge (1967) and further developed by Anderson (1972). With this technique, continuous and discrete variables, jointly or separately, can be handled with equal facility. There is no restriction on the number of variables that can be used other than those imposed by sample size and computer time. To date, 20 variables have been used and there is no doubt that larger problems could be handled - they have just not arisen.

2. *Formulation of the problem*

Suppose that the discrimination between the k populations $H_1, H_2, \ldots, H_k$ is to be based on the vector of observations $\underset{\sim}{x}^T = (1, x_1, x_2, \ldots, x_p)$, where all the components of $\underset{\sim}{x}$ are real but some represent discrete variables and some represent continuous variables. $x_0 \equiv 1$ has been introduced for notational simplicity. Sample points are available from each population and the discrimination problem is to allocate further sample points $\underset{\sim}{x}$ to populations. Suppose further that the likelihood of $\underset{\sim}{x}$ given H_s is $f_s(\underset{\sim}{x})$ $(s=1,\ldots,k)$ and that the points to be allocated are from the mixture of the distributions of $H_1, H_2, \ldots, H_k$ in the proportions $\underset{\sim}{\Pi}^T = (\Pi_1, \Pi_2, \ldots, \Pi_k)$, where $\sum_{s=1}^{k} \Pi_s = 1$. Then the simplest optimizing method of discrimination is to maximize the probability of correct allocation. This implies that the sample point $\underset{\sim}{x}$ is allocated to H_s if

$$\Pi_s f_s(\underset{\sim}{x}) \geq \Pi_t f_t(\underset{\sim}{x}) \qquad (t=1,\ldots,k;\ t\neq s) \tag{1}$$

or

$$\mathrm{pr}(H_s|\underset{\sim}{x}) \geq \mathrm{pr}(H_t|\underset{\sim}{x}) \qquad (t=1,\ldots,k;\ t\neq s) \tag{2}$$

This is proved by many authors, including Rao (1965, p.375). This approach will be taken here but the logistic methods to be developed can be used to advantage with other allocation strategies (see §9).

Since the allocation rules depend so fundamentally on the posterior probabilities $\{\mathrm{pr}(H_s|\underset{\sim}{x})\}$, Cox (1966) and Day and Kerridge (1967) postulated the logistic form for them, giving

$$\mathrm{pr}(H_1|\underset{\sim}{x}) = \exp(\alpha_0+\alpha_1 x_1+\alpha_2 x_2+\ldots+\alpha_p x_p)\mathrm{pr}(H_2|\underset{\sim}{x}) \tag{3}$$

$$\mathrm{pr}(H_2|\underset{\sim}{x}) = 1/\{1+\exp(\alpha_0+\alpha_1 x_1+\ldots+\alpha_p x_p)\}. \tag{4}$$

They estimated the $\{\alpha_i\}$ directly in contrast to the standard discrimination approach with multivariate normal distributions which effectively estimates the $\{\alpha_i\}$ as functions of the estimates of the means and covariances.

There is an obvious extension of the logistic model to k populations (Cox, 1970, p.104),

$$\left.\begin{aligned} \mathrm{pr}(H_s|\underset{\sim}{x}) &= \exp(\underset{\sim}{x}^T\underset{\sim}{\alpha}_s)\mathrm{pr}(H_k|\underset{\sim}{x}) \qquad (s=1,\ldots,k-1) \\ \mathrm{pr}(H_k|\underset{\sim}{x}) &= 1/\{1+\sum_{s=1}^{k-1}\exp(\underset{\sim}{x}^T\underset{\sim}{\alpha}_s)\}, \end{aligned}\right. \tag{5}$$

where $\underset{\sim}{\alpha}_s^T = (\alpha_{s0},\alpha_{s1},\ldots,\alpha_{sp})$ for $(s=1,2,\ldots,k-1)$. Clearly the success of the logistic approach to discrimination depends on the extent to which equations (5) are satisfied. The equations hold exactly in many situations for example if the variables are:

(i) multivariate normal with equal dispersion matrices
(ii) multivariate independent dichotomous
(iii) multivariate dichotomous following the log-linear model (Birch, 1963) with equal second and higher order effects
(iv) a combination of (i) and (iii).

Thus equations (5) are satisfied exactly in the very situations that are usually considered in discrimination problems and they also hold, exactly or approximately, in many other cases. The equations hold over a wider class of distributions if second order terms are included in the logistic model (5), but at the cost of introducing many more parameters to estimate.

It should be emphasized that once the $\{\underset{\sim}{\alpha}_s\}$ have been estimated the allocation rule defined in equations (2) and (3) is extremely easy to use since it is equivalent to the rule: $\underset{\sim}{x}$ is allocated to H_s if $(\underset{\sim}{\alpha}_s-\underset{\sim}{\alpha}_t)^T\underset{\sim}{x} > 0$ for $t = 1,\ldots,k;\ t \neq s$, where for convenience, $\underset{\sim}{\alpha}_k^T = (0,0,\ldots,0)$. Thus the logistic approach yields linenar discriminant functions.

3. *Estimation when sampling the mixture*

Cox (1966) and Day and Kerridge (1967) assumed that sample points $\underset{\sim}{x}$ were available from the mixture of the two distributions. Generalising this to k populations, suppose that of a sample size n, $n_{\underset{\sim}{x}}$ are observed at $\underset{\sim}{x}$ and of these $n_{s\underset{\sim}{x}}$ are from H_s for all $\underset{\sim}{x}$; $s = 1,\ldots,k$. $\sum_{s=1}^{k} n_{s\underset{\sim}{x}} = n_{\underset{\sim}{x}}$. If $\sum_{\underset{\sim}{x}} n_{s\underset{\sim}{x}} = n_s$, then here the $\{n_s\}$ are random variables.

Let $\phi_{\underset{\sim}{x}}$ be the likelihood of the mixture distribution at $\underset{\sim}{x}$; then $\phi_{\underset{\sim}{x}} = \sum_{s=1}^{k} \Pi_s f_s(\underset{\sim}{x})$, where the $\{\Pi_s\}$ are the unknown mixing proportions. The log-likelihood of the sample can be written as

$$\log L = \text{constant} + \sum_{\tilde{x}}\sum_{s} n_{s\tilde{x}} \log(p_{s\tilde{x}}\phi_{\tilde{x}}).$$

It follows after some reduction, Anderson (1972), that the maximum likelihood equations for the $\{\alpha_s\}$ are

$$\frac{\partial \log L}{\partial \alpha_{sj}} = \sum_{\tilde{x}} (n_{s\tilde{x}} - n_{\tilde{x}} p_{s\tilde{x}}) x_j = 0 \quad (s=1,\ldots,k-1;\ j=0,1,\ldots,p). \tag{6}$$

The iterative solution of the equations for $k = 2$ is discussed at length by Cox (1970, p.87) and will be discussed more generally in §6.

4. *Estimation when each population is sampled separately*

The above sampling scheme is not usually associated with discrimination problems. More often, samples are taken from each population separately, so that the $\{n_s\}$ are fixed. It is then necessary to define the mixture of distributions from which the sample points to be allocated are drawn. This is done by specifying the mixing proportions, $\tilde{\Pi}^T = (\Pi_1, \Pi_2, \ldots, \Pi_k)$ of $H_1, H_2, \ldots, H_k$, where $\sum_{s=1}^{k} \Pi_s = 1$. These might be given, or estimated from other data. Where mixing proportions are not appropriate, likelihood ratios can be used (see §5).

In the above notation, the joint likelihood of the samples from $H_1, H_2, \ldots$ and H_k is given by

$$\log L = \text{constant} + \sum_{s=1}^{k} \sum_{\tilde{x}} n_{s\tilde{x}} \log\{L(\tilde{x}|H_s)\}. \tag{7}$$

Suppose, for the moment, that x_j is dichotomous with values 0 or 1 ($j=1,\ldots,p$), then

$$L(\tilde{x}|H_s) = \text{pr}(\tilde{x}|H_s) = \text{pr}(H_s|\tilde{x})\text{pr}(\tilde{x})/\text{pr}(H_s).$$

Again, let $\phi_{\tilde{x}} = \text{pr}(\tilde{x})$, then

$$L(\tilde{x}|H_s) = p_{s\tilde{x}}\phi_{\tilde{x}}/\Pi_s, \tag{8}$$

giving

$$\log L = \text{constant} + \sum_{s=1}^{k} \sum_{\tilde{x}} n_{s\tilde{x}} \log(p_{s\tilde{x}}\phi_{\tilde{x}}), \tag{9}$$

the constant including all terms independent of the $\{\tilde{\alpha}_s\}$

and the $\{\phi_{\underset{\sim}{x}}\}$. This likelihood is apparently identical to the one given by the Cox-Day-Kerridge formulation but it must be remembered that the unknown parameters are related by the functionally independent conditions

$$\sum_{\underset{\sim}{x}} \phi_{\underset{\sim}{x}} = 1, \quad \sum_{\underset{\sim}{x}} p_{s\underset{\sim}{x}} \phi_{\underset{\sim}{x}} = \Pi_s \quad (s=1,\ldots,k-1), \tag{10}$$

where the summations are over all $\underset{\sim}{x}$ values, so that the estimation problems in the two cases are not the same.

It is possible to estimate the parameters by maximizing the likelihood (9) subject to the constraints (10) in the straigthforward way, using Aitchison and Silvey's (1958) results on constrained maximum likelihood estimation. However, the algebra is heavy and the asymptotic dispersion matrix appears to have a very unwieldy form. Fortunately, a slightly less direct approach gives considerable simplifications.

It must be remembered that the proportions $\underset{\sim}{\Pi}$ can be fixed at arbitrary levels, subject to the usual conditions on probabilities. Suppose that there is interest in two sets of proportions, $\underset{\sim}{\Pi}$, to which all the preceding equations refer, and $\underset{\sim}{\Pi}'$, for which

$$\mathrm{pr}'(H_s|\underset{\sim}{x}) = p'_{s\underset{\sim}{x}} = \exp(\underset{\sim}{x}^T\underset{\sim}{\alpha}')\mathrm{pr}'(H_k|\underset{\sim}{x}) \quad (s=1,\ldots,k-1) \tag{11}$$

$$\mathrm{pr}'(H_k|\underset{\sim}{x}) = p'_{k\underset{\sim}{x}} = 1/\{1+\sum_{s=1}^{k-1}\exp(\underset{\sim}{x}^T\underset{\sim}{\alpha}'_s)\} \tag{12}$$

and $\quad \mathrm{pr}'(\underset{\sim}{x}) = \phi'_{\underset{\sim}{x}} \quad$ for all $\underset{\sim}{x}$.

The samples from $L(\underset{\sim}{x}|H_s)$ do not depend on the choice of $\underset{\sim}{\Pi}$, so

$$L(\underset{\sim}{x}|H_s) = p_{s\underset{\sim}{x}}\phi_{\underset{\sim}{x}}/\Pi_s = p'_{s\underset{\sim}{x}}\phi'_{\underset{\sim}{x}}/\Pi'_s \quad (s=1,\ldots,k). \tag{13}$$

Dividing the sth equation of (13) by the kth, it follows that

$$p'_{s\underset{\sim}{x}} = \exp(\beta_s+\underset{\sim}{x}^T\underset{\sim}{\alpha}_s)p'_k(\underset{\sim}{x}) \quad (s=1,\ldots,k-1) \tag{14}$$

$$p'_{k\underset{\sim}{x}} = 1/\{1+\sum_{s=1}^{k-1}\exp(\beta_s+\underset{\sim}{x}^T\underset{\sim}{\alpha}_s)\}, \tag{15}$$

where $\beta_s = \log\{\Pi_k\Pi'_s/(\Pi'_k\Pi_s)\}$. Thus

$$\left.\begin{array}{ll} \alpha'_{s0} = \alpha_{s0} + \beta_s & (s=1,\ldots,k-1) \\ \alpha'_{sj} = \alpha_{sj} & (j=1,\ldots,p) \end{array}\right) \qquad (16)$$

Thus only the $\{\alpha_{s0}\}$ are changed when $\underset{\sim}{\Pi}$ is varied, all the other logistic coefficients $\{\alpha_{sj}\}$ $(j\neq 0)$ remaining invariant. Thus the maximum likelihood estimates of the $\{\underset{\sim}{\alpha}'_s\}$ for any choice of $\underset{\sim}{\Pi}'$ can be derived immediately from those of $\{\underset{\sim}{\alpha}_s\}$ and vice versa. It became apparent when working on the constrained maximization problem that the algebra simplified considerably when $\underset{\sim}{\Pi} = \underset{\sim}{\Pi}^*$, where $\Pi^*_s = n_s/n$ $(s=1,\ldots,k)$. The obvious procedure is to carry out the constrained likelihood maximization with this choice of $\underset{\sim}{\Pi}^*$ and to adjust the $\{\alpha_{s0}\}$ afterwards for the desired $\underset{\sim}{\Pi}$.

It was shown by Anderson (1972) that the estimates of the $\{\underset{\sim}{\alpha}_s\}$ with $\Pi = \Pi^*$ given by maximizing the likelihood (9) subject to the constraints (10) are given by the equations:

$$\sum_{\underset{\sim}{x}} (n_{s\underset{\sim}{x}} - n_{\underset{\sim}{x}} p_{s\underset{\sim}{x}}) x_j = 0 \qquad (s=1,\ldots,k-1;\ j=0,1,\ldots,p). \qquad (17)$$

The result is obtained after a considerable amount of algebra, using Lagrange's undetermined multipliers.

It can be seen that the equations (6) and (17) are identical so the same computer procedures can be used to estimate the $\{\underset{\sim}{\alpha}_s\}$ in the two systems, mixture and separate sampling, despite their logical differences. In fact, the use of equations (6) for estimating the $\{\underset{\sim}{\alpha}_s\}$ with separate sampling, adjusting the $\{\alpha_{s0}\}$ afterwards to give the required proportions, first suggested itself to the author as an approximation. However, it can now be seen to be a constrained maximum likelihood solution with corresponding desirable properties (Aitchison and Silvey, 1958).

Let the asymptotic dispersion matrices of the estimates of the $\{\underset{\sim}{\alpha}_s\}$ for separate and mixture sampling be $\underset{\sim}{D}_S$ and $\underset{\sim}{D}_M$, respectively. Anderson (1972) used Aitchison and Silvey's (1958) result on asymptotic dispersion matrices with constrained maximum likelihood to show that $\underset{\sim}{D}_M$ and $\underset{\sim}{D}_S$ differ only in the variances and covariances of the $\{\alpha_{s0}\}$.

So far the discussion in this section has been restricted to dichotomous variables but the approach can easily be extended to polychotomous variables. For example, without any loss of generality, several variables with three unordered classes could be coded as pairs of dichotomous variables. This approach may not be realistic if there are a lot of unordered polychotomous variables with many classes because of the number of extra parameters. However, where there is an element of ordering in the classes, it should be possible to introduce a linear or quadratic trend term and thus economize on parameters.

Anderson (1972) discussed the extension of the methods of this section to continuous variables and concluded that this was permissible.

5. *Unknown mixing proportions*

There are some situations where mixing proportions $\underset{\sim}{\Pi}$ cannot be used. It may be that there is an underlying mixture but $\underset{\sim}{\Pi}$ is unknown and cannot be estimated. Alternatively, logical considerations may preclude a mixture and posterior probabilities. Perhaps the most obvious criterion for discrimination in these situations is the likelihood ratio and it is pleasing that estimation equations (6) or (17) can still be used.

Let the likelihood ratio for H_s and H_t be R_{st}; then from equations (5) and (8), assuming hypothetical proportions $\underset{\sim}{\Pi}$,

$$R_{st} = \frac{L(\underset{\sim}{x}|H_s)}{L(\underset{\sim}{x}|H_t)} = \frac{p_{s\underset{\sim}{x}}}{p_{t\underset{\sim}{x}}} \frac{\Pi_t}{\Pi_s}$$

so

$$\left.\begin{array}{ll} \log R_{st} = \underset{\sim}{x}^T(\underset{\sim}{\alpha}_s - \underset{\sim}{\alpha}_t) - r_{st} & (s,t=1,\ldots,k-1) \\ \log R_{sk} = \underset{\sim}{x}^T\underset{\sim}{\alpha}_s - r_{sk} & (s=1,\ldots,k-1), \end{array}\right) \quad (18)$$

where $r_{st} = \log(\Pi_s/\Pi_t)$, $(s,t=1,\ldots,k)$.

Thus, if the $\{\underset{\sim}{\alpha}_s\}$ are estimated with $\underset{\sim}{\Pi}^*$ defining a hypothetical mixture, estimates of the likelihood ratios can be obtained as indicated in equation (18).

6. *Iterative estimation of the logistic parameters*

The equations (6) and (17) are identical so there

is no point in distinguishing between them in the solution of the maximum likelihood equations. Cox (1970, p.87) and Day and Kerridge (1967) discuss the solution of these equations for $k = 2$ using a Newton-Raphson procedure. Anderson (1972) gave the appropriate equations for the Newton-Raphson procedure for k populations, which has been programmed as an Algol procedure. Cox (1970) suggested taking starting values from the least squares estimates on the linearized logistic function. However, as Cox noted, this does not work well when the samples are well separated and the author prefers starting values of zero for all components of the $\{\underset{\sim}{\alpha}_s\}$. This has worked well in many applications although in a few cases, one restart from an intermediate set of coefficients has been necessary.

One difficulty that can arise in the above iterative procedure is that non-unique maxima of the likelihood occur if the sample points from each population can be separated by hyperplanes. For example, if $k = 2$, and the sample points from H_1 lie to one side of the hyperplane L and the points from H_2 lie on the other side of L then values of $\underset{\sim}{\alpha}_1 = r\underset{\sim}{\alpha}_1^+$ exist for which $P(H_1|\underset{\sim}{x}) \to 1$ at all sample points from H_1 and $P(H_2|\underset{\sim}{x}) \to 1$ at all sample points from H_2, as $r \to \infty$. From the purely discriminatory point of view, this does not matter too much since good sample discrimination is obtained with any of the separating hyperplanes but the corresponding estimates of the $\{\underset{\sim}{\alpha}_s\}$ will not be reliable. If this complete separation of the sample points does exist, a convergent procedure for finding the maximum likelihood estimates of the $\{\underset{\sim}{\alpha}_s\}$ must find a set of logistic coefficients which give complete separation. These results are proved and discussed further by Anderson (1972).

7. *Logistic discrimination in medical diagnosis*

Differential diagnosis is one of the obvious situations where discrimination between groups is required to be based on qualitative variables, sometimes together with quantitative variables. An example of the use of logistic discrimination in the field of rheumatoid arthritis will be given.

There is a risk that people suffering from rheumatoid arthritis will also contract eye disease, keratoconjunctivitis sicca. This complaint can be diagnosed reliably and treated successfully by an eye specialist but his services are not available to screen all patients with rheumatoid arthritis. The question is whether a simple screening system can be set up to enable the medical staff of a Rheumatic Centre, who are not eye specialists, to decide which patients to refer to the eye hospital.

It was decided that the diagnostic system was to be based on 10 symptoms of the presence or absence type, so logistic discrimination was obviously appropriate. The ith variable, x_i, was taken to be 0 if the symptom was absent and 1 otherwise. The 10 symptoms were observed on 40 rheumatoid patients with keratoconjunctivitis sicca and on 37 patients who had no eye complaint ('normals'). This set of patients will be called Series I. The maximum likelihood estimate of $\underset{\sim}{\alpha}_1$, $\hat{\underset{\sim}{\alpha}}_1$ was found using the method described in the last section. Let $z = \hat{\underset{\sim}{\alpha}}_1^T \underset{\sim}{x}$, then

$$z = 4.0 - 4.4x_1 - 2.1x_2 - 1.1x_3 - 4.7x_4 - 3.5x_5 - 0.8x_6 + 0.8x_7 - 2.4x_8 + 1.8x_9 - 0.9x_{10} \quad (19)$$

The z-scores of all the patients in Series I were calculated using equation (18) and are shown in fig. 1a. Remembering the interpretation of z as a posterior probability {equation (5)}, keratoconjunctivitis sicca is more likely if $z < 0$ and it is less likely if $z > 0$. Complete separation did not occur but quite good differentiation between the two conditions was obtained. To test the method estimated from the Series I patients, the ten symptoms were observed on a further set of 41 patients (Series II) of whom 24 had keratoconjunctivitis sicca. The z-scores of all these patients were calculated using equation (19) and are plotted in fig. 1b. It can be seen that the results for the two Series are quite comparable.

For the application of the method to further patients, Series I and II were amalgamated and the logistic coefficients $\underset{\sim}{\alpha}_1$ re-estimated to be $\hat{\underset{\sim}{\alpha}}_1^{(2)}$. Let $z_2 = \underset{\sim}{x}^T \hat{\underset{\sim}{\alpha}}_1^{(2)}$; so

$$z_2 = 4.7 - 5.2x_1 - 3.0x_2 - 1.3x_3 - 5.4x_4 - 4.0x_5 + 1.1x_6 + 0.8x_7 - 1.9x_8 + 2.1x_9 - 2.0x_{10}. \quad (20)$$

The scores of all patients in Series I and II were recalculated using equation (20) and these are shown in fig. 1c.

The estimated coefficients in (19) and (20) are very similar, as are the plots of points in fig. 1a, b and c. In addition, the error rates are acceptable, so it is concluded that the system is stable and satisfactory.

This study is described in greater detail by Anderson, Whaley, Williamson and Buchanan (1972).

8. Logistic methods in epidemiology

Another application of the logistic methods developed here is in epidemiology, investigating factors related to diseases with low incidences. Retrospective studies have many advantages in this field and are used extensively (Mantel and Haenszel, 1959). The procedure followed is exactly the separate sampling scheme introduced in §4: subjects are sampled independently from the diseased group (D) and the normal or disease-free group (N). Thus the methods of §4 and §6 could be used to estimate $pr(D|\underset{\sim}{x})$, provided that this has the logistic form (3), where $\underset{\sim}{x}$ represents the levels of the factors under consideration. However, the overall incidence of D is often difficult to specify exactly so epidemiologists prefer to use relative risks instead of posterior probabilities. The relative risk, R, of a single factor (X: present; $\overline{X}$: absent) can be defined as

$$R = \frac{pr(D|X)}{pr(D|\overline{X})} .$$

This can be extended to the relative risk of two sets of factors $\underset{\sim}{x}$ and $\underset{\sim}{y}$,

$$R(x,y) = \frac{pr(D|\underset{\sim}{x})}{pr(D|\underset{\sim}{y})} .$$

If D has a low incidence then both these relative risks will be independent of the actual level of the incidence. Fortunately, the logistic parameters $\underset{\sim}{\alpha}$ of $pr(D|\underset{\sim}{x})$ have a direct interpretation in terms of relative risks.

$$pr(D|\underset{\sim}{x}) = \exp(\underset{\sim}{\alpha}^T\underset{\sim}{x})/\{1+\exp(\underset{\sim}{\alpha}^T\underset{\sim}{x})\} \simeq \exp(\underset{\sim}{\alpha}^T\underset{\sim}{x}),$$

since the incidence of D is low. Thus

$$R(\underset{\sim}{x},\underset{\sim}{y}) = \exp\{\underset{\sim}{\alpha}^T(\underset{\sim}{x}-\underset{\sim}{y})\} = \exp\{\sum_{i=1}^{p} \alpha_i(x_i-y_i)\},$$

since $x_0 = 1$ for all $\underset{\sim}{x}$.

One advantage of this approach is that interactions between the factors can be investigated and estimated quite simply by including terms like $\alpha_{ij}x_ix_j$ in the logistic function for $pr(D|\underset{\sim}{x})$. The same estimation procedure can be used. Another advantage is that continuous factors like age can be used as such without splitting their ranges up into intervals. Thirdly, competing risks between diseases D_1 and D_2 can be examined by using the logistic functions (5) for the posterior probabilities of more than two populations.

Retrospective studies in epidemiology can, then, be analysed in terms of the logistic models introduced in earlier sections. Although there are other ways of looking at dependent factors and competing risks, the logistic method does give a flexible and concise approach that satisfies many demands. If the sample from N is matched to the sample from D on certain factors, complications do arise but these can be handled by extending the analysis of §4. It is hoped to develop these ideas in a further communication.

9. *Discussion*

The chief advantages of the logistic approach to discrimination are, firstly, that it can be used with equal facility whether the variables are discrete or continuous and, secondly, that the estimation procedure is efficient under many different assumptions about the underlying distributions. Thus it rivals the generality of the distribution-free methods of allocation, for example, Fisher's linear discriminant function, but unlike the latter, yields estimates of posterior probabilities and likelihood ratios, which may be the major objective of a study as in epidemiology (§8). Because the parameterization is specific and the estimation is by maximum likelihood, estimates of the standard errors of the parameters are available and likelihood ratio hypothesis testing is possible, comparing maximized log-likelihoods under various assumptions.

When planning the sort of studies described here, it is sometimes possible to use either mixture or separate sampling. In the author's opinion, the latter is to be preferred, particularly when one or more of the mixing proportions are small, since it is possible to achieve balance between the sample sizes from the various populations when sampling separately but not when sampling the mixture.

The discrimination method of this paper has been concerned with applying logistic estimates of posterior probabilities or likelihood ratios to the simple allocation rules of equations (1) and (2) but these same estimates could be used in any system which requires estimates of the $\{pr(H_s|\underset{\sim}{x})\}$. For example, logistic procedures could be used in decision problems with utilities (Anderson, 1972) or in the constrained discrimination methods of Marshall and Olkin (1968) and Anderson (1969).

It is concluded that the method of estimating the logistic form for posterior probabilities discussed in this paper furnishes a valuable tool for a wide range of problems in addition to discrimination.

The author is very grateful to Professor Buchanan of the Centre for Rheumatic Diseases, Glasgow, and to Dr. Whaley and Dr. Williamson of the Western Infirmary, Glasgow, for providing the data on rheumatic patients for the application in §7.

REFERENCES

Aitchison, J. & Silvey, S.D. (1958), Maximum likelihood estimation of parameters subject to restraints, *Ann. Math. Stat.*, 29, 813-829.

Anderson, J.A. (1969), Discrimination between k populations with constraints on the probabilities of misclassification, *J.R. Statist. Soc.* B, 31, 123-139.

Anderson, J.A. (1972), Separate sample logistic discrimination, *Biometrika*, 59, 19-35.

Anderson, J.A. and Boyle, J.A. (1968), Computer diagnosis: statistical aspects, *Brit. Med. Bull.*, 24, 230-235.

Anderson, J.A., Whaley, K., Williamson, J. & Buchanan, W.W. (1972), A statistical aid to the diagnosis of keratoconjunctivitis sicca, *Quart. J. Med.*, 41, 175-189.

Birch, M.W. (1963), Maximum likelihood in three-way contingency tables, *J.R. Statist. Soc.* B, 25, 220-233.

Cochran, W.G. & Hopkins, C.E. (1961), Some classification problems with multivariate qualitative data, *Biometrics*, 17, 10-32.

Cox, D.R. (1966), Some procedures associated with the logistic qualitative response curve, *Research Papers in Statistics: Festschrift for J. Neyman*, (F.N. David, Ed.), Wiley, London, 55-71.

Cox, D.R. (1970), *The Analysis of Binary Data*, Methuen: London.

Day, N.E. & Kerridge, D.F. (1967), A general maximum likelihood discriminant, *Biometrics*, 23, 313-323.

Johns, M.V. Jr. (1961), An empirical Bayes approach to non-parametric two-way classification, *Studies in Item Analysis and Prediction*, (H. Solomon, Ed.) Stanford University Press, 221-232.

Linhart, H. (1959), Techniques for discriminant analysis with discrete variables, *Metrika*, 2, 138-149.

Mantel, N. & Haenszel, W. (1959), Statistical aspects of the analysis of data from retrospective studies of disease, *J. Nat. Cancer Inst.*, 22, 719-748.

Marshall, A.W. & Olkin, I. (1968), A general approach to some screening and classification procedures, *J.R. Statist. Soc.* B, 30, 407-443.

Martin, D.C. & Bradley, R.A. (1972), Probability models, estimation and classification for multivariate dichotomous populations, *Biometrics*, 28, 203-221.

Rao, C.R. (1965), *Linear Statistical Inference and its Applications*, Wiley: New York.

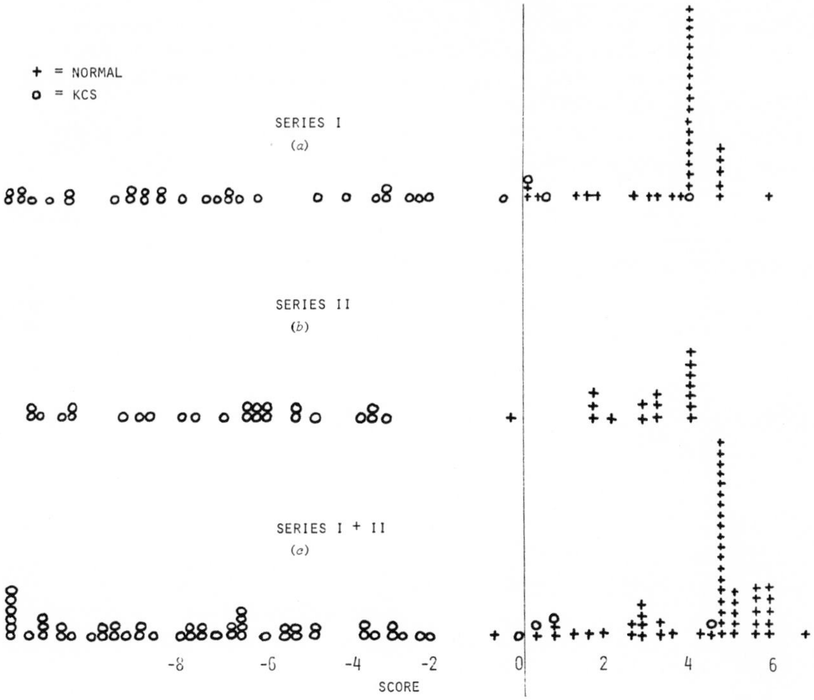

Fig. 1 The distribution of z-scores in rheumatoid arthritis; patients with keratoconjunctivitis sicca (o) and those without the eye complaint (+).

(a) gives the scores of Series I patients estimated from Series I;

(b) gives the scores of Series II patients estimated from Series II;

(c) gives the scores of Series I and II patients estimated from Series I and II.

ASYMPTOTIC EVALUATION OF THE PROBABILITIES OF MISCLASSIFICATION BY LINEAR DISCRIMINANT FUNCTIONS[1]

T. W. Anderson
Stanford University

1. Introduction

The problem of classifying an observation into one of two multivariate normal populations with a common covariance matrix might be called the classical classification problem. Fisher's linear discriminant function (Fisher, 1936) serves as a criterion when samples are used to estimate the parameters of the two distributions. The exact probabilities of misclassifications when using this criterion are difficult to compute because the distribution of the criterion is virtually intractable. Wald (1944) made considerable progress towards finding the distribution, but only managed to express the criterion as a function of three angles whose distribution he gave. T. W. Anderson (1951) and Rosedith Sitgreaves (1952) continued with the problem. For further references see T. W. Anderson, Das Gupta, and Styan (1972), Subject Matter Code 6.2.

If the parameters are known, the Neyman-Pearson Fundamental Lemma can be applied to the classical classification problem (as done by Wald, 1944) to obtain a discriminant function that is linear in the components of the vector to be classified. The distribution of this statistic is normal; the mean and variance depend only on the Mahalanobis distance between the two populations. Since the procedure for classification is to classify into one population or the other depending on whether this statistic is greater or less than a constant, the probabilities

[1]Prepared under the auspices of Office of Naval Research contract N00014-67-A-0112-0030.

of misclassification are found directly from the normal distribution. If the constant is 0, the probabilities are equal and the procedure is minimax.

When the parameters are unknown and there is available a sample from each population, the mean of each population is estimated by the mean of the respective sample and the common covariance matrix of the populations is estimated by using deviations from the respective means in the two samples. The classification function W, proposed by T. W. Anderson (1951), is obtained by replacing the parameters in the linear function resulting from the Neyman-Pearson Fundamental Lemma by the estimates; the substitution for parameters has been called "plugging in" estimates. This criterion differs from Fisher's discriminant function by subtraction of the average of the Fisher discriminant function at the two sample means. Then the distribution depends only on the population distance, and this fact makes the distribution problem simpler (T. W. Anderson, 1951 and Sitgreaves, 1952), though it is still rather intractable.

When the sizes of the two samples increase, the limiting distribution of W approaches a normal distribution, whose mean and variance depend on the Mahalanobis distance; if the limiting mean is subtracted from W and the difference is divided by the limiting standard deviation, the statistic has the standard normal distribution as its limiting distribution. Bowker and Sitgreaves (1961) and Okamoto (1963) with correction (1968) have given asymptotic expansions of the distributions to the order of the reciprocal of the square of the sample sizes. The approximate probability depends on the unknown parameter (the distance).

The "Studentized" W statistic is W less the estimate of its limiting mean divided by the estimate of its limiting standard deviation. It, too, has the standard normal distribution as its limiting distribution. If a statistician wants to set his cut-off point to achieve a specified probability of misclassification, he can use this Studentized W. An asymptotic expansion of the distribution of this statistic has been given by T. W. Anderson (1972).

In this paper we compare these two approximations

to the probabilities of misclassification and their uses. For further discussion of the classification problem see Anderson (1958), Chapter 6.

2. *The asymptotic expansion of the distribution of the classification statistic* W

Let the two populations be $N(\mu^{(1)}, \Sigma)$ and $N(\mu^{(2)}, \Sigma)$, and let the two samples be $x_1^{(1)},\ldots,x_{N_1}^{(1)}$ and $x_1^{(2)},\ldots,x_{N_2}^{(2)}$, respectively. The observation to be classified is x, which has the distribution $N(\mu, \Sigma)$, where $\mu = \mu^{(1)}$ or $\mu = \mu_1^{(2)}$. The classification statistic W is

$$W = (\bar{x}^{(1)} - \bar{x}^{(2)})' S^{-1} [x - \tfrac{1}{2}(\bar{x}^{(1)} + \bar{x}^{(2)})], \tag{1}$$

where

$$\bar{x}^{(1)} = \frac{1}{N_1} \sum_{j=1}^{N_1} x_j^{(1)}, \quad \bar{x}^{(2)} = \frac{1}{N_2} \sum_{j=1}^{N_2} x_j^{(2)}, \tag{2}$$

$$\begin{aligned} nS = &\sum_{j=1}^{N_1} (x_j^{(1)}-\bar{x}^{(1)})(x_j^{(1)}-\bar{x}^{(1)})' \\ &+ \sum_{j=1}^{N_2} (x_j^{(2)}-\bar{x}^{(2)})(x_j^{(2)}-\bar{x}^{(2)})', \end{aligned} \tag{3}$$

and $n = N_1 + N_2 - 2$. The rule is to classify x as coming from $N(\mu^{(1)}, \Sigma)$ if $W > c$ and from $N(\mu^{(2)}, \Sigma)$ if $W \leq c$, where c may be a constant, particularly 0, or a function of $\bar{x}^{(1)}$, $\bar{x}^{(2)}$, and S.

The squared Mahalanobis distance is

$$\alpha = (\mu^{(1)} - \mu^{(2)})' \Sigma^{-1} (\mu^{(1)} - \mu^{(2)}), \tag{4}$$

which can be estimated by

$$a = (\bar{x}^{(1)} - \bar{x}^{(2)})' S^{-1} (\bar{x}^{(1)} - \bar{x}^{(2)}). \tag{5}$$

The limiting distribution of W as $N_1 \to \infty$ and $N_2 \to \infty$ is normal with variance α and mean $\frac{1}{2}\alpha$ if $\underset{\sim}{x}$ is from $N(\underset{\sim}{\mu}^{(1)}, \underset{\sim}{\Sigma})$ and mean $-\frac{1}{2}\alpha$ if $\underset{\sim}{x}$ is from $N(\underset{\sim}{\mu}^{(2)}, \underset{\sim}{\Sigma})$; that is, the standard normal distribution $N(0,1)$ is the limiting distribution of $(W - \frac{1}{2}\alpha)/\sqrt{\alpha}$ for $\underset{\sim}{x}$ coming from $N(\underset{\sim}{\mu}^{(1)}, \underset{\sim}{\Sigma})$ and of $(W + \frac{1}{2}\alpha)/\sqrt{\alpha}$ for $\underset{\sim}{x}$ coming from $N(\underset{\sim}{\mu}^{(2)}, \underset{\sim}{\Sigma})$.

Okamoto's expansion of the probability distribution [(1963), Corollary 1] to terms of order n^{-1} is

$$
\begin{aligned}
&\Pr\left\{\frac{W - \frac{1}{2}\Delta^2}{\Delta} \le u \,\middle|\, \underset{\sim}{\mu} = \underset{\sim}{\mu}^{(1)}\right\} \\
&= \Phi(u) + \frac{1}{n}\phi(u)\left\{\Delta\left[(1 + \frac{p}{2}k - \frac{p-2}{2}\frac{1}{k})\frac{1}{\Delta^2} - \frac{p-1}{2}\right]\right. \\
&\quad - \left[(1 + \frac{1}{2}k + \frac{1}{2}\frac{1}{k})\frac{p-3}{\Delta^2} + \frac{3p-2}{2} + \frac{1}{2}\frac{1}{k} + \frac{\Delta^2}{4}\right] . u \qquad (6) \\
&\quad \left. - \Delta\left[(1 + \frac{1}{k})\frac{1}{\Delta^2} + 1\right]u^2 - \left[(1 + \frac{1}{2}k + \frac{1}{2}\frac{1}{k})\frac{1}{\Delta^2} + 1\right]u^3\right\} \\
&\quad + O(n^{-2}) ,
\end{aligned}
$$

where $k = \lim_{n\to\infty} N_1/N_2$ as $N_1 \to \infty$ and $N_2 \to \infty$, $\Delta^2 = \alpha$, and $\Phi(\ )$ and $\phi(\ )$ are the cumulative distribution function and density of $N(0,1)$, respectively. If $\lim_{n\to 1} N_1/N_2 = 1$, then

$$
\begin{aligned}
&\Pr\left\{\frac{W - \frac{1}{2}\Delta^2}{\Delta} \le u \,\middle|\, \underset{\sim}{\mu} = \underset{\sim}{\mu}^{(1)},\quad \lim_{n\to\infty}\frac{N_1}{N_2} = 1\right\} \\
&= \Phi(u) + \frac{1}{n}\phi(u)\left\{\Delta\left[\frac{2}{\Delta^2} - \frac{p-1}{2}\right]\right. \qquad (7) \\
&\left. - 2\left[\frac{p-3}{\Delta^2} + \frac{3p-1}{2} + \frac{\Delta^2}{4}\right]u - \Delta\left[\frac{2}{\Delta^2} + 1\right]u^2 - \left[\frac{2}{\Delta^2} + 1\right]u^3\right\} + O(n^{-2})
\end{aligned}
$$

$$= \Phi(u) + \frac{1}{n}\,\phi(u)\left\{\left[\frac{2}{\Delta^2} + 1\right](\Delta+u)(1-u^2) - \frac{p+1}{2}\,\Delta - \left[2\,\frac{p-2}{\Delta^2} + \frac{3p+1}{2} + \frac{\Delta^2}{4}\right]u\right\} + O(n^{-2}) .$$

The relation between the cut-off point c and the argument u is

$$c = u\Delta + \frac{1}{2}\,\Delta^2 , \qquad u = \frac{c - \frac{1}{2}\,\Delta^2}{\Delta} . \tag{8}$$

The probability of misclassification when $\underset{\sim}{x}$ is from $N(\underset{\sim}{\mu}^{(1)}, \underset{\sim}{\Sigma})$ is (6) [or (7)] with u given by (8); the probability depends importantly on the parameter α.

A cut-off point of particular interest is $c = 0$, which corresponds to $u = -\frac{1}{2}\,\Delta$. If $N_1 = N_2$, this defines a minimax procedure. In this case the probability of misclassification is

$$\begin{aligned} &\Pr\left\{W \le 0 \,\middle|\, \underset{\sim}{\mu} = \underset{\sim}{\mu}^{(1)},\ \lim_{n\to\infty}\frac{N_1}{N_2} = 1\right\} \\ &= \Phi\left(-\frac{\Delta}{2}\right) + \frac{1}{n}\,\phi\left(\frac{\Delta}{2}\right)\left\{\frac{p-1}{\Delta} + \frac{p}{4}\,\Delta\right\} + O(n^{-2}) . \end{aligned} \tag{9}$$

As far as this approximation goes, the correction term is positive; that is, the probability of a misclassification error is greater than the value of the normal approximation. For a given value of Δ the correction term and hence the probability (to order n^{-1}) increases with p. For a given value of p the probability (to order n^{-1}) decreases with Δ.

Okamoto (as well as Bowker and Sitgreaves) expanded the characteristic function. The method of Anderson (1972) could be used to obtain the result.

3. *The asymptotic expansion of the distribution of the Studentized* W

To use the approximate probability given by (6) one must know the parameter $\alpha = \Delta^2$, but this is generally unknown; then the statistician cannot achieve, even approximately, a desired probability. However, he can use the fact that a is a consistant estimate of α and therefore $(W - \frac{1}{2} a)/\sqrt{a}$ and $(W + \frac{1}{2} a)/\sqrt{a}$ have $N(0,1)$ as the limiting distribution in cases $\underset{\sim}{\mu} = \underset{\sim}{\mu}^{(1)}$ and $\underset{\sim}{\mu} = \underset{\sim}{\mu}^{(2)}$, respectively.

We can write

$$W - \frac{1}{2} a = (\bar{\underset{\sim}{x}}^{(1)} - \bar{\underset{\sim}{x}}^{(2)})' \underset{\sim}{S}^{-1} (\underset{\sim}{x} - \bar{\underset{\sim}{x}}^{(1)}). \tag{10}$$

Then

$$\Pr\left\{ \frac{W - \frac{1}{2} a}{\sqrt{a}} \le u \right\} = \Pr\left\{ (\bar{\underset{\sim}{x}}^{(1)} - \bar{\underset{\sim}{x}}^{(2)})' \underset{\sim}{S}^{-1} (\underset{\sim}{x}-\underset{\sim}{\mu}) \le u \sqrt{(\bar{\underset{\sim}{x}}^{(1)}-\bar{\underset{\sim}{x}}^{(2)})'\underset{\sim}{S}^{-1}(\bar{\underset{\sim}{x}}^{(1)}-\bar{\underset{\sim}{x}}^{(2)})} + (\bar{\underset{\sim}{x}}^{(1)}-\bar{\underset{\sim}{x}}^{(2)})'\underset{\sim}{S}^{-1}(\bar{\underset{\sim}{x}}^{(1)}-\underset{\sim}{\mu}) \right\}. \tag{11}$$

Since $\underset{\sim}{x}$ has the distribution $N(\underset{\sim}{\mu}, \underset{\sim}{\Sigma})$ independently of $\bar{\underset{\sim}{x}}^{(1)}$, $\bar{\underset{\sim}{x}}^{(2)}$ and $\underset{\sim}{S}$, the conditional distribution of $(\bar{\underset{\sim}{x}}^{(1)}-\bar{\underset{\sim}{x}}^{(2)})'\underset{\sim}{S}^{-1}(\underset{\sim}{x}-\underset{\sim}{\mu})$ is

$$N[0, (\bar{\underset{\sim}{x}}^{(1)}-\bar{\underset{\sim}{x}}^{(2)})'\underset{\sim}{S}^{-1}\underset{\sim}{\Sigma}\underset{\sim}{S}^{-1}(\bar{\underset{\sim}{x}}^{(1)}-\bar{\underset{\sim}{x}}^{(2)})],$$

and

$$r = \frac{(\bar{\underset{\sim}{x}}^{(1)}-\bar{\underset{\sim}{x}}^{(2)})'\underset{\sim}{S}^{-1}(\underset{\sim}{x}-\underset{\sim}{\mu})}{\sqrt{(\bar{\underset{\sim}{x}}^{(1)}-\bar{\underset{\sim}{x}}^{(2)})'\underset{\sim}{S}^{-1}\underset{\sim}{\Sigma}\underset{\sim}{S}^{-1}(\bar{\underset{\sim}{x}}^{(1)}-\bar{\underset{\sim}{x}}^{(2)})}} \tag{12}$$

has the distribution $N(0,1)$. Then (11) is

$$\Pr\left\{\frac{W-\frac{1}{2}a}{\sqrt{a}} \le u\right\}$$

$$= \Pr\left\{ r \le \frac{u\sqrt{(\bar{x}^{(1)}-\bar{x}^{(2)})'S^{-1}(\bar{x}^{(1)}-\bar{x}^{(2)})}+(\bar{x}^{(1)}-\bar{x}^{(2)})'S^{-1}(\bar{x}^{(1)}-\mu)}{\sqrt{(\bar{x}^{(1)}-\bar{x}^{(2)})'S^{-1}\Sigma S^{-1}(\bar{x}^{(1)}-\bar{x}^{(2)})}} \right\}$$

$$= \mathcal{E}\Phi\left[\frac{u\sqrt{(\bar{x}^{(1)}-\bar{x}^{(2)})'S^{-1}(\bar{x}^{(1)}-\bar{x}^{(2)})}+(\bar{x}^{(1)}-\bar{x}^{(2)})'S^{-1}(\bar{x}^{(1)}-\mu)}{\sqrt{(\bar{x}^{(1)}-\bar{x}^{(2)})'S^{-1}\Sigma S^{-1}(\bar{x}^{(1)}-\bar{x}^{(2)})}}\right], \tag{13}$$

where the expectation is with respect to $\bar{x}^{(1)}$, $\bar{x}^{(2)}$, and S.

When $\mu = \mu^{(1)}$, $\bar{x}^{(1)} - \bar{x}^{(2)}$, $\bar{x}^{(1)} - \mu$, and S converge in probability to $\mu^{(1)} - \mu^{(2)}$, 0, and Σ, respectively. We can expand the argument of $\Phi(\ )$ in a Taylor's series in terms of $\sqrt{n}$ times the differences between the estimates and their probability limits. When the expansion includes third degree terms and the expectations computed, the result is

$$\Pr\left\{\frac{W-a}{\sqrt{a}} \le u \,\middle|\, \mu=\mu^{(1)}\right\} = \Phi(u) + \frac{1}{n}\phi(u)\left[\frac{(p-1)}{\Delta}(1+k)-(p-\frac{1}{4}+\frac{1}{2}k)u-\frac{1}{4}u^3\right]+O(n^{-2}). \tag{14}$$

Interchanging N_1 and N_2 gives

$$\Pr\left\{\frac{W+\frac{1}{2}a}{\sqrt{a}} \le v \,\middle|\, \mu=\mu^{(2)}\right\} = \Phi(v) - \frac{1}{n}\phi(v)\left[\frac{p-1}{\Delta}(1+\frac{1}{k})+(p-\frac{1}{4}+\frac{1}{2k})v+\frac{1}{4}v^3\right]+O(n^{-2}) . \tag{15}$$

The proof of these results was given by T. W. Anderson (1972). If $\lim_{n\to\infty} N_1/N_2 = k = 1$,

$$\Pr\left\{ \frac{W-a}{\sqrt{a}} \le u \,\middle|\, \underset{\sim}{\mu}=\underset{\sim}{\mu}^{(1)},\ \lim_{n\to\infty} \frac{N_1}{N_2} = 1 \right\}$$
$$= \Phi(u) + \frac{1}{n}\,\phi(u)\left\{ 2\,\frac{p-1}{\Delta} - (p + \tfrac{1}{4})u - \frac{1}{4}\,u^3 \right\} + O(n^{-2}). \tag{16}$$

The correction term in (14) [(15) or (16)] is positive for $u < 0$. If $p = 1$, the correction term does not depend on Δ; if $p > 1$, the correction term decreases with Δ. For $u < 0$, the correction term increases with p.

For $u = -\frac{1}{2}\Delta$ (which is not $c = 0$)

$$\Pr\left\{ \frac{W-a}{\sqrt{a}} \le -\frac{\Delta}{2} \,\middle|\, \underset{\sim}{\mu}=\underset{\sim}{\mu}^{(1)},\ \lim_{n\to\infty} \frac{N_1}{N_2} = 1 \right\}$$
$$= \Phi(-\tfrac{\Delta}{2}) + \frac{1}{n}\,\phi(\tfrac{\Delta}{2})\left\{ 2\,\frac{p-1}{\Delta} + \frac{4p+1}{8}\,\Delta + \frac{\Delta^3}{32} \right\} + O(n^{-2}). \tag{17}$$

4. *Numerical values of the correction term for the Studentized* W *when* $N_1 = N_2$

We can obtain an idea of the importance of the term of order 1/n by studying numerical values of it. We consider the second term in (16), which is the error to order n^{-1} of using $\Phi(u)$ for the probability of misclassification. The correction relative to the nominal probability of misclassification is

$$\frac{1}{n}\,\frac{\phi(u)}{\Phi(u)}\left[2\,\frac{p-1}{\Delta} - (p + \tfrac{1}{4})u - \frac{u^3}{4}\right]. \tag{18}$$

Table 1 gives values of the term in brackets for the five values of u corresponding to values of $\Phi(u)$ of .1, .05, .025, .01, and .005, and various values of p and Δ. It is 4.0893 for $u = -1.28155$ [$\Phi(u) = .1$], $p = 2$, and $\Delta = 2$.

The correction relative to the nominal probability of misclassification is the value in the table multiplied by the ratio $\phi(u)/\Phi(u)$ divided by $n = N_1 + N_2 - 2$. In the example above it is $4.0893 \times 1.755 = 7.1767$ divided by n. If $N_1 = N_2 = 25$, then $n = 48$ and the correction relative to the nominal probability of misclassification is about .15. Here the correction would be rather small. For values of N_1 and N_2 somewhat larger, one might be willing to neglect the correction. One would hope that for these values of N_1 and N_2 the error when using this correction term would be rather small.

We might also be interested in the correction at $u = -\frac{1}{2}\Delta$. Table 2 gives the information. For example, for $\Delta = 4$ $\Phi(-\frac{1}{2}\Delta) = .022750$ (which would be the minimax probability if the parameters were known) and the correction is the appropriate number in the fourth column multiplied by .053991 divided by n. If $N_1 = N_2 = 25$ and $p = 2$, then $n = 48$ and the correction relative to the nominal probability is $7 \times 2.383/48 = .3475$.

5. *Comparison of the expansions of the distributions of* W *and the Studentized* W

It is striking that the asymptotic expansion of the distribution of the Studentized W is much simpler than that of W itself [the comparison of (6) with (14) and (7) with (16)], except for the particular case of $u = -\frac{1}{2}\Delta$ [(9) with (17)] which has special meaning for W (c = 0), but not for the Studentized W.

It is of interest to compare the correction terms of the two asymptotic expansions. The difference is

$$\Pr\left\{\frac{W - \frac{1}{2}a}{\sqrt{a}} \leq u \,\middle|\, \underset{\sim}{\mu} = \underset{\sim}{\mu}^{(1)}\right\} - \Pr\left\{\frac{W - \frac{1}{2}\alpha}{\sqrt{\alpha}} \leq u \,\middle|\, \underset{\sim}{\mu} = \underset{\sim}{\mu}^{(1)}\right\} \tag{19}$$

$$= \frac{1}{n}\phi(u)\left\{\frac{p-2}{2}\,\frac{2+k+1/k}{\Delta} + \frac{p-1}{2}\Delta + [(1 + \frac{1}{2}k + \frac{1}{2}\frac{1}{k})\frac{p-3}{\Delta^2}\right.$$

$$+ \frac{2p-3}{4} - \frac{1}{2} k + \frac{1}{2}\frac{1}{k} + \frac{\Delta^2}{4}] u + [(1 + \frac{1}{k})\frac{1}{\Delta} + \Delta] u^2$$

$$+ [\frac{2+k+1/k}{2\Delta^2} + \frac{3}{4}] u^3 \Big\} + O(n^{-2}) .$$

If $\lim_{n\to\infty} N_1/N_2 = k = 1$, the expression simplifies to

$$\Pr\left\{\frac{W - \frac{1}{2} a}{\sqrt{a}} \leq u \,|\, \underset{\sim}{\mu}=\underset{\sim}{\mu}^{(1)},\ \lim_{n\to\infty}\frac{N_1}{N_2} = 1\right\}$$

$$- \Pr\left\{\frac{W - \frac{1}{2}\alpha}{\sqrt{\alpha}} \leq u \,|\, \underset{\sim}{\mu}=\underset{\sim}{\mu}_1 ,\ \lim_{n\to\infty}\frac{N_1}{N_2} = 1\right\} \tag{20}$$

$$= \frac{1}{n}\,\phi(v)\left\{2\,\frac{p-2}{\Delta} + \frac{p-1}{2}\,\Delta + 2\left[\frac{p-3}{\Delta^2} + \frac{2p-3}{4} + \frac{\Delta^2}{4}\right] u\right.$$

$$\left. + \ [\frac{2}{\Delta} + \Delta]\, u^2 - [\frac{2}{\Delta^2} + \frac{3}{4}]\, u^3\right\} + O(n^{-2}) .$$

In particular, for $u = -\frac{1}{2}\Delta$ the difference is

$$\Pr\left\{\frac{W - \frac{1}{2} a}{\sqrt{a}} \leq -\frac{\Delta}{2} \,|\, \underset{\sim}{\mu}=\underset{\sim}{\mu}^{(1)},\ \lim_{n\to\infty}\frac{N_1}{N_2} = 1\right\}$$

$$- \Pr\left\{W \leq 0 \,|\, \underset{\sim}{\mu}=\underset{\sim}{\mu}^{(1)},\ \lim_{n\to\infty}\frac{N_1}{N_2} = 1\right\} \tag{21}$$

$$= \frac{1}{n}\,\phi(\frac{\Delta}{2})\left\{\frac{p-1}{\Delta} + (\frac{p}{4} + \frac{1}{8})\Delta + \frac{1}{32}\,\Delta^3\right\} + O(n^{-2}) .$$

Put another way, the correction term for $\Pr\{(W-a)/\sqrt{a} \leq -\frac{1}{2}\Delta\}$ is twice the correction term for $\Pr\{W \leq 0\}$ plus $\phi(\frac{1}{2}\Delta)\{\Delta/8 + \Delta^3/32\}/n$. The latter term, which does not depend on p, is usually small; values of $\Delta/8 + \Delta^3/32$ are given in Table 3. Comparison with Table 2 shows that for $p > 1$ this term is small except for large Δ. Thus, roughly speaking, the correction for the Studentized W is about that of W itself.

Okamoto (1963) has given numerical values of the term of order $1/n$ and the term of order $1/n^2$ in the expansion of $\Pr\{W \leq 0 | \mu = \mu^{(1)}\}$ for $N_1 = N_2 = 100$ $(n = 198)$ for various values of p and Δ. His values for $1/n$ are about twice the values we can compute from Table 2. In his table for small values of p and Δ the ratio of the term of order $1/n^2$ to the term of order $1/n$ is very roughly $1/n$. The maximum of the $1/n^2$ term over Δ increases with p. At $p = 7$, for example, it is about .0008. The table suggests that for small or moderate values of p the second correction term can be safely ignored for moderately large values of N_1 and N_2.

6. *Comparison of approximate densities and moments*

Corresponding to the approximate distributions of $(W-\alpha)/\sqrt{\alpha}$ and $(W-a)/\sqrt{a}$ (for $\mu=\mu^{(1)}$) are densities and moments. It is of some interest to compare these.

The approximate density of $(W - \frac{1}{2}\Delta^2)/\Delta$ is

$$\begin{aligned}\phi(u)\Bigg\{1 &- \frac{1}{n}\Bigg[\left(1 + \frac{1}{2}k + \frac{1}{2}\frac{1}{k}\right)\frac{p-3}{\Delta^2} + \frac{3p-2}{2} + \frac{1}{2}\frac{1}{k} + \frac{\Delta^2}{4} \\ &+ \left(\frac{3 + \frac{1}{2}pk - \frac{1}{2}(p-6)/k}{\Delta} - \frac{p-5}{2}\Delta\right)u \\ &- \left(\frac{p-6}{\Delta^2}\left(1 + \frac{1}{2}k + \frac{1}{2}\frac{1}{k}\right) + \frac{3p-8}{2} + \frac{1}{2}\frac{1}{k} + \frac{\Delta^2}{4}\right)u^2 \\ &+ \left(\frac{1 + 1/k}{\Delta} + \Delta\right)u^3 + \left(\frac{1 + \frac{1}{2}k + \frac{1}{2}\frac{1}{k}}{\Delta^2} + 1\right)u^4\Bigg]\Bigg\},\end{aligned} \tag{22}$$

which for $k = 1$ is

$$\phi(u)\left\{1-\frac{1}{n}\left[2\,\frac{p-3}{\Delta^2}+\frac{3p-1}{2}+\frac{\Delta^2}{4}+\left(\frac{6}{\Delta}-\frac{p-5}{2}\Delta\right)u\right.\right.$$

$$\left.\left.-\left(2\,\frac{p-6}{\Delta^2}+\frac{3p-7}{2}+\frac{\Delta^2}{4}\right)u^2+\left(\frac{2}{\Delta}+\Delta\right)u^3+\left(\frac{2}{\Delta^2}+1\right)u^4\right]\right\}. \tag{23}$$

The approximate density of $(W-\frac{1}{2}a)/\sqrt{a}$ is

$$\phi(u)\left\{1-\frac{1}{n}\left[p-\frac{1}{4}+\frac{1}{2}k+\frac{(p-1)(1+k)}{\Delta}u\right.\right.$$

$$\left.\left.-(p-1+\frac{1}{2}k)u^2-\frac{u^4}{4}\right]\right\}, \tag{24}$$

which for $k = 1$ is

$$\phi(u)\left\{1-\frac{1}{n}\left[p+\frac{1}{4}+2\,\frac{p-1}{\Delta}u-(p-\frac{1}{2})u^2-\frac{u^4}{4}\right]\right\}. \tag{25}$$

The approximate mean of $(W-\frac{1}{2}\Delta^2)/\Delta$ is

$$-\frac{1}{n}\left[\frac{6+\frac{1}{2}pk-\frac{1}{2}(p-12)/k}{\Delta}-\frac{p-11}{2}\Delta\right], \tag{26}$$

which for $k = 1$ is

$$-\frac{1}{n}\left[\frac{12}{\Delta}-\frac{p-11}{2}\Delta\right]; \tag{27}$$

the approximate second-order moment is

$$1+\frac{1}{n}\left[\frac{(2p-30)(1+\frac{1}{2}k+\frac{1}{2}\frac{1}{k})}{\Delta^2}+3p+26-\frac{1}{k}+\frac{1}{2}\Delta^2\right], \tag{28}$$

which for $k = 1$ is

$$1 + \frac{1}{n}\left[\frac{4p - 60}{\Delta^2} + 3p - 25 + \frac{1}{2}\Delta^2\right] . \tag{29}$$

The approximate mean of $(W - \frac{1}{2} a)/\sqrt{a}$ is

$$-\frac{1}{n}\frac{(p-1)(1+k)}{\Delta} , \tag{30}$$

which for $k = 1$ is

$$-\frac{1}{n} 2 \frac{p-1}{\Delta} ; \tag{31}$$

the approximate second-order moment is

$$1 + \frac{1}{n}(2p + 1 + k) , \tag{32}$$

which for $k = 1$ is

$$1 + \frac{1}{n}(2p + 2) . \tag{33}$$

In each case the "approximate" moment is the moment of the approximate density. The approximate second-order moment is also the approximate variance. For $(W - \frac{1}{2} a)/\sqrt{a}$ the approximate mean is negative for $p > 1$ (while it is 0 for the standard normal distribution); its numerical value increases with p and decreases with Δ. The approximate variances are greater than 1 (the value for the standard normal distribution); it increases with p, but does not depend on Δ.

7. *Achieving a given probability of misclassification*

Suppose one wants to achieve a given probability p of misclassification when $\underset{\sim}{\mu}=\underset{\sim}{\mu}^{(1)}$, say. How should one choose the cut-off point $c = u\sqrt{a} + \frac{1}{2} a$ for W or equivalently u for $(W - \frac{1}{2} a)/\sqrt{a}$?

Let u_0 be the number such that $\Phi(u_0) = p$. Then the probability of misclassification is

$$p + \frac{1}{n}\,\phi(u_0)\left[\frac{(p-1)(1+k)}{\Delta} - (p - \frac{1}{4} + \frac{1}{2}\,k)u_0 - \frac{1}{4}\,u_0^3\right] + O(n^{-2}) \,. \tag{34}$$

The correction term of order n^{-1} contains the unknown parameter Δ (if $p > 1$). However, Δ can be estimated by $\sqrt{a}$. These facts suggest taking

$$u = u_0 - \frac{1}{n}\left[\frac{(p-1)(1+k)}{\sqrt{a}} - (p - \frac{1}{4} + \frac{1}{2}\,k)u_0 - \frac{1}{4}\,u_0^3\right]. \tag{35}$$

Then

$$\begin{aligned} &\Pr\left\{\frac{W - \frac{1}{2}\,a}{\sqrt{a}} \le u \,\middle|\, \mu=\mu^{(1)}\right\} \\ &= \Pr\left\{\frac{W - \frac{1}{2}\,a}{\sqrt{a}} + \frac{1}{n}\,\frac{(p-1)(1+k)}{\sqrt{a}} \le u^*\right\} \,, \end{aligned} \tag{36}$$

where

$$u^* = u_0 + \frac{1}{n}\,[(p - \frac{1}{4} + \frac{1}{2}\,k)u_0 + \frac{1}{4}\,u_0^3] \,. \tag{37}$$

If $p = 1$, this probability is (14) with $u = u^*$, which is $p + O(n^{-2})$. When $p > 1$, we calculate the probability of misclassification as

$$\begin{aligned} \Pr\{W - \frac{1}{2}\,a \le u^*\sqrt{a} - \frac{1}{n}(p-1)(1+k)\} &= \Pr\{(\bar{x}^{(1)}-\bar{x}^{(2)})'S^{-1}(x-\mu) \\ &\le u^* \sqrt{(\bar{x}^{(1)}-\bar{x}^{(2)})'S^{-1}(\bar{x}^{(1)}-\bar{x}^{(2)})} \\ &+ (\bar{x}^{(1)}-\bar{x}^{(2)})'S^{-1}(\bar{x}^{(1)}-\mu) - \frac{1}{n}(p-1)(1+k)\} \end{aligned} \tag{38}$$

$$= \Phi\left[\frac{u^*\sqrt{(\bar{x}^{(1)}-\bar{x}^{(2)})'S^{-1}(\bar{x}^{(1)}-\bar{x}^{(2)})}+(\bar{x}^{(1)}-\bar{x}^{(2)})'S^{-1}(\bar{x}^{(1)}-\mu)-\frac{1}{n}(p-1)(1+k)}{\sqrt{(\bar{x}^{(1)}-\bar{x}^{(2)})'S^{-2}(\bar{x}^{(1)}-\bar{x}^{(2)})}}\right],$$

where $\bar{x}^{(1)}-\bar{x}^{(2)}$, $\bar{x}^{(1)}-\mu$ and S have the joint distribution given in Anderson (1972). Then the expansion of $\Phi(\ )$ is

$$\Phi\left\{u^* + \frac{1}{\sqrt{n}} C^*(Z,V) + \frac{1}{n} D^*(Y,Z,V) + r^*_{7n}(Y,Z,V)\right.$$
$$\left. - \frac{1}{n}(p-1)(1+k)\left[\frac{1}{\Delta} - \frac{1}{\Delta^3\sqrt{n}}(\delta'Y - \delta'V\delta) + r^*(Y,Z,V)\right]\right\}$$

$$= \Phi(u^*) + \phi(u^*)\left\{\frac{1}{\sqrt{n}} C^*(Z,V) + \frac{1}{n}\left[D^*(Y,Z,V)\right.\right.$$
$$\left. - \frac{1}{2} u^* C^{*2}(Z,V) - \frac{1}{n}(p-1)(1+k)\frac{1}{\Delta}\right] \tag{39}$$
$$\left. + \frac{1}{\Delta^3 n^{3/2}}(p-1)(1+k)(\delta'Y-\delta'V\delta)\right\} + \frac{1}{n^{3/2}} r^*_8(Y,Z,V)$$
$$+ \frac{1}{n^2} r^*_9(Y,Z,V) + r^*_{10n}(Y,Z,V),$$

where $C^*(Z,V)$, $D^*(Y,Z,V)$, and $r^*_{7n}(Y,Z,V)$ are $C(Z,V)$, $D(Y,Z,V)$ and $r_{7n}(Y,Z,V)$ of Anderson (1972), with u replaced by u^* and $r^*(Y,Z,V)$ in the remainder term in (19) of Anderson (1972). The expected value of $\Phi(\ )$ is

$$\Phi(u^*) + \frac{1}{n}\phi(u^*)\left[-\left(p - \frac{1}{4} + \frac{1}{2k}\right)u^* - \frac{1}{4}u^{*3}\right] + O(n^{-2})$$

$$= \Phi(u_0) + O(n^{-2}) \tag{40}$$

$$= p + O(n^{-2}).$$

REFERENCES

Anderson, T. W. (1951), Classification by multivariate analysis, *Psychometrika*, 16, 31-50.

Anderson, T. W. (1958), *An Introduction to Multivariate Statistical Analysis*, John Wiley & Sons, Inc., New York.

Anderson, T. W. (1972), An asymptotic expansion of the distribution of the "Studentized" classification statistic W, Technical Report No.9, Stanford University. To appear *Ann. Statist.*

Anderson, T. W., Somesh Das Gupta and George P. H. Styan (1972), *A Bibliography of Multivariate Statistical Analysis*, Oliver & Boyd, Ltd., Edinburgh.

Bowker, Albert H. and Rosedith Sitgreaves (1961), An asymptotic expansion for the distribution function of the W-classification statistic, *Stud. Item Anal. Predict.* (H. Solomon, Ed.), 285-292.

Fisher, Ronald A. (1936), The use of multiple measurements in taxonomic problems, *Ann. Eugenics*, 7, 179-188.

Okamoto, Mashashi (1963), An asymptotic expansion for the distribution of linear discriminant function, *Ann. Math. Statist.*, 34, 1286-1301.

Sitgreaves, Rosedith (1952), On the distribution of two random matrices used in classification procedures, *Ann. Math. Statist.*, 23, 263-270.

Wald, Abraham (1944), On a statistical problem arising in the classification of an individual into one of two groups, *Ann. Math. Statist.*, 15, 145-162.

TABLE 1

$$2\,\frac{p-1}{\Delta} - (p + \tfrac{1}{4})u - \frac{u^3}{4}$$

$u = -1.28155$
$\Phi(u) = .100$
$\phi(u) = .17550$
$\phi(u)/\Phi(u) = 1.755$

p \ Δ	1	2	3	4	6	∞
1	2.13	2.13	2.13	2.13	2.13	2.13
2	5.41	4.41	4.08	3.91	3.74	3.41
4	11.97	8.97	7.97	7.47	6.97	5.97
8	25.10	18.10	15.77	14.60	13.43	11.10

$u = -1.64485$
$\Phi(u) = .05$
$\phi(u) = .10314$
$\phi(u)/\Phi(u) = 2.063$

p \ Δ	1	2	3	4	6	∞
1	3.17	3.17	3.17	3.17	3.17	3.17
2	6.81	5.81	5.48	5.31	5.15	4.81
4	14.10	11.10	10.10	9.50	9.10	8.10
8	28.68	21.68	19.35	18.18	17.06	14.68

TABLE 1 (Cont.)

$u = -1.95996$
$\Phi(u) = .025$
$\phi(u) = .05844$
$\phi(u)/\Phi(u) = 2.338$

p \ Δ	1	2	3	4	6	∞
1	4.33	4.33	4.33	4.33	4.33	4.33
2	8.29	7.29	6.96	6.79	6.63	6.29
4	16.21	13.21	12.21	11.71	11.21	10.21
8	32.05	25.05	22.72	21.55	20.39	18.05

$u = -2.32635$
$\Phi(u) = .01$
$\phi(u) = .02665$
$\phi(u)/\Phi(u) = 2.665$

p \ Δ	1	2	3	4	6	∞
1	6.06	6.06	6.06	6.06	6.06	6.06
2	10.38	9.38	9.05	8.88	8.72	8.38
4	19.03	16.03	15.03	14.53	14.03	13.03
8	36.34	29.34	27.01	25.84	24.67	22.34

$u = -2.57583$
$\Phi(u) = .005$
$\phi(u) = .01446$
$\phi(u)/\Phi(u) = 2.892$

p \ Δ	1	2	3	4	6	∞
1	7.49	7.49	7.49	7.49	7.49	7.49
2	11.07	11.07	10.73	10.57	10.40	10.07
4	18.22	18.22	17.22	16.72	16.22	15.22
8	32.52	32.52	30.19	29.02	27.86	25.52

TABLE 2

$$2\,\frac{p-1}{\Delta} + \left(\frac{p}{2} + \frac{1}{8}\right) + \frac{\Delta^3}{32}$$

p \ Δ	1	2	3	4	6
1	.65625	1.50000	2.71875	4.50000	10.50000
2	3.15625	3.50000	4.88542	7.00000	13.83333
4	8.15125	7.50000	9.21875	12.00000	20.00000
8	18.15625	15.50000	17.88542	22.00000	40.83333
$\phi(-\frac{1}{2}\Delta)$	.35206	.24197	.129518	.053991	.0044318
$\Phi(-\frac{1}{2}\Delta)$	.30854	.15866	.066807	.022750	.0013499
$\phi(-\frac{1}{2}\Delta)/\Phi(-\frac{1}{2}\Delta)$	1.141	1.525	1.939	2.383	3.283

TABLE 3

$$\frac{\Delta}{8} + \frac{\Delta^3}{32}$$

Δ	1	2	3	4	6
$\frac{\Delta}{8} + \frac{\Delta^2}{32}$	.15626	.50000	1.21875	2.50000	7.50000

GRAPHICAL TECHNIQUES FOR HIGH DIMENSIONAL DATA

D. F. Andrews
University of Toronto

1. *Summary and Introduction*

Pictures of data serve many important functions in statistical analysis. They are useful in the early stages of analysis, in the initial screening and cleaning of data. Pictures are also useful in the subjective art of model formulation, where they assist in the selection of variables or effects to be included in a model, and help in checking the assumptions of a model. Probability plots, and other plots of residuals are used extensively in the analysis of linear regression models; see for example Draper and Smith (1966) or Daniel and Wood (1971).

The need for pictures of multivariate data is even greater than for univariate data. Typically the amounts of multivariate data are larger and the relationships, though frequently geometrical cannot be readily assimilated from a listing of the data. The use of colour displays and the dimension of time in interactive displays offer new possibilities. Some techniques for producing two-dimensional plots are available. Among these are the use of 'glyphs' by Anderson (1960) and probability plots of distances by Gnanadesikan and Wilk (1969).

In this paper a method is described for mapping k-dimensional points into functions which may then be plotted. In Section 2, the method is described and its properties are briefly described. In Section 3, the function plots are used in connection with a problem to which clus-

---

Preparation of this paper was supported in part by the National Research Council of Canada.

ter analysis was applied. In Section 4, the use of function plots for comparing covariance matrices of several populations is outlined.

2. *Function Plots of High-Dimensional Data*

The method is applicable to quantitative data in k dimensions, metric data. A data point may be represented as a vector $\underset{\sim}{x} = (x_1, \ldots, x_n)$. A way of plotting such a point is to map it into a space of functions and plot the resulting functions. There are of course many possible mappings. The following method has many useful geometrical and statistical properties and uses a family of functions which are widely understood.

For each k variate observation x, define a function

$$f_{\underset{\sim}{x}}(t) = \frac{1}{\sqrt{2}} x_1 + x_2 \sin(t) + x_3 \cos(t)$$

$$+ x_4 \sin(2t) + x_5 \cos(2t) + \ldots .$$

This function may be plotted over the range $-\pi \leq t \leq \pi$. Each data point will thus produce a curve drawn across the page.

2.1 *Properties of Function plots*

The properties of such plots are discussed in Andrews (1972). Some of these properties are summarized here.

(i) The mapping is linear. If $\underset{\sim}{x}, \underset{\sim}{y}$ and $\underset{\sim}{z}$ are points in k-dimensional space and if

$$\underset{\sim}{x} = a\underset{\sim}{y} + b\underset{\sim}{z} ,$$

then the functions share the same linear relation

$$f_{\underset{\sim}{x}}(t) = a\, f_{\underset{\sim}{y}}(t) + b\, f_{\underset{\sim}{z}}(t)$$

and so the function of the average satisfies

$$f_{\underset{\sim}{x}}(t) = n^{-1} \sum_{i=1}^{n} f_{\underset{\sim}{x}_i}(t) .$$

(ii) The mapping preserves distances. If the distance between two functions f and g is defined by

$$||f - g||_{L_2} = \int_{-\pi}^{\pi} |f(t) - g(t)|^2 \, dt \ ,$$

then the distance between the functions corresponding to $\underset{\sim}{x}$ and $\underset{\sim}{y}$,

$$||f_{\underset{\sim}{x}}(t) - f_{\underset{\sim}{y}}(t)||_{L_2} = \pi \, \Sigma(x_i - y_i)^2$$

is proportional to the Euclidean distance between these points. Thus, since the eye tends to appreciate the L_2 function norm, functions will look close together if and only if the corresponding data points are close together. This property is particularly useful in cluster related problems.

(iii) The mapping yields one-dimensional projections. For a particular value of t say t_0, the function value $f_{\underset{\sim}{x}}(t_0)$ is proportional to the projection of $\underset{\sim}{x}$ on the vector $f_{\underset{\sim}{1}}$

$$f_{\underset{\sim}{1}}(t_0) = (1/\sqrt{2}, \sin(t_0), \cos(t_0), \sin(2t_0), \ldots) \ .$$

As t_0 changes the projections of the data on a continuum of directions are recorded on the plot. If the data form different clusters in different orthogonal subspaces, these different clusterings will be reflected in the projections of $f_1(t)$ as it passes through or near these spaces. This if more than one clustering is appropriate this will be shown in the plot.

(iv) The mapping preserves variances. If the original data have been transformed so that the components are approximately independent with equal variance σ^2, then the variance of $f_{\underset{\sim}{x}}(t)$ is given by

$$\text{var}(f_{\underset{\sim}{x}}(t)) = \sigma^2(1/2 + \sin^2(t_0) + \cos^2(t_0) + \sin^2(2t_0) + \ldots)$$

$$= \sigma^2 \begin{cases} k/2 \ , \text{ if } k \text{ is odd} \ , \\ k/2 + \varepsilon, \ |\varepsilon| \leq 1/2, \text{ if } k \text{ is even} \ . \end{cases}$$

Thus the variance of the function values is constant or very nearly so. This variance may be used to construct critical regions for tests or confidence sets for location parameters at $t=t_0$. These sets will appear as bands of approximately equal width drawn across the page.

(v) The mapping generates over all tests. Since the function value $f_{\underset{\sim}{x}}(t)$ proportional to the projection of $\underset{\sim}{x}$ on any vector is less than the length of $\underset{\sim}{x}$, tests may be constructed for any value of t. Thus a test of the hypothesis $E(\underset{\sim}{x}) = \underset{\sim}{\mu}$ may be based on the inequality

$$||f_{\underset{\sim}{x}}(t) - f_{\underset{\sim}{\mu}}(t)||^2 \leq \frac{1}{2}(k+1)\ \sigma^2\ \chi_k^2\ (\alpha)$$

which is true with probability $1 - \alpha$ for *all* values of t.

The properties discussed above also hold with obvious slight changes for functions of the form

$$f_{\underset{\sim}{x}}(t) = x_1\sin(t) + x_2\cos(t) + x_3\sin(2t) + x_4\cos(2t) + \ldots$$

or similar functions.

3. *An Example Involving Clusters*

In this section we give an application of the function plots to a problem involving clustering of four-dimensional data. Moser and Scott (1961) investigated socio-economic data on a large number of British towns. They started with 57 variables. Many of these were highly related. They calculated principal components calculated from the correlation matrix of the 57 variables. The first four components accounted for 60.4% of the original variation. The values of these components for 155 towns are given in Table 1.

Moser and Scott clustered the data partly 'by hand' using two-dimensional projections of subsets of the first four principal components. Their objective was to produce clusters containing about 10 towns each. Table 1 also gives the composition of each cluster.

Figures 1.1-1.15 are function plots of the 14 clusters and of all the data. These were reduced from ordinary printer plots. A function is plotted as a character string made up of units of length ten. The first three characters give the sequence number of the observation in the previous table. The next 3 characters indicate the group or cluster. The last 4 characters are just dots.

The functions plotted were based on the mapping

$$\underset{\sim}{x} \to f_{\underset{\sim}{x}}(t) = x_1 \sin(t) + x_2 \cos(t) + x_3 \sin(2t) + \ldots \quad .$$

The clusters appear as relatively tight bands, the thinner the band, the smaller the size of the cluster. Some clusters, notably those in Figures 1.8, 1.11, 1.12 appear less dense than many of the others and the presence of some holes in the center of these cluster bands suggests that they might be better divided and assigned to two or more different clusters.

H. Andrews (1971) re-examined this data and produced another clustering using a clustering algorithm on the same four principal components. The resulting clusters are given in Table 2 and plotted in Figure 2.1-2.15. The clusters appear somewhat more homogeneous. Figure 2.7, a relatively tight cluster contains parts of Figures 1.5, 1.7, 1.8, all somewhat fuzzy clusters. Similarly, Figure 2.1 is composed of parts of Figures 1.11 and 1.12; Andrews' clustering yields in these instances much smaller clusters.

Despite these many improvements there are many similarities between the 'hand' clustering of Moser and Scott and that of Andrews. It must be noted that some of the improvement in the latter approach may be due to dropping the constraint that the clusters contain approximately equal numbers of towns.

A real question remains however. To what extent is the division into clusters artificial?

3.1 Is there more than one cluster?

The plot of all the towns repeated in Figures 1.15 and 2.15 is a structure-free mass of points. If the data consisted of small, widely separated clusters this plot would appear as a superposition of several narrow ribbons of functions. This does not appear to be the case. There may be a weaker cluster structure obscured by the poor resolution of the printer plot or there may be only one cluster. A better method of displaying all the data is required for this purpose.

Histograms or preferably Chi-squared probability plots of the distances from a particular point to the remaining n-1 points may be made for each point. Clustering will produce anomalies in these plots. The following procedure gives some idea of the number of clusters, their size and composition. This procedure uses the structure of function plots to map each data point and its neighbours onto the plotting surface in a continuous manner and then to plot a measure of local density of each point.

Each town corresponds to a data vector $\underset{\sim}{x}_0$ and for each $\underset{\sim}{x}_0$ there is a value of t_0 such that $\underset{\sim}{x}_0$ is closer to the point $\pm f_{\underset{\sim}{1}}(t_0)$ than to any other point on the curve $\{\pm f_{\underset{\sim}{1}}(t): -\pi \leq t \leq \pi\}$. At $t = t_0$ the ordinate $f_{\underset{\sim}{x}}(t_0)$ is plotted with a character giving a measure of local density.

This measure of local density at $\underset{\sim}{x}_0$ is an integer (modulo 10) representing the number of data points within a distance d_0 of and including $\underset{\sim}{x}_0$. These neighbouring points $\underset{\sim}{x}_i$ are plotted with a '.' at $(t_0, f_{\underset{\sim}{x}_i}(t_0))$.

Without understanding the nature of the critical distance d_0 and hence of the printed integers, it is clear that Figures 3.1-3.3 represent non-homogeneous batches of dots and integers. These batches are suggestive of clusters.

More objective evidence may be inferred once the critical distance d_0 is specified. This distance was found by calculating the Euclidean norms of the vector of the first four principal components for each town.

$$||\underset{\sim}{x}_i|| \qquad (i = 1,\ldots,155) ,$$

sorting these numbers to get

$$||x||_{(i)} , \quad \text{where} \quad x_{(i)} \leq \cdots \leq x_{(155)} ,$$

and choosing d_0 equal to

$$||x||_{(2)} \quad \text{in Figure 3.1 ,}$$

$$||x||_{(4)} \quad \text{in Figure 3.2 ,}$$

$$||x||_{(8)} \quad \text{in Figure 3.3 .}$$

Now if the points $\underset{\sim}{x}_i$ were uniformly distributed in R^4 and if $\underset{\sim}{x}_i$ is sufficiently far from the origin so that the distance to the rth nearest neighbours of $\underset{\sim}{x}_i$ may be assumed to be independent of the distance to the sth nearest neighbour of $\underset{\sim}{o}$ then the probability

P (exactly r points are closer to $\underset{\sim}{x}_i$ than the sth closest point is to $\underset{\sim}{o}$)

$$= \frac{n!\; n!\; (r+s-1)!\; (2n-r-s)!}{r!\; (n-r)!\; (s-1)!\; (n-s)!\; 2n!}$$

which for large n is approximately equal to

$$\frac{(r+s-1)!}{r!\; (s-1)!} \left(\frac{1}{2}\right)^{r+s} .$$

Now for most non-uniform distributions the density of points decreases monotonically away from $\underset{\sim}{o}$ and thus the variable r is stochastically less than the distribution derived above would indicate. If an obscured value of r seems large relative to this distribution it will be even more significant under a distribution which decreases monotonically away from the origin.

In Figure 3.1 we put the integer $1 + (r \bmod 10)$ for values of $s = 2$ (Figure 3.1), $s = 4$ (Figure 3.2), $s = 8$ (Figure 3.3).

The probability distribution r and hence of $r + 1$ is given in Table A for the case $s = 2$.

Table A

Probability distribution of integers in Figure 3.1 calculated under the assumption that there is only one cluster.

1 + (r mod 10)	Probability
1	0.2527
2	0.2515
3	0.1883
4	0.1254
5	0.0783
6	0.0470
7	0.0273
8	0.0156
9	0.0088
0	0.0049

Thus the occurrence of large numbers of the digits '7', '8' in Figure 3.1 represents highly significant evidence of the presence of clusters.

Some idea of the number and composition of clusters can be obtained by a careful examination of this plot.

4. *Plotting Covariance Matrices*

Frequently in discrimination and classification problems data is available from known populations. It is often convenient to act as if these populations have a common covariance matrix. It is important to examine this assumption. In addition to some numerical basis for comparison it would be useful to have a method of displaying graphically the component covariance matrices. Function plots provide one method of doing this.

Let C_i (i=1,...,p) be a set of p, k×k covariance matrices and let C_0 be some estimate of their common value. The C_i (i=1,...,p) may be obtained from data but we will assume they are positive semi-definite and that C_0 is positive definite. Let $C_{1/2}$ be a square root of C_0 so that $C_{1/2}C_{1/2} = C_0$. Then define

$$C_i^* = C_{1/2}^{-1} \, C_i \, C_{1/2}^{-1}$$

If $C_1 = \ldots = C_p = C_o$, then $C_i^* = I$ the identity matrix.

Now for each value of t we can solve the equation in $\alpha(t)$

$$\alpha^2(t)\ \underset{\sim}{f_1}(t)\ C_i^*\ \underset{\sim}{f_1}(t) = K\ ,$$

where K is some constant. There will be two solutions.

$$\text{If}\quad C_1 = \ldots = C_p = C_o\ ,\quad \text{then}$$

$$\alpha_i(t) = \pm\ \{K/(\underset{\sim}{f_1}(t)\underset{\sim}{f_1}(t))\}^{1/2}$$

$$= \pm\ \{K/(\tfrac{1}{2}\ k)\}^{1/2}\quad .$$

For each covariance matrix C_i, the two solutions for $\alpha_i(t)$ are two functions for t which may be plotted on the range $-\pi \leq t \leq \pi$. Under the hypothesis $C_1 = \ldots = C_p = C_o$, these functions will appear as p copies of the same pair of horizontal straight lines. However, if the covariance matrices differ the functions $\alpha_i(t)$ will typically be different.

The curve $\underset{\sim}{f_1}(t)$ lies on a hyperplane and a plot based on $\underset{\sim}{f_1}(t)$ will be insensitive to differences in the covariance matrices that occur solely in the space orthogonal to this hyperplane. Since this hyperplane has dimension greater than $\frac{1}{2}$ k, two such hyperplanes span the space and two plots will be sufficient. The second plot may be obtained by permuting the components of $\underset{\sim}{f_1}(t)$.

In most situations the covariance matrices C_i will be determined from data and hence will be different. The graphical representation of these matrices may be used to determine, among other things, whether or not

(i) the matrices are similar and may be considered equal in the subsequent analysis;
(ii) a few sub-populations must be treated differently;
(iii) the populations form several distinct groups, each group having ansimilar covariance structure (clustering by

covariance);
(iv) some transformation of the data lessens the differences in the covariance matrices.

5. *Concluding Remarks*

The graphical techniques described here share with other multivariate techniques a strong dependence on the choice and scaling of the original data. Cormack (1971) comments on the indiscriminant use of some common classification procedures. The need for 'clear thinking' remains. Graphical tools are useful if not essential in this process.

ACKNOWLEDGEMENT

Conversation with Drs. A. M. Herzberg and J. R. Kettenring were helpful in the preparation of this paper.

REFERENCES

Anderson, E. (1960), A semigraphical method for the analysis of complex problems, *Technometrics*, 2, 387-392.

Andrews, D. F. (1972), Plots of high-dimensional data, *Biometrics*, 28, 125-136.

Andrews, H. F. (1971), A cluster analysis of British towns, *Urban Studies*, 8, 271-284.

Cormack, R. M. (1971), A review of classification, *J. R. Statist. Soc.* A, 321-367.

Daniel, C. and Wood, F. S. (1971), *Fitting Equations to data*, Wiley: New York.

Draper. N. R. and Smith, H. (1966), *Applied Regression Analysis*, Wiley: New York.

Gnanadesikan, R. and Wilk, M. B. (1969), Data analytic methods in multivariate statistical analysis, *Multivariate Analysis* II (P. R. Krishnaiah, Ed.), 593-638, Academic Press: New York.

Moser, C. A. and Scott, W. (1961), *British Towns*, Oliver and Boyd: Edinburgh.

Table 1

155 British Towns
Clustered by Moser and Scott
First Four Principal Components

N =155K = 4SCALE = 0.0300SHIFT = 0.0

	1	9.550	-4.370	3.090	1.520
	2	8.720	-5.700	-0.800	3.230
	3	5.900	-5.410	1.810	1.680
	4	7.580	-4.310	1.340	-0.270
1.1	5	7.240	-5.810	1.300	2.850
	6	6.510	-4.840	1.240	1.980
	7	6.740	-3.850	-0.840	1.060
	8	7.050	-4.490	1.350	1.400
	9	3.490	-4.890	1.130	2.280
	10	5.240	-3.480	-0.100	0.740
	11	3.760	-2.130	0.710	-0.320
	12	2.050	-3.740	3.440	0.810
	13	2.680	-0.150	2.700	-0.800
	14	2.190	-1.870	0.400	-0.890
1.2	15	3.670	-1.950	1.980	-1.090
	16	2.440	-2.130	2.170	-0.690
	17	2.070	-1.440	2.350	-0.960
	18	1.930	-1.160	2.310	-1.730
	19	1.930	-1.340	3.770	-2.540
	20	4.790	-1.340	1.220	1.020
	21	-0.860	-0.690	3.170	0.050
	22	0.280	-1.580	2.220	-0.980
	23	-0.800	-1.290	2.440	-0.280
	24	0.400	-0.700	0.420	-0.530
	25	-0.360	-0.990	2.270	-1.170
	26	0.200	-2.620	2.800	-2.450
	27	0.490	-2.570	1.690	-2.050
1.3	28	0.240	-0.780	2.890	-2.520
	29	-0.730	-1.810	2.610	-2.430
	30	-0.030	-1.650	2.670	-2.100
	31	0.420	-1.550	1.420	-0.660
	32	0.640	-1.220	-0.300	-2.760
	33	-0.520	-2.030	0.990	-0.620
	34	-0.090	-1.690	1.360	-0.660
	35	0.440	-0.720	0.890	-2.340
	36	-1.020	-0.710	1.260	1.250
	37	-2.610	0.470	0.280	-3.190
	38	-1.480	-0.470	0.660	-0.700
	39	-1.680	-0.430	3.360	-2.210
	40	-1.630	-0.060	1.980	-2.340
	41	-2.790	-1.340	-0.040	-2.740
	42	-1.910	-1.070	1.730	-0.070
1.4	43	-2.140	0.420	1.400	-1.560
	44	-2.440	0.450	0.940	-1.470
	45	-2.430	-0.320	1.800	-2.160
	46	-3.040	1.140	1.440	-2.200
	47	-2.540	-0.150	-1.380	-0.770
	48	-2.270	-1.310	1.160	-1.020
	49	-1.030	-0.790	-1.030	-1.430
	50	-2.580	1.880	1.600	-1.520
	51	-3.220	0.110	0.740	1.280
	52	-3.790	-1.800	0.910	4.500
	53	-3.660	-1.780	3.240	0.440
	54	-4.260	-1.430	1.230	1.520
1.5	55	-3.290	-0.550	1.540	3.170
	56	-3.270	-0.280	0.870	1.200
	57	-2.570	-0.210	0.660	0.790
	58	-3.670	-1.510	-0.980	4.260
	59	-3.070	-0.380	-0.270	-0.700
	60	-3.130	-1.290	-1.320	0.770
	61	-1.260	-2.290	-1.670	-0.600
	62	-2.010	-3.860	-1.910	0.890
	63	-2.290	-3.350	-1.550	1.520
	64	-4.600	-3.040	-0.070	1.380
	65	-2.590	-3.370	-1.190	0.810
	66	-1.640	-1.700	-1.710	-2.130
	67	-1.420	-1.050	-1.940	-2.170
1.6	68	-2.640	-2.040	-1.640	-3.490
	69	-0.920	-1.860	-1.860	-3.430
	70	-0.980	-1.840	-1.010	-2.580
	71	-2.130	-2.000	-1.200	0.810
	72	-3.470	-2.400	-2.440	-0.890
	73	-2.760	-1.410	-0.960	-1.310
	74	-3.050	-3.080	-2.340	-2.070
	75	-0.980	-1.730	-0.400	-2.140
	76	-2.800	-1.770	0.190	-1.390
	77	-7.510	-0.440	-1.590	5.530
	78	-5.980	-2.440	0.970	5.700
	79	-6.170	-1.410	1.920	5.820
	80	-5.620	-0.500	3.180	2.160

Table 1 (Continued)

	[illegible]	[illegible]	[illegible]	[illegible]	[illegible]
	81	-6.200	-1.400	2.090	1.040
	82	-5.640	-0.630	-3.680	2.010
	83	-6.110	0.800	-0.200	1.440
1.7	84	-4.600	-0.820	-3.220	0.320
	85	-6.160	-1.380	-0.820	0.290
	86	-4.660	-0.890	0.650	-0.160
	87	-5.790	-0.950	-0.810	-0.200
	88	-4.650	1.120	3.340	0.530
	89	-4.310	2.490	4.400	-2.310
	90	-5.070	-0.080	3.520	1.520
	91	-3.820	2.420	2.720	-1.130
	92	-4.190	3.170	1.900	-2.330
	93	-4.270	1.180	1.640	-1.770
1.8	94	-7.350	2.920	-1.280	3.010
	95	-5.570	1.070	-0.240	1.390
	96	-4.880	0.210	0.670	1.480
	97	-5.920	0.460	-0.910	-0.720
	98	-5.740	1.800	0.320	0.690
	99	-4.480	0.730	-0.460	-0.280
	100	-3.850	-0.340	-3.190	-2.310
	101	-3.820	3.190	-2.510	-0.610
	102	10.300	3.440	1.730	0.620
	103	9.060	2.030	1.080	0.400
	104	6.360	2.550	1.170	0.370
	105	6.240	0.560	0.450	0.830
1.9	106	7.130	2.270	-1.250	1.030
	107	7.290	2.150	-0.710	0.240
	108	8.060	3.320	-1.870	-0.200
	109	6.940	0.660	-2.480	1.130
	110	9.340	1.570	-2.980	1.020
	111	2.550	-0.330	-3.890	0.760
	112	4.730	-0.800	-3.660	2.000
	113	2.600	1.590	-2.960	-0.180
	114	4.570	1.430	-3.010	0.830
	115	6.190	-1.210	-2.380	1.360
	116	4.290	2.580	-2.260	-0.500
1.10	117	3.030	2.240	-2.470	-1.160
	118	4.760	1.130	-1.580	0.340
	119	3.510	-0.190	-0.800	-0.310
	120	5.380	2.880	-0.960	0.600
	121	4.570	0.240	0.920	2.340
	122	3.110	0.320	1.900	1.240
	123	2.100	6.040	2.600	1.360
	124	4.750	4.240	3.440	0.620
	125	4.260	5.520	3.040	-0.470
	126	1.930	5.850	2.180	-0.720
	127	1.280	5.110	0.520	0.890
1.11	128	5.100	6.030	1.410	0.660
	129	3.830	4.600	0.280	0.310
	130	4.690	3.960	-2.120	0.250
	131	2.720	4.460	-1.960	2.230
	132	4.000	4.220	-4.180	-1.550
	133	2.800	4.370	-4.670	-1.730
	134	0.620	2.190	4.430	-1.120
	135	0.770	0.840	3.100	-1.670
	136	1.280	2.770	2.960	0.900
	137	-1.400	1.600	0.760	-2.840
	138	0.940	3.300	1.410	-0.960
1.12	139	0.940	0.180	-0.210	-1.460
	140	-1.120	2.200	0.050	-1.690
	141	0.550	2.430	-1.760	-1.830
	142	0.340	3.300	-3.020	-2.090
	143	-1.060	-0.670	-5.700	2.410
	144	-2.260	-0.930	-5.380	1.200
	145	-1.080	0.650	-2.990	0.310
	146	0.720	-0.360	-4.010	-0.270
	147	0.210	-1.500	-4.930	-0.040
1.13	148	-0.300	1.810	-2.460	-0.870
	149	-0.360	2.340	-4.020	-0.660
	150	-0.540	0.690	-4.120	-0.670
	151	-0.620	-0.700	-5.340	-1.690
	152	-0.420	1.140	-2.940	-2.850
	153	-2.820	5.950	-2.630	-0.610
1.14	154	-4.090	4.220	-3.760	0.320
	155	-6.320	4.760	-3.910	1.460

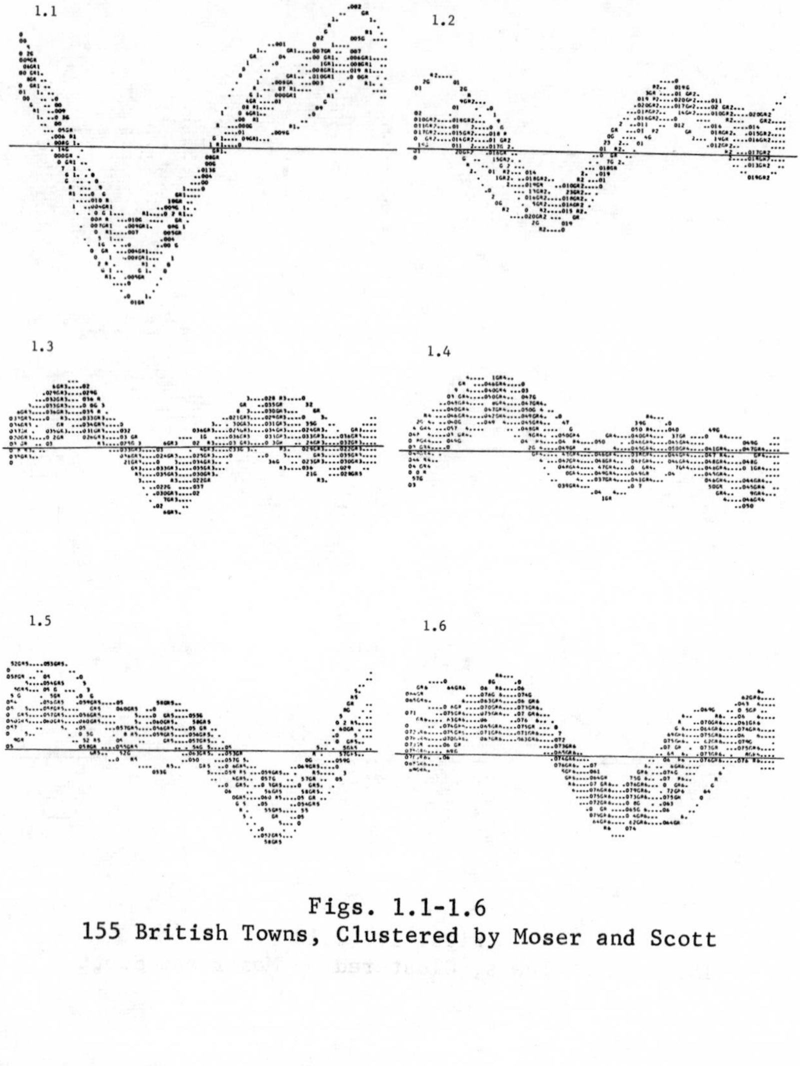

Figs. 1.1-1.6
155 British Towns, Clustered by Moser and Scott

1.7

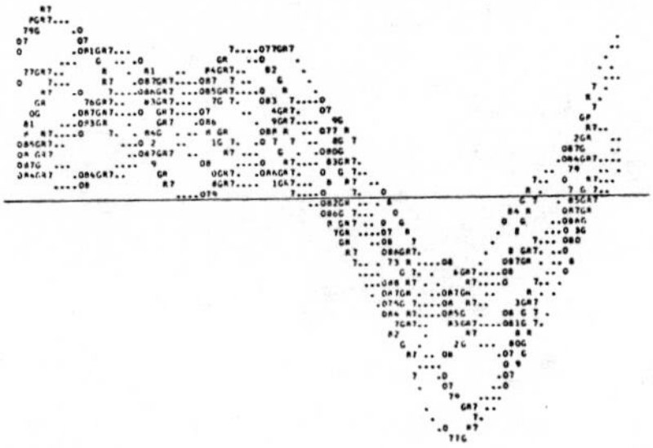

1.8

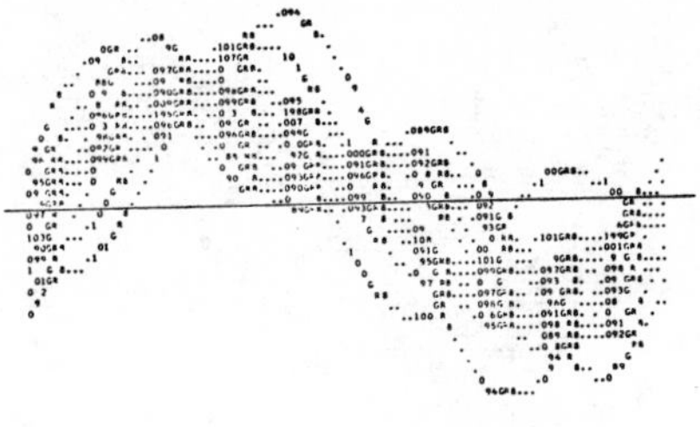

1.9

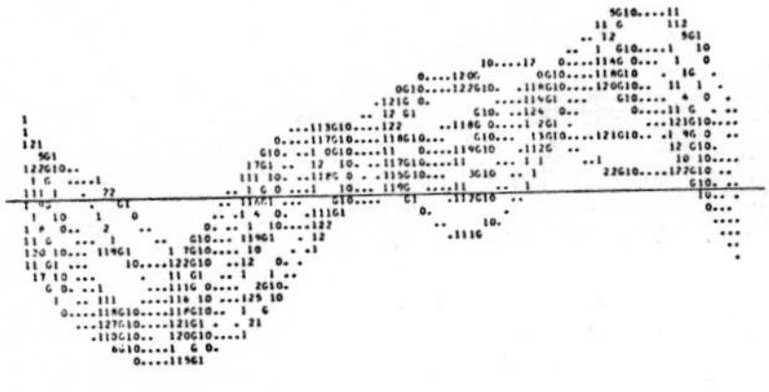

1.10

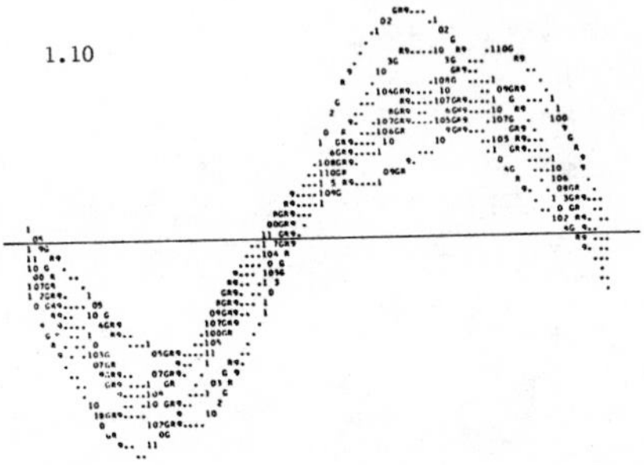

1.11

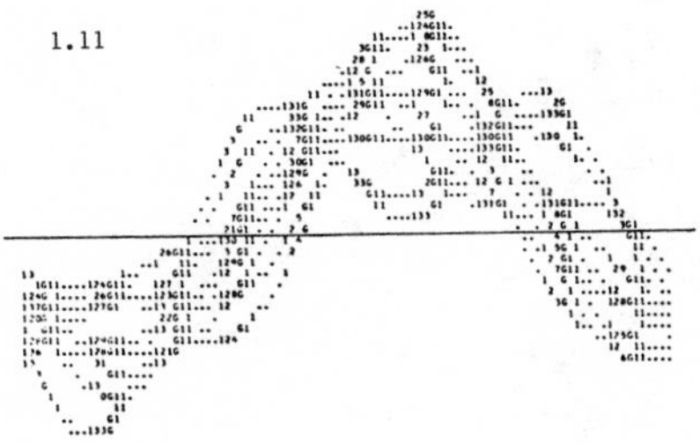

1.12

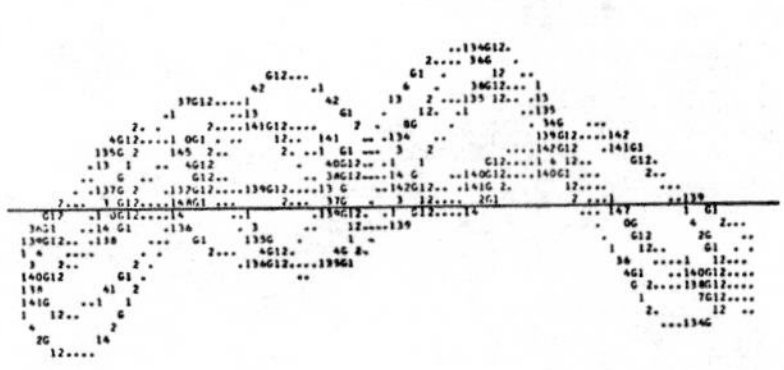

Figs. 1.7-1.12
155 British Towns, Clustered by Moser and Scott

1.13

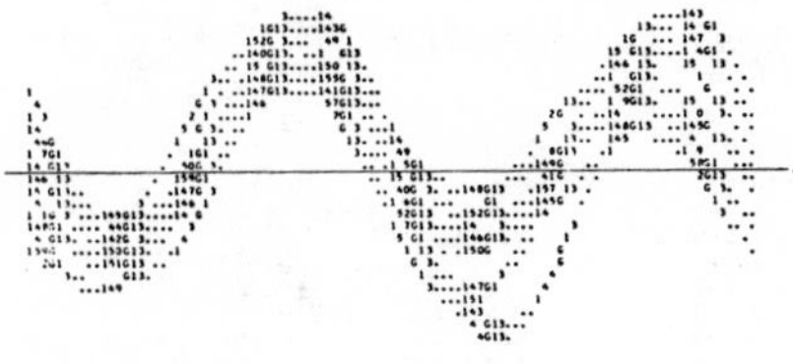

1.14

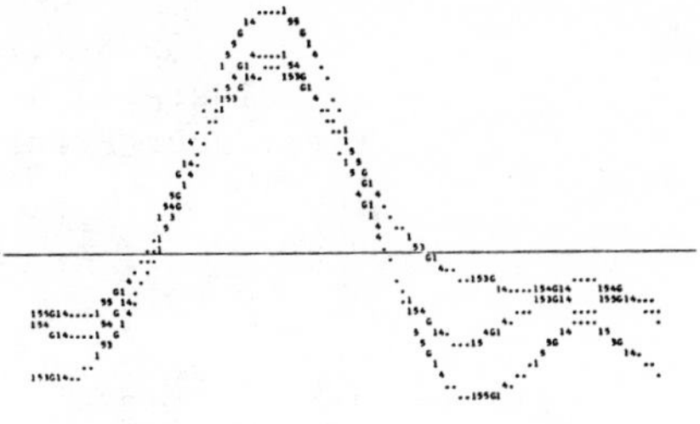

1.15

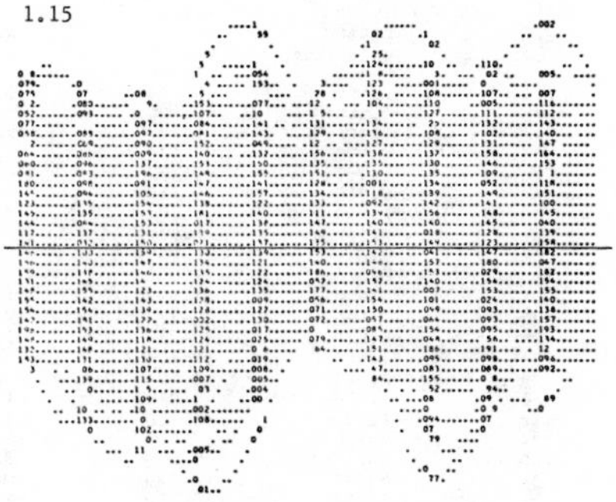

Figs. 1.13-1.15
155 British Towns, Clustered by Moser and Scott

Table 2

155 British Towns
Clustered by H. Andrews
First Four Principal Components

N =155K = 4SCALE = 0.03000SHIFT = 0.0

	123	2.100	6.040	2.600	1.360
	124	4.750	4.240	3.440	0.620
	125	4.260	5.520	3.040	-0.470
	126	1.930	5.850	2.180	-0.720
2.1	127	1.280	5.110	0.520	0.890
	128	5.100	6.030	1.410	0.660
	129	3.830	4.600	0.280	0.310
	136	1.280	2.770	2.960	0.900
	138	0.940	3.300	1.410	-0.960
	18	1.930	-1.160	2.310	-1.730
	19	1.930	-1.340	3.770	-2.540
	21	-0.860	-0.690	3.170	0.050
	22	0.280	-1.580	2.220	-0.980
	23	-0.800	-1.290	2.440	-0.280
	24	0.400	-0.700	0.420	-0.530
	25	-0.360	-0.990	2.270	-1.170
	26	0.200	-2.620	2.800	-2.450
	27	0.490	-2.570	1.690	-2.050
	28	0.240	-0.780	2.890	-2.520
	29	-0.730	-1.810	2.610	-2.430
	30	-0.030	-1.650	2.670	-2.100
2.2	31	0.420	-1.550	1.420	-0.660
	32	0.640	-1.220	-0.300	-2.760
	33	-0.520	-2.030	0.990	-0.620
	34	-0.090	-1.690	1.360	-0.660
	35	0.440	-0.720	0.890	-2.340
	36	-1.020	-0.710	1.260	1.250
	38	-1.480	-0.470	0.660	-0.700
	39	-1.680	-0.430	3.360	-2.210
	42	-1.910	-1.070	1.730	-0.070
	48	-2.270	-1.310	1.160	-1.020
	134	0.620	2.190	4.430	-1.120
	135	0.770	0.840	3.100	-1.670
	139	0.940	0.180	-0.210	-1.460
	111	2.550	-0.330	-3.890	0.760
	141	0.550	2.430	-1.760	-1.830
	142	0.340	3.300	-3.020	-2.090
	143	-1.060	-0.670	-5.700	2.410
	144	-2.260	-0.930	-5.380	1.200
	145	-1.080	0.650	-2.990	0.310
2.3	146	0.720	-0.360	-4.010	-0.270
	147	0.210	-1.500	-4.930	-0.040
	148	-0.300	1.810	-2.460	-0.870
	149	-0.360	2.340	-4.020	-0.660
	150	-0.540	0.690	-4.120	-0.670
	151	-0.620	-0.700	-5.340	-1.690
	152	-0.420	1.140	-2.940	-2.850
	11	3.760	-2.130	0.710	-0.320
	12	2.050	-3.740	3.440	0.810
	13	2.680	-0.150	2.700	-0.800
	14	2.190	-1.870	0.400	-0.890
	15	3.670	-1.950	1.980	-1.090
2.4	16	2.440	-2.130	2.170	-0.690
	17	2.070	-1.440	2.350	-0.960
	20	4.790	-1.340	1.220	1.020
	119	3.510	-0.190	-0.800	-0.310
	121	4.570	0.240	0.920	2.340
	122	3.110	0.320	1.900	1.240
	7	6.740	-3.850	-0.840	1.060
	10	5.240	-3.480	-0.100	0.740
2.5	112	4.730	-0.800	-3.660	2.000
	115	6.190	-1.210	-2.380	1.360
	94	-7.350	2.920	-1.280	3.010
	101	-3.820	3.190	-2.510	-0.610
2.6	153	-2.820	5.950	-2.630	-0.610
	154	-4.090	4.220	-3.760	0.320
	155	-6.320	4.760	-3.910	1.460
	37	-2.610	0.470	0.280	-3.190
	40	-1.630	-0.060	1.980	-2.340
	43	-2.140	0.420	1.400	-1.560
	44	-2.440	0.450	0.940	-1.470
	45	-2.430	-0.320	1.800	-2.160
	46	-3.040	1.140	1.440	-2.200
2.7	50	-2.580	1.880	1.600	-1.520
	89	-4.310	2.490	4.400	-2.310
	91	-3.820	2.420	2.720	-1.130
	92	-4.190	3.170	1.900	-2.330
	93	-4.270	1.180	1.640	-1.770
	137	-1.400	1.600	0.760	-2.840
	140	-1.120	2.200	0.050	-1.690

Table 2 (Continued)

	1	9.550	-4.370	3.090	1.520
	2	8.720	-5.700	-0.800	3.230
	3	5.900	-5.410	1.810	1.680
	4	7.580	-4.310	1.340	-0.270
2.8	5	7.240	-5.810	1.300	2.850
	6	6.510	-4.840	1.240	1.980
	8	7.050	-4.490	1.350	1.400
	9	3.490	-4.890	1.130	2.280
	51	-3.220	0.110	0.740	1.280
	53	-3.660	-1.780	3.240	0.440
	54	-4.260	-1.430	1.230	1.520
	55	-3.290	-0.550	1.540	3.170
	56	-3.270	-0.280	0.870	1.200
	57	-2.570	-0.210	0.660	0.790
	80	-5.620	-0.500	3.180	2.160
	81	-6.200	-1.400	2.090	1.040
	83	-6.110	0.800	-0.200	1.440
2.9	85	-6.160	-1.380	-0.820	0.290
	86	-4.660	-0.890	0.650	-0.160
	87	-5.790	-0.950	-0.810	-0.200
	88	-4.650	1.120	3.340	0.530
	90	-5.070	-0.080	3.520	1.520
	95	-5.570	1.070	-0.240	1.390
	96	-4.880	0.210	0.670	1.480
	97	-5.920	0.460	-0.910	-0.720
	98	-5.740	1.800	0.320	0.690
	99	-4.480	0.730	-0.460	-0.280
	60	-3.130	-1.290	-1.320	0.770
	62	-2.010	-3.860	-1.910	0.890
	63	-2.290	-3.350	-1.550	1.520
	64	-4.600	-3.040	-0.070	1.380
2.10	65	-2.590	-3.370	-1.190	0.810
	71	-2.130	-2.000	-1.200	0.810
	72	-3.470	-2.400	-2.440	-0.890
	82	-5.640	-0.630	-3.680	2.010
	84	-4.600	-0.820	-3.220	0.320
	102	10.300	3.440	1.730	0.620
	103	9.060	2.030	1.080	0.400
	104	6.360	2.550	1.170	0.370
	105	6.240	0.560	0.450	0.830
2.11	106	7.130	2.270	-1.250	1.030
	107	7.290	2.150	-0.710	0.240
	108	8.060	3.320	-1.870	-0.200
	109	6.940	0.660	-2.480	1.130
	110	9.340	1.570	-2.980	1.020
	113	2.600	1.590	-2.960	-0.180
	114	4.570	1.430	-3.010	0.830
	116	4.290	2.580	-2.260	-0.500
	117	3.030	2.240	-2.470	-1.160
2.12	118	4.760	1.130	-1.580	0.340
	120	5.380	2.880	-0.960	0.600
	130	4.690	3.960	-2.120	0.250
	131	2.720	4.460	-1.960	2.230
	132	4.000	4.220	-4.180	-1.550
	133	2.800	4.370	-4.670	-1.730
	52	-3.790	-1.800	0.910	4.500
	58	-3.670	-1.510	-0.980	4.260
2.13	77	-7.510	-0.440	-1.590	5.530
	78	-5.980	-2.440	0.970	5.700
	79	-6.170	-1.410	1.920	5.820
	41	-2.790	-1.340	-0.040	-2.740
	47	-2.540	-0.150	-1.380	-0.770
	49	-1.030	-0.790	-1.030	-1.430
	59	-3.070	-0.380	-0.270	-0.700
	61	-1.260	-2.290	-1.670	-0.600
	66	-1.640	-1.700	-1.710	-2.130
	67	-1.420	-1.050	-1.940	-2.170
2.14	68	-2.640	-2.040	-1.640	-3.490
	69	-0.920	-1.860	-1.860	-3.430
	70	-0.980	-1.840	-1.010	-2.580
	73	-2.760	-1.410	-0.960	-1.310
	74	-3.050	-3.080	-2.340	-2.070
	75	-0.980	-1.730	-0.400	-2.140
	76	-2.800	-1.770	0.190	-1.390
	100	-3.850	-0.340	-3.190	-2.310

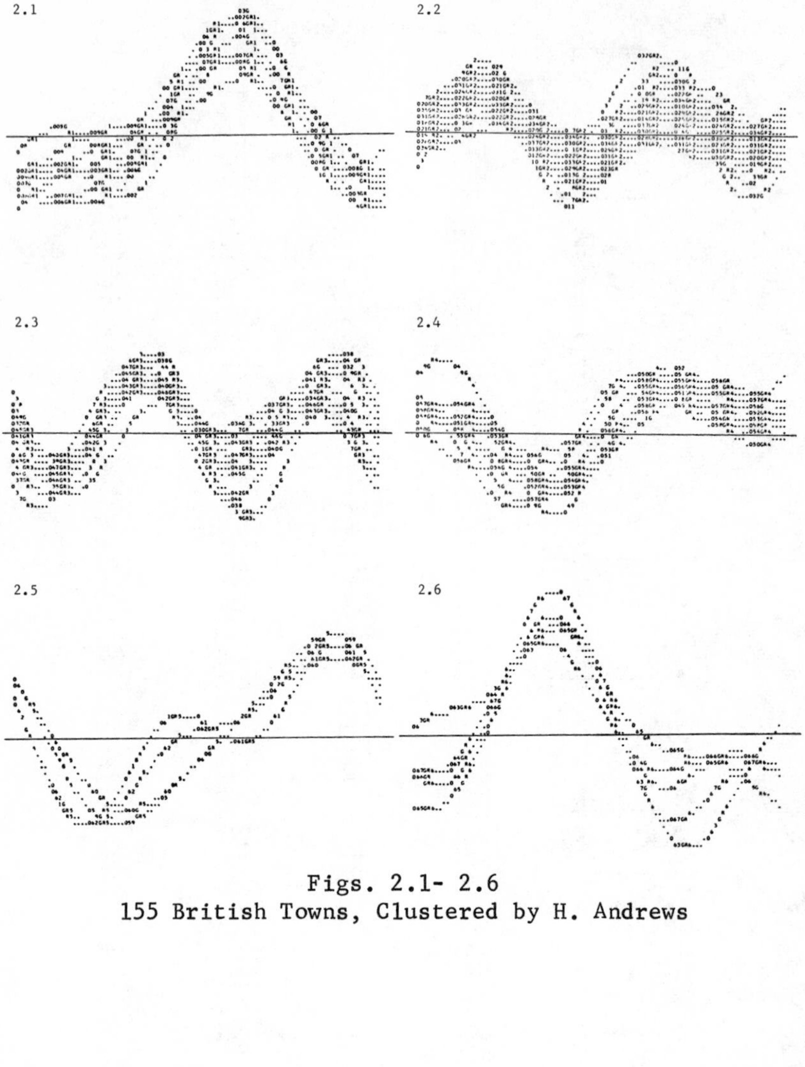

Figs. 2.1- 2.6
155 British Towns, Clustered by H. Andrews

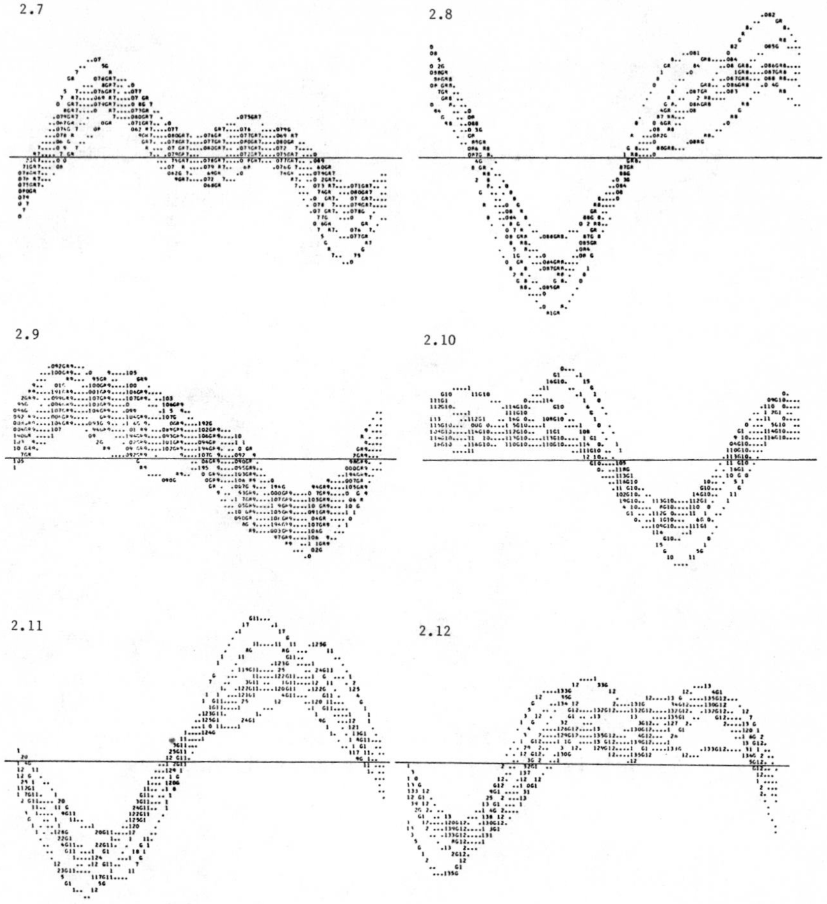

Figs. 2.7-2.12
155 British Towns, Clustered by H. Andrews

2.13

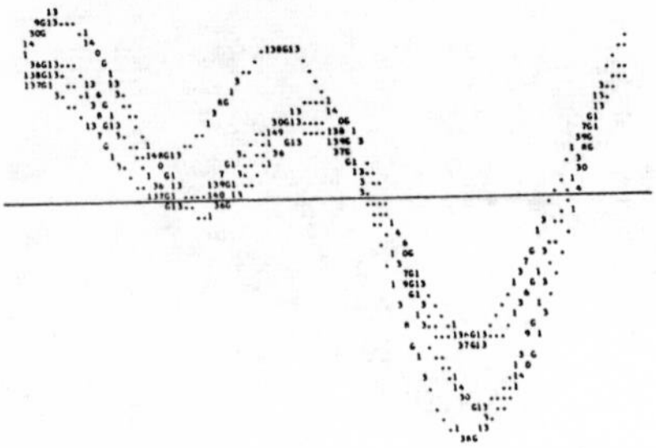

2.14

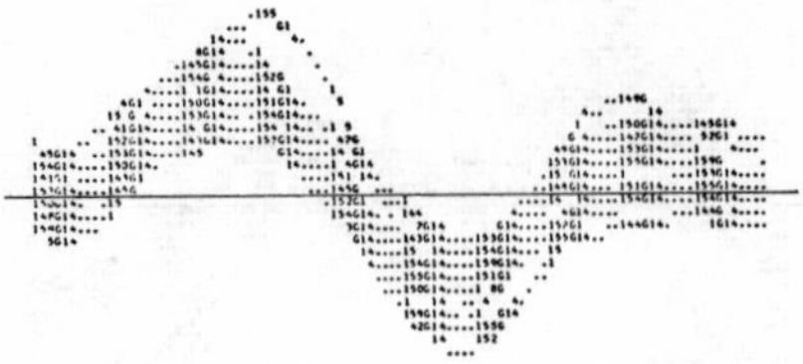

2.15

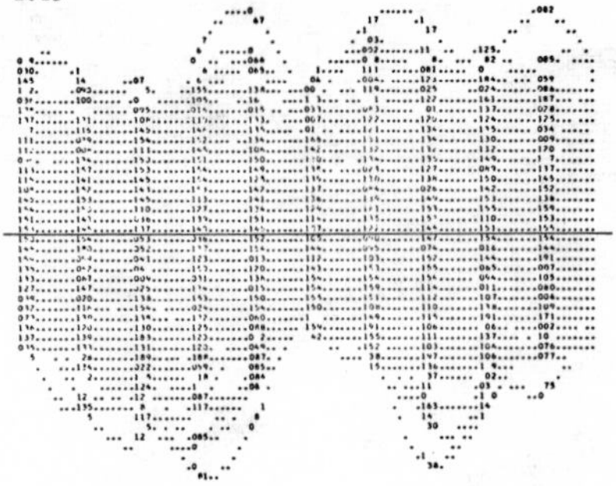

Figs. 2.13-2.15
155 British Towns, Clustered by H. Andrews

3.1

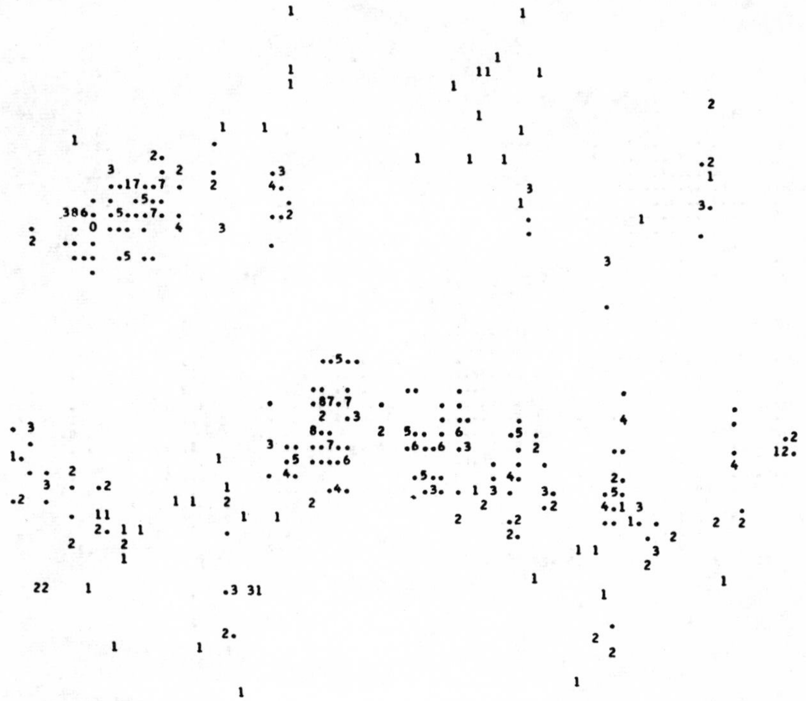

Fig. 3.1
Plot of Local Densities and Nearest Neighbours
(See Section 3.1)

3.2

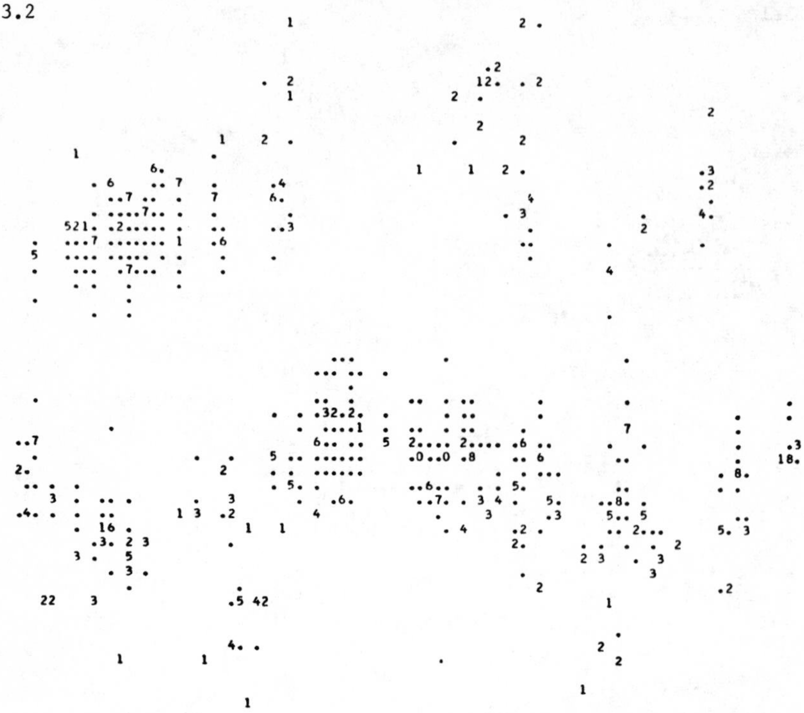

Fig. 3.2
Plot of Local Densities and Nearest Neighbours
(See Section 3.1)

3.3

Fig. 3.3
Plot of Local Densities and Nearest Neighbours
(See Section 3.1)

DISTANCE, DISCRIMINATION AND ERROR*

T. Cacoullos
University of Athens

Introduction

Optimum procedures in discriminant analysis for normal alternative populations with identical dispersion matrices are intimately related to decision rules for choosing, among several populations, the nearest to another population in the sense of the Mahalanobis (squared) distance Δ_{ij}^2 between $\Pi_i:N(\mu_i,M)$ and $\Pi_j:N(\mu_j,M)$, defined by $\Delta_{ij}^2 = (\mu_i-\mu_j)'M^{-1}(\mu_i-\mu_j)$. For example, it was shown by Cacoullos (1965a) that under certain conditions on the configuration of the points $\mu_1,\ldots,\mu_k$ in the p-space of the normal distributions, the admissible minimax classification procedure for assigning a p-component vector observation x to one of the alternative populations $\Pi_i:N(\mu_i,M)$, $i=1,\ldots,k$, is also the admissible invariant minimax procedure for the problem of selecting the nearest (called *topothetical*) of the Π_i to $\Pi:N(\mu,M)$, from which the observation was taken. Of course, the classification or discrimination (also referred to as the identification) problem is the special case of the topothetical problem in which the unknown mean μ is restricted to one of the μ_i, $i=1,\ldots,k$ (see Theorem 2.1 below). Admissible invariant Bayes topothetical procedures were given by Cacoullos (1965b) when the μ_i are unknown and μ and M are known.

Desirable properties of decision rules based on sample analogues of the Mahalanobis distance result from the fact that it emerges as the natural measure of dissimilarity (distance) between homoscedastic normal populations. Thus, it is equivalent, in this case, to the

---

*This work was partially supported by NATO Grant No.417.

Kullback-Leibler information measure, to Jeffreys divergence and to Bhattacharya's measure of divergence $\rho(f_1,f_2)$ between two densities f_1,f_2. This is known as the affinity between distributions and is defined by $\rho(f_1,f_2) = \int\sqrt{f_1(x)}\,\sqrt{f_2(x)}\,dx$. It has been extended to several distributions and been used extensively by Matusita for the study of several statistical inference problems, including a discrete distribution classification problem (Matusita, 1956). For normal distributions with the same dispersion matrix, the Mahalanobis squared distance Δ^2 is equal to $-8 \log \rho$ (see, e.g., Matusita, 1966).

Analogues of the Mahalanobis distance also appear in the discrimination problem for infinite-dimensional normal distributions, i.e., Gaussian processes (see Rao and Varadarajan, 1963). A special type of stochastic process which admits the same treatment as the finite-dimensional normal case is the normal p-dimensional diffusion process or Wiener process with drift. The reduction of the corresponding topothetical or identification problem to the standard p-variate normal case is indicated in the next section.

The main purpose of this paper is to give ways of evaluating lower bounds for the probability of correct decision under the minimum-distance classification or topothetical rule and the assumption that the populations are apart by certain given minimum distance. Equivalently, it is shown how sample sizes can be chosen in advance to guarantee a minimum given probability P^* of correct decision, or control the misclassification error rate.

1. *The problem for diffusion processes*

Let $X_i(t)$, $0 \leq t \leq t_i$, $i=1,\ldots,n$, be n independent realizations of a p-dimensional normal diffusion process (Wiener process with drift), $X(t)$ with $X(0)=0$, mean vector $t\mu$ and dispersion matrix tM. $X(t)$ will be identified with a normal population $\Pi:N(t\mu,tM)$, where the drift parameter μ per unit of time is unknown. Given k $(k\geq 2)$ diffusion processes $\Pi_i:N(t\mu_i,tM)$, $i=1,\ldots,k$, with common dispersion matrix M, we wish to select the nearest Π_i to Π in the sense of the Mahalanobis distance, that is, such that

$\Delta_i^2 = \min_{1\leq j\leq k} \Delta_j^2$ where the Mahalanobis (squared) distance Δ_i^2 between Π_i and Π is defined by $\Delta_i^2 = (\mu_i-\mu)'M^{-1}(\mu_i-\mu)$. The corresponding classification or discrimination problem would be the special case of identifying Π with one of the Π_i, i=1,...,k, i.e., assigning X(t) to one of the processes $\Pi_1,\ldots,\Pi_k$.

Suppose the only unknown parameter is the mean vector μ. The problem can be simplified by sufficiency as follows.

First, note that each realization of X(t) in a t-interval $[0,\tau]$ can be replaced by the value of $X(\tau)$ only at $t = \tau$. This is so because X(t) is a process with independent increments, i.e., for every $t < t'$, $X(t')-X(t)$ is independent of X(t). Hence, for any $0 = t_0 < t_1 < t_2 <\ldots< t_r$, the random variables $X(t_1),X(t_2)-X(t_1),\ldots, X(t_r)-X(t_{r-1})$ are independently normally distributed with means $t_1\mu,(t_2-t_1)\mu,\ldots,(t_r-t_{r-1})\mu$ and dispersion matrices $t_1M,(t_2-t_1)M,\ldots,(t_r-t_{r-1})M$, respectively; however, the sum

$$\sum_{i=1}^{r}[X(t_i)-X(t_{i-1})] = X(t_r)$$

is a sufficient statistic for μ. Thus, as far as inferences on μ are concerned, a realization $X(t_1),\ldots,X(t_r)$ of X(t) with $t_1 < t_2 < \ldots < t_r$ can be replaced by the value $X(t_r)$ of X(t) at the largest (last) paremeter point t_r. The argument extends to the case in which X(t) is observed over the entire interval $[0,\tau]$ when the above sum is replaced by

$$\int_0^\tau X(t)dt = X(\tau)-X(0) = X(\tau) .$$

Second, if n independent realizations $X_i(t)$, $0 \leq t \leq t_i$, of X(t) are given, i = 1,...,n, then the sum $X_1(t_1) +\ldots+ X_n(t_n)$ or the sample mean

$$\overline{X} = \frac{1}{n}\sum_{i=1}^{n} X_i(t_i)$$

is a sufficient statistic for μ. $\overline{X}$ is distributed as $N(\overline{t}\mu, \overline{t}/n\ M)$, where $n\overline{t} = t_1 +\ldots+ t_n$. Clearly, changing the scale of t, we can take $\overline{t} = 1$.

Moreover, under the assumption that the variance-covariance matrix M is known, it can be taken equal to the identity matrix I, since there exists a nonsingular matrix C such that $CMC' = I$ and the problem is invariant under the group of affine transformations

$$X(t) \to AX(t) + b, \tag{1.1}$$

where A is any nonsingular (constant) p×p matrix and b a constant p-component vector (see Cacoullos, 1965a).

2. Case of known M and $\mu_1,\ldots,\mu_k$

The preceding discussion shows that the problem for diffusion processes can be cast into the usual form for p-variate normal populations: On the basis of a random sample $X_1,\ldots,X_n$ from $\Pi:N(\mu,I)$, choose that $\Pi_i:N(\mu_i,I)$ which is closest to Π in the sense of the Mahalanobis distance, which, in this case (M=I), becomes the usual Euclidean distance $d(\mu_i,\mu) = |\mu_i-\mu|$. Thus we want to choose an i such that

$$d(\mu_i,\mu) \leq d(\mu_j,\mu), \quad j \neq i. \tag{2.1}$$

This problem was studied in detail by Cacoullos (1965a) from the point of view of optimal procedures (admissible minimax and Bayes), mainly under the assumption of maximum dimensionality of the linear space spanned by the points $\mu_1,\ldots,\mu_k$. In this case, the k points $\mu_1, \ldots,\mu_k$ can be regarded as the vertices of a (k-1)-simplex, σ_{k-1} say; for k=3, σ_2 is a triangle, for k=4, σ_3 is a tetrahedron, etc.; moreover, there is a unique point $\mu = \mu_0$ in the parameter space Ω of μ equidistant from the k population means $\mu_1,\ldots,\mu_k$, i.e., with $d(\mu_i,\mu_0) = d(\mu_j,\mu_0)$ for all $i,j = 1,\ldots,k$. Then the parametric subset of Ω

$$\omega_i = \{\mu:\delta_{ij}(\mu) \equiv d^2(\mu,\mu_i)-d^2(\mu,\mu_j)<0, \text{ all } j\neq i\}, \tag{2.2}$$

where decision d_i, that Π_i is the nearest to Π_0, is the correct decision, has the boundary point μ_0 common with every other ω_j, $j\neq i$. This makes impossible the existence of a minimax procedure (cf., the type I and II error probabilities of the test of $\mu \leq 0$ versus $\mu > 0$) with minimum probability of correct decision bigger than the lower bound 1/k, which is guaranteed by the trivial (but, of course, inadmissible) completely randomized procedure which chooses each d_i with probability 1/k, motivated the introduction of an indifference zone

$$\omega_o(\lambda) = \{\mu:\delta_{ij}(\mu) < -\lambda d^2(\mu_i,\mu_j),\ j\neq i\} \tag{2.3}$$

where $\lambda > 0$ determines the "size" of the indifference zone.

Thus Ω is partitioned as follows:

$$\Omega = \omega_o(\lambda) + \omega_1(\lambda) + \ldots + \omega_k(\lambda),$$

where now decision d_i is appropriate in the parametric subset

$$\omega_i(\lambda) = \{\mu:\delta_{ij}(\mu) \leq -\lambda d^2(\mu_i,\mu_j),\ j\neq i,\ i=1,\ldots,k\} \tag{2.4}$$

instead of in ω_i defined in (2.2).

The results of the following theorem (Cacoullos, 1965a) are quite pertinent in the evaluation of the performance of the minimum distance rule (see (2.8) below).

Theorem 2.1. The class of all translations T_ω of the system $\omega = (\omega_1,\ldots,\omega_k)$ in the (k-1)-space E_{k-1} of the simplex σ_{k-1}, which by invariance under the transformations (1.1) can be identified with the sample space of $\overline{X}$ (projected on E_{k-1}), is an admissible class of invariant partitions (procedures) for the topothetical problem of locating μ into one of the regions $\omega_1(\lambda),\ldots,\omega_k(\lambda)$. Moreover, the equalizer partition $R(\lambda) \in T_\omega$, where $R(\lambda) = (R_1(\lambda),\ldots,R_k(\lambda))$ is such that

$$P[\overline{X} \in R_i(\lambda)|\mu=\mu_i(\lambda)] = \text{constant} \quad \text{for} \quad i=1,\ldots,k$$

gives a) the admissible minimax procedure for the discrimination problem with alternative normal populations centered at $\mu_1(\lambda),\ldots,\mu_k(\lambda)$ with $\mu_i(\lambda)$ as the solution in μ of the equations which define boundaries (hyperplanes) of the indifference region in (2.3), that is of

$$\delta_{ij}(\mu) = -\lambda d^2(\mu_i,\mu_j),\quad j\neq i\quad (i=1,\ldots,k) \tag{2.5}$$

b) the admissible invariant minimax procedure for the topothetical problem with indifference region $\omega_0(\lambda)$.

It should be recalled that a partition $R = (R_1, \ldots, R_k) \in T_\omega$ can also be defined by

$$R_i = \{\overline{x}_n : \delta_{ij}(\overline{x}) \leq c_i - c_j,\ j \neq i\} \quad (i=1,\ldots,k) \qquad (2.6)$$

where the sample analogue $\delta_{ij}(\overline{X})$ of the distance differences $\delta_{ij}(\mu)$,

$$\delta_{ij}(\overline{X}) = (2\overline{X} - \mu_i - \mu_j)'(\mu_j - \mu_i),$$

is distributed as $N(\delta_{ij}(\mu), \frac{4}{n} d^2(\mu_i, \mu_j))$ and for each i the joint distribution of the $\delta_{ij}(\overline{X}_n)$, $j \neq i$, is a (k-1)-variate normal. The $\delta_{ij}(x)$ are usually referred to as the linear discriminant functions. Thus the minimax procedure requires the use of tables of a (k-1)-variate normal distribution function and moreover trial and error iterative procedures (see, e.g., Anderson, 1958, Chapter 6). On the other hand, the lower bound $P_i(R)$, say, of the probability of correctly taking decision d_i that Π_i is the nearest population to Π_0 under the indifference region $\omega_0(\lambda)$ and when using procedure R in (2.6) is (cf. Theorem 2.1b)

$$\begin{aligned} P_i(R) &= \inf_{\mu \varepsilon \omega_i(\lambda)} P[\overline{X}_n \varepsilon R_i | \mu] = P[\overline{X}_n \varepsilon R_i | \mu = \mu_i(\lambda)] \\ &= P[\delta_{ij}(\overline{X}) \leq c_i - c_j,\ j \neq i | \mu = \mu_i(\lambda)] \ . \end{aligned} \qquad (2.7)$$

Thus again tables of the (k-1)-variate normal distribution function are required for its evaluation. The same difficulty arises if one uses the minimum-distance procedure $R^o = (R_1^o, \ldots, R_k^o)$, defined by

$$R_i^o : \delta_{ij}(\overline{x}) = d^2(\overline{x}, \mu_i) - d^2(\overline{x}, \mu_j) < 0 \ , \ (j \neq i) \quad (2.8)$$

This procedure maximizes the average probability of correct decision (the Bayes procedure under equal a priori probabilities and 0-1 loss function).

If our goal is to control the error rate or, equivalently, achieve a preassigned bound $P^*(1/k < P^* < 1)$ for the minimum probability P_0 of correct decision under R^o, that is,

$$P_o = \min_{1 \le i \le k} P_i(R^o) \ge P^* , \tag{2.9}$$

then the question arises as to how large a sample size n should be chosen to satisfy (2.9) for given $\lambda > 0$ (indifference region $\omega_o(\lambda)$). It was suggested in Cacoullos (1970) that n be chosen so that

$$P[|\overline{X}-\mu_i(\lambda)| \le \frac{\delta(\lambda)}{2} | \mu=\mu_i(\lambda)] \ge P^* \tag{2.10}$$

where

$$\delta(\lambda) = \min_{i \ne j} |\mu_i(\lambda)-\mu_j(\lambda)| = \lambda \min_{i \ne j} |\mu_i-\mu_j| = \lambda\delta \quad \text{(say)}, \tag{2.11}$$

since it can be shown (Cacoullos, 1965a) that, by (2.5),

$$\mu_i(\lambda) - \mu_j(\lambda) = \lambda(\mu_i-\mu_j) \quad (i,j=1,2,\ldots,k). \tag{2.12}$$

Indeed we have

$$\begin{aligned} P_i(R^o) &= P[\delta_{ij}(\overline{X})<0,\ j\ne i | \mu=\mu_i(\lambda)] \\ &= P[d(\overline{X},\mu_i)<d(\overline{X},\mu_j), j\ne i | \mu=\mu_i(\lambda)] \qquad (2.13) \\ &\ge P[|\overline{X}-\mu_i(\lambda)| < \frac{\lambda\delta}{2} | \mu=\mu_i(\lambda)] \quad (i=1,\ldots,k) \end{aligned}$$

because $|\overline{x}-\mu_i(\lambda)| < \frac{\lambda\delta}{2}$ implies $\overline{x} \in R_i^o$ since, by the triangle inequality and (2.11),

$$|\overline{x}-\mu_j(\lambda)| \ge |\mu_i(\lambda)-\mu_j(\lambda)|-|\overline{x}-\mu_i(\lambda)| > \lambda\delta - \frac{\lambda\delta}{2} = \frac{\lambda\delta}{2} \quad (j\ne i).$$

Since $\overline{X}$ is $N(\mu, \frac{1}{n} I_{k-1})$, the probability requirement (2.9) will be satisfied if (2.10) is satisfied, that is, if

$$P[\chi^2_{k-1} \le \frac{1}{4} n^2\delta^2\lambda^2] \ge P^* \tag{2.14}$$

where χ^2_ν denotes a chi-square random variable with ν degrees of freedom. Thus we have shown the following

Proposition 2.1. If $X_1,\ldots,X_{n_o}$ observations are taken from $\Pi:N(\mu,I)$ where n_o is the smallest value of n satisfying (2.14), that is,

$$G_{k-1}(\frac{1}{4}n^2\delta^2\lambda^2) \geq P^* ,$$

where G_ν denotes the distribution function of a chi-square with ν degrees of freedom, then the minimum distance rule R^o chooses the nearest neighbor of Π_o among the $\Pi_i:N(\mu_i,I)$ with probability of correct selection at least P^*, provided the indifference zone is given by (2.3) and

$$\min_{i\neq j} \Delta_{ij} = \min_{i\neq j}|\mu_i-\mu_j| = \delta .$$

The solution for the discriminant problem with alternative populations Π_i (i.e., μ is one of the $\mu_1,\ldots,\mu_k$) is also given by (2.14) with $\lambda = 1$.

It is worth noting that for the discrimination problem the preceding arguments are still valid under a general configuration of the mean vector points $\mu_1,\ldots,\mu_k$; that is, if the dimensionality of the linear space spanned by these points is $s < \min(k-1,p)$ (e.g., three collinear points, four coplanar points, etc.), then again (2.14) gives the required sample size n if we take $\lambda = 1$ and $k-1 = s$. This becomes very useful in view of the fact that the exact probabilities P_i are very difficult to evaluate in this case because of the shape of the corresponding classification regions R_i^o. When the points $\mu_1,\ldots,\mu_k$ define a (k-1)-simplex, the regions R_i^o are convex polyhedral cones with common vertex the equidistant point μ_o.

An alternative way of obtaining a lower bound for P_o is by using Bonferoni-type inequalities. Thus we have, e.g.,

$$P_1(R^o) = P[\delta_{1j}(\bar{X})<0,\ j\neq 1|\mu=\mu_1] \geq 1 - \sum_{j=2}^{k} P[\delta_{1j}(\bar{X})\geq 0|\mu=\mu_1]$$

$$= 1 - \sum_{j=2}^{k} \Phi(-\frac{\sqrt{n}}{2}\Delta_{1j}) \geq 1 - (k-1)\Phi(-\frac{\sqrt{n}}{2}\delta_1)$$

where $\delta_1 = \min(\Delta_{12},\Delta_{13},\ldots,\Delta_{1k})$. Thus, taking $\delta = \min(\delta_i,\ldots,\delta_k)$, we obtain (cf. Lachenbruch, 1973)

$$P_o \geq 1 - (k-1)\Phi(-\frac{\sqrt{n}}{2}\delta). \qquad (2.15)$$

This bound is better for certain k and δ than the corresponding one provided by (2.14) with $\lambda = 1$; for example, for $k = 2$, (2.15) gives the exact minimax probability $\Phi(c)$ with $c = \sqrt{n}\,\delta/2$, whereas (2.13) gives $P_o = \Phi(c) - \Phi(-c)$; for $k = 3$ we have

$$P[X_2^2 \leq c^2] < P[X_1^2 \leq c^2] = \Phi(c) - \Phi(-c) = 1 - 2\Phi(c) ,$$

that is, again (2.15) is better than (2.14). However, for larger values of k, (2.14) will, in general, give better bounds than (2.15).

3. *The case of unknown means and known dispersion matrix*

When, in addition to μ, the mean vectors $\mu_1,\ldots,\mu_k$ are also unknown, we assume that a sample of size n_i is available from population Π_i so that $\overline{x}_i$ will be the corresponding estimate of μ_i, $i=1,\ldots,k$. The minimum-distance procedure

$$R' = (R_1',\ldots,R_k')$$

in this case both for the discrimination problem of assigning $\overline{x}$ to one of the $\Pi_i:N(\mu_i,I)$ and the problem of the nearest neighbor to $\Pi:N(\mu,I)$ is defined by

$$R_i' : d(\overline{x},\overline{x}_i) \leq d(\overline{x},\overline{x}_j), \quad j\neq i \quad (i=1,\ldots,k). \qquad (3.1)$$

The evaluation of the probabilities of correct selection requires the distribution of the analogue $d_{ij}(\overline{X})$ of $\delta_{ij}(\overline{X})$, i.e., of

$$d_{ij}(\overline{X}) = d^2(\overline{X},\overline{X}_i) - d^2(\overline{X},\overline{X}_j) = (2\overline{X}-\overline{X}_i-\overline{X}_j)'(\overline{X}_j-\overline{X}_i) \qquad (3.2)$$

which, as a difference of two non-central χ^2 variates, is exceedingly complicated. Therefore, to guarantee a given P^* as in the previous section, we will choose the sample sizes n and n_i by using a similar argument to the one employed for the choice of n when the means $\mu_1,\ldots,\mu_k$ were known.

We will treat the discrimination problem assuming that the minimum value of $\Delta_{ij} = |\mu_i-\mu_j|$ is at least δ, where δ is given. The corresponding topothetical problem does not lend itself to a similar treatment when the means are unknown.

For any β, $0 < \beta < 1$, let N_i be the smallest value of n_i which satisfies

$$P[d(\overline{X}_i,\mu_i) < \delta/4] = P[\chi_p^2 < \frac{n_i\delta^2}{16}] \geq \beta, \quad i=1,\ldots,k, \tag{3.3}$$

where we used the fact that $\overline{X}_i$ is $N(\mu_i, \frac{1}{n_i} I)$ and hence

$$n_i d^2(\overline{X}_i,\mu_i) = n_i(\overline{X}_i-\mu_i)'(\overline{X}_i-\mu_i) \tag{3.4}$$

has the χ^2 distribution with p degrees of freedom. Similarly, let N_0 be the smallest value of n which satisfies

$$P[d(\overline{X},\mu) < \frac{\delta}{4}] = P[\chi_p^2 < \frac{n\delta^2}{16}] \geq \beta . \tag{3.5}$$

From (3.3) and (3.4), it follows that $N_i = N_0$, $i=1,\ldots,k$.

Take a sample of size N_0 from each of the k+1 populations $\Pi_0,\Pi_1,\ldots,\Pi_k$, where N_0 corresponds to $\beta^{k+1} = P^*$. We will show that

$$P_i(R') = P\,[\text{correctly taking decision } d_i \text{ under } R'] \geq P^* \quad i = 1,\ldots,k . \tag{3.6}$$

For this it is enough to show that the region

$$S_i = \{(\overline{x},\overline{x}_1,\ldots,\overline{x}_k) : d(\overline{x},\mu_i) < \frac{\delta}{4}, \; d(\overline{x}_i,\mu_i) < \frac{\delta}{4}, \; i=1,\ldots,k\} \subset R_i' \tag{3.7}$$

so that, by the independence of $\overline{X},\overline{X}_1,\ldots,\overline{X}_k$ and in virtue of (3.4), (3.5), we will have

$$P_i(R') = P[R_i'|\mu=\mu_i] \geq P[S_i|\mu=\mu_i] = P[d(\overline{X},\mu_i) < \frac{\delta}{4}|\mu=\mu_i] \times \prod_{i=1}^{k} P[d(\overline{X}_i,\mu_i) < \frac{\delta}{4}] \geq \beta^{k+1} = P^* . \tag{3.8}$$

In order to show (3.7), note that from $d(\overline{x},\mu_i) < \frac{\delta}{4}$,

$d(\bar{x}_i,\mu_i) < \frac{\delta}{4}$, we get by the triangle inequality

$$d(\bar{x},\bar{x}_i) \leq d(\bar{x}_i,\mu_i) + d(\mu_i,\bar{x}) < \frac{\delta}{2}$$

and from $d(\bar{x}_j,\mu_j) < \frac{\delta}{4}$, $d(\bar{x},\mu_i) < \frac{\delta}{4}$ we get

$$d(\bar{x},\bar{x}_j) \geq \Delta_{ij} - d(\bar{x}_j,\mu_j) - d(\bar{x},\mu_i) > \delta - \frac{\delta}{4} - \frac{\delta}{4} = \frac{\delta}{2} .$$

Hence every sample point in S_i satisfies (3.1) and belongs to R_i'.

If the sample sizes n_i are given for $i=1,\ldots,k$, then the corresponding probabilities in (3.3) are

$$\beta_i = G_p(\frac{1}{16} n_i \delta^2) , \quad i=1,\ldots,k ,$$

and the sample size n from Π should be chosen as the smallest value of n satisfying

$$G_p(\frac{1}{16} n\delta^2) \geq \frac{P^*}{\beta_i \ldots \beta_k} .$$

It should be noted that when the β_i and P^* are such that $P^*/(\beta_1,\ldots,\beta_k)$ is not less than one, there is no n to achieve the given P^*. A similar remark applies, when n is given, as in the usual discrimination problem with $n = 1$, and it is required to choose the n_i.

4. *The case of unknown parameters*

When, in addition to the mean vectors $\mu_o,\mu_1,\ldots,\mu_k$, the common dispersion matrix M of the k+1 populations is unknown, this is estimated by the pooled sample covariance matrix S where

$$NS = \sum_{i=0}^{k} \sum_{j=1}^{n_i} (x_{ij}-\bar{x}_i)(x_{ij}-\bar{x}_i)' , \qquad (4.1)$$

$$N = n_o + n_1 + \ldots + n_k - (k+1) \quad (N \geq p)$$

and $\bar{x}_i$ the sample mean from $\Pi_i:N(\mu_i,M)$, $i=0,1,\ldots,k$.

The minimum distance classification procedure is given by (cf. (3.1))

$$R_i^* : D_{oi} \leq D_{oj} , \quad j \neq i \quad (i=1,\ldots,k)$$

where D_{ij}^2 denotes the sample Mahalanobis (squared) distance, i.e.

$$D_{ij}^2 = (\overline{x}_i - \overline{x}_j)' S^{-1} (\overline{x}_i - \overline{x}_j) \quad (i,j=0,1,\ldots,k) .$$

The analogues of the $d(\overline{X}_i, \mu_i)$ and $d(\overline{X}_o, \mu_i)$ are no longer independent because they involve S; more precisely (see, e.g., Anderson, 1958), the statistics

$$D_i^2(\overline{X}_i) = (\overline{X}_i - \mu_i)' S^{-1} (\overline{X}_i - \mu_i) \qquad i=1,\ldots,k$$

and $D_i^2(\overline{X}_o)$, under the hypothesis $\mu_o = \mu_i$, are distributed as the ratios

$$\frac{N}{n_i} \frac{\chi^2_{\nu_i}}{\chi^2_{\nu}} \qquad i=0,1,\ldots,k$$

respectively, where all the chi-square chance variables are independent with $\nu_i = p$, $i=0,1,\ldots,k$ and $\nu = N-p+1$ degrees of freedom. Hence the basic probability P^* requirement for R^*, by the analogues of (3.7) and (3.8), becomes

$$P_i(R^*) = P[D_i(\overline{X}_o) < \frac{\delta}{4} , D_j(\overline{X}_j) < \frac{\delta}{4} , j=1,\ldots,k | \mu=\mu_i]$$

$$= P[\frac{N}{n_i} \frac{\chi^2_{\nu_i}}{\chi^2_{\nu}} < \frac{\delta^2}{16} , i=0,1,\ldots,k] \geq P^* . \tag{4.2}$$

Thus, for $n_i = n$, $i=0,1,\ldots,k$, we require the distribution function of the maximum of several correlated F statistics which, inter alia, appears in several ranking and selection problems (Gupta and Sobel, 1962; Gupta, 1963; Krishnaiah and Armitage, 1964). Actually, these authors have provided related tables for the special case of equal degrees of freedom for all the χ^2, i.e., $\nu_i = p$, $i=0,1,\ldots,k$. The tables by Krishnaiah and Armitage give cut-off points for $P^* = 0.90, 0.95, 0.025, 0.01$, and by trial and error one can use these tables for the solution of (4.2), only under the restriction of equal sample sizes (cf. the remark at the end of Section 3 for the case of n or n_i fixed). Therefore, to avoid further complications, suppose all the sample sizes are equal to n. Then we have

$$N = (k+1)(n-1) \ , \quad \nu = N - p + 1 \ . \tag{4.3}$$

The distribution function of

$$Y = \max_{i=0,1,\ldots,k} \left\{ \frac{N}{n_i} \frac{\chi^2_{\nu_i}}{\chi^2_{\nu}} \right\}$$

is readily seen to be

$$H(y) = P[\chi^2_{\nu_i} \leq \chi^2_{\nu} y n_i/N \ , \ i=0,1,\ldots,k]$$

$$= \int_0^{\infty} g_{\nu}(x) \prod_{i=0}^{k} G_{\nu_i}(xyn_i/N)dx \quad ,$$

where $G_m(x)$ and $g_m(x)$ are the distribution function and density of a (central) χ^2_m, respectively. Hence, the probability requirement in (4.2) for $n_i = n$, $i=0,1,\ldots,k$ becomes

$$\int_0^{\infty} g_{\nu}(x)[G_p(xyn/N)]^{k+1}dx = P^* \tag{4.4}$$

with $y = \delta^2/16$, $\delta = \min \Delta_{ij}$ $(i \neq j)$ and ν, N as given by (4.3).

An approximation, to get around the computational difficulty of the solution of (4.4), is obtained by using Bonferoni's inequality, as in (2.15). We can write (4.2)

$$P[\frac{N}{n_i} \frac{\chi^2_{\nu_i}}{\chi^2_{\nu}} < \frac{\delta^2}{16}, \ i=0,1,\ldots,k] = P[F_{\nu_i,\nu} < \frac{\nu n_i}{\nu_i N} \frac{\delta^2}{16}, \ i=0,1,\ldots,k]$$

$$\geq 1 - \sum_{i=0}^{k} P[F_{\nu_i,\nu} > \frac{\nu n_i}{\nu_i N} \frac{\delta^2}{16}] \geq P^* \tag{4.5}$$

where $F_{s,m}$ denotes an F chance variable with s and m degrees of freedom and $\nu_i = p$, $i=0,1,\ldots,k$. Then the required common sample n^* is the smallest value of n which satisfies

$$G_{p,\nu}\left(\frac{\nu n}{pN}\frac{\delta^2}{16}\right) \geq \frac{k+P^*}{k+1}, \quad N = (k+1)(n-1), \ \nu = N-p+1, \qquad (4.6)$$

where $G_{s,m}$ denotes the distribution function of the $F_{s,m}$ chance variable. As intuitively expected, (4.6) has a solution in n for every given set of values for p, k, δ and P^*. Actually, it is easily seen that the left-hand side is an increasing function of n. This follows from the fact that as n_i gets larger $\overline{X}_i$ approaches μ_i and the corresponding D_i gets (stochastically) smaller, so that the probability $P_i(R^*)$ in (4.2), for every fixed δ, gets larger; hence the Bonferoni bounds in (4.5) get larger and the resulting $G_{p,\nu}$ in (4.6) for $n_i = n$ also gets larger.

Remark 1. In the usual classification problem, when an observation x is to be assigned to one of the k populations $\Pi_i,\ldots,\Pi_k$, we have $n_0 = 1$ and, therefore, S will be estimated as in (4.1) where the summation will be over $i=1,\ldots,k$; then $N = n_1 +\ldots+ n_k - k$. If the same sample size n is to be chosen from each Π_i, then, to achieve a P^*, one has to solve for n the analogue of (4.4), i.e.

$$\int_0^\infty g_\nu(x)G_p(xy/N)[G_p(xyn/N)]^k = P^*, \ N = k(n-1), \ \nu = N-p+1.$$

As already remarked at the end of section 3, such an n may not exist.

Remark 2. We saw that the common sample size n required to achieve the desired probability P^*, say 0.90 or 0.95, can be determined, provided we are given a lower bound δ of the minimum of Δ_{ij}. If the experimenter cannot choose in advance a δ (a measure of the indifference zone introduced in the parameter space), then one can use the estimate $d = \min D_{ij}$ obtained on the basis of the samples from $\Pi_1,\ldots,\Pi_k$. Using d instead of δ one can proceed as above to determine the required sample size n from Π_0; in this respect, the remark at the end of Section 3.1 applies. Of course, the effects of such an approximation, both on the resulting n and the lower bounds of the probability of correct decision as exhibited above, are, no doubt, hard to evaluate (cf. T.W. Anderson, 1973).

REFERENCES

Anderson, T. W. (1973), Asymptotic evaluation of the probabilities of misclassification by linear discriminant functions, this Volume.

Anderson, T. W. (1958), *An Introduction to Multivariate Statistical Analysis*, Wiley: New York.

Armitage, J. V. and Krishnaiah, P. R. (1964), Tables for the studentized largest chi-square distribution and their applications, ARL64-188, Aerospace Research Laboratories, Wright-Patterson Air-Force Base, Dayton, Ohio.

Cacoullos, T. (1965a), Comparing Mahalanobis distances I: Comparing distances betweek k known normal populations and other unknown, *Sankhyā* A, 27, 1-22.

Cacoullos, T. (1965b), Comparing Mahalanobis distances II: Bayes procedures when the mean vectors are unknown. *Sankhyā* A, 27, 23-32.

Cacoullos, T. (1970), Some remarks on topothetical procedures, Proceedings ISI 38th Session, Washington, D. C., *Bull. Intern. Stat. Instit.*, I, 128-131.

Gupta, S. S. (1963), On a selection and ranking procedure for gamma populations, *Ann. Inst. Stat. Math.*, 14, 199-216.

Gupta, S. S. and Sobel, M. (1962), On the smallest of several correlated F statistics, *Biometrika*, 49, 509-523.

Lachenbruch, P. A. (1973), Some results on the multiple group discriminant problem, this Volume.

Matusita, K. (1956), Decision rule, based on the distance for the classification problem, *Ann. Inst. Stat. Math.*, 8, 67-77.

Matusita, K. (1966), A distance and related statistics in multivariate analysis, *Multivariate Analysis* (P. R. Krishnaiah, Ed.), Academic Press: New York, 187-200.

Rao, C. R. and Varadarajan, V. S. (1963), Discrimination of Gaussian processes, *Sankhyā* A, 25, 303-330.

THEORIES AND METHODS IN CLASSIFICATION: A REVIEW

*Somesh Das Gupta**
University of Minnesota

1A. Introduction

In this review paper I have restricted my attention only to major theoretical papers. However I have tried to be objective, as far as possible, in selecting papers from the enormous bulk of literature in this area. The recently published bibliography on multivariate analysis (Anderson, Das Gupta, and Styan, 1972) lists over 400 papers published before 1967 in the area of classification and discrimination. Moreover, some results are available in the well-known textbooks by Anderson (1958) and by Rao (1952), besides few books (listed in the references) completely devoted to this and allied fields. Anyway, I apologize for omitting many papers, especially many important applied papers and useful computer programs.

In the literature we find many names for this general area of problems; for example, allocation, identification, prediction, pattern recognition, selection, besides the standard terms, such as, classification and discrimination. Whatever names may be attached, it is clear that this branch has attracted many researchers from different disciplines. From the existing literature, I have extracted the main formulations of the classification problem and reviewed almost all the important resultssunder different broad categories of problems.

1B. Early History (up to 1950)

In the first survey of discriminatory analysis, Hodges (1950) aptly mentioned the following.

*Supported by U.S. Army Research Grant DA-ARO-D-31-124-70-G-102.

> In his invited address at the meeting of the Institute of Mathematical Statistics in Berkeley, California, June 16, 1949, Professor M. A. Girshick pointed out that the development of discriminatory analysis reflects the same broad phases as does the general history of statistical inference. We may distinguish a Pearsonian stage,..., followed by a Fisherian stage. Professor Girshick further notes a Neyman-Pearson stage and a contemporary Waldian stage....

In the early work, the classification problem was not precisely formulated and often confounded with the problem of testing the equality of two or more distributions; the term "discriminatory analysis" was used for both. In practice, the following scheme was generally followed for the two-population classification problem. Suppose we have three distributions F, F_1, F_2 and T_i is a test statistic designed to test the hypothesis $F = F_i$ $(i=1,2)$. The decision $F = F_i$ is taken if T_i is the smaller of T_1 and T_2; sometimes the critical values of T_i's are compared in order to take the decision. Thus statistics for testing the equality of two distributions played an important role. Generally, such a test statistic may be considered as a measure of divergence between the two distributions. Karl Pearson (in a paper by Tildesley (1921)) proposed one such measure, termed as the "coefficient of racial likeness (CRL)." This was modified by Morant in 1928 and by Mahalanobis in 1927 and 1930. Mahalanobis called his measure D^2 and suggested (1930) also some measures of divergence in variability, skewness and kurtosis and studied the distributions of these measures. In 1926, Pearson published the first considerable theoretical work on the CRL and suggested the following form for the coefficient when the variables are dependent:

$$\frac{n_1 n_2}{n_1+n_2}(\bar{x}_1-\bar{x}_2)' S^{-1}(\bar{x}_1-\bar{x}_2) ,$$

where $\bar{x}_i$ is the sample mean vector based on a sample of size n_i from the ith population $(i=1,2)$ and S is the pooled sample covariance matrix. In 1935 and 1936, Mahalanobis gave the dependent-variate versions of his D^2-statistic in the classical and the studentized forms. The distributions of these statistics were studied by Bose (1936a,b), Bose (1936), Bose and Roy (1938), and Bhattacharya and Naryana (1941). In 1931 Hotelling suggested a test statistic T^2 which is a constant multiple of the studentized Mahalanobis D^2 and obtained its null distribution.

Hodges remarked that "the first clear statement of the problem of discrimination, and the first proposed solution to that problem were given by Fisher in the middle of the 1930's...the ideas of Fisher first appeared in print in papers by other people (Barnard, 1935; Martin, 1936)." Earlier than this, Morant (1926) considered the problem of classifying a skull into Eskimo or modern English groups by two sets of tests. Fisher's own first work on the subject appeared in his paper in 1936. For the univariate two-population problem Fisher suggested a rule which classifies the observation x into the ith population if $|x-\overline{x}_i|$ is the smaller of $|x-\overline{x}_1|$ and $|x-\overline{x}_2|$. For a p-component observation vector (p>1), Fisher reduced the problem to the univariate one by considering an "optimum" linear combination (called the "linear discriminant function") of the p components. For a given linear combination Y of the p components, Fisher considered the ratio between the difference in the sample means of the Y-values and the standard error within samples of the Y-values and maximized this ratio in order to define the optimum linear combination. It turns out that the coefficients of this optimum linear combination are proportional to $S^{-1}(\overline{x}_1-\overline{x}_2)$. Incidentally, Fisher (1936) suggested a test for the equality of two normal distributions with the same unknown covariance matrix and this test is the same as the one proposed by Hotelling (1931).

The next development was influenced by the pioneering fundamental work by Neyman and Pearson (1933, 1936). For the two-population problem, Welch (1939) derived the forms of Bayes rules and the minimax Bayes rule when the distributions are known; he illustrated the theory with multivariate normal distributions with the same covariance matrix. This example was also considered by Wald (1944), who further proposed some heuristic rules by replacing the unknown parameters by their respective (maximum likelihood) estimates. Wald studied the distribution of the proposed classification statistic. Von Mises (1944) obtained the rule which maximizes the minimum probability of correct classification. The problem of classification into two normal distributions with different covariance matrices was treated by Cavalli (1945) and Penrose (1947) when p = 1 and by Smith (1947) for general p. In a series of papers, Rao (1946, 1947a, 1947b, 1948, 1949a, 1949b, 1950) suggested different methods of classification into two or more populations following the ideas of Neyman-Pearson and Wald; in particular, Rao suggested a measure of distance between two groups and considered the possibility of withholding decision (through "doubtful" regions) and preferential decisions. Rao's development is for the case when the distributions are all known. General theoretical results on the classification problem (as a special case) in the framework of decision theory are given in the book by Wald (1950) and in a paper by Wald and Wolfowitz (1950).

References (1)

Tildesley, M. L. (1921), A first study of the Burmese skull, *Biometrika*, 13, 247-251.

Pearson, K. (1926), On the coefficient of racial likeness, *Biometrika*, 18, 105-117.

Morant, G. M. (1926), A first study of crainology of England and Scotland from neolithic to early historic times, with special reference to Anglo-Saxon skulls in London museums, *Biometrika*, 18, 56- .

Mahalanobis, P. C. (1927), Analysis of race mixture in Bengal, *Jour. and Proc. Asiatic Soc. of Bengal*, 23, No.3.

Morant, G. M. (1928), A preliminary classification of European races based on cranial measurements, *Biometrika*, 20(B), 301-375.

Mahalanobis, P. C. (1930), On tests and measurements of group divergence, *Jour. and Proc. Asiatic Soc. Bengal*, 26, 541-588.

Hotelling, H. (1931), The generalization of Student's ratio, *Ann. Math. Statist.*, 2, 360-378.

Neyman, J. and Pearson, E. S. (1933a), On the problem of the most efficient tests of statistical hypotheses, *Phil. Trans. Roy. Soc.* A, 231, 281- .

Neyman, J. and Pearson, E. S. (1933b), On the testing of statistical hypotheses in relation to probability a priori, *Proc. Camb. Phil. Soc.*, 9, 492- .

Barnard, M. M. (1935), The secular variations of skull characters in four series of Egyptian skulls, *Ann. Eug.*, 6, 352-371.

Mahalanobis, P. C. (1936), On the generalized distance in statistics, *Proc. Nat. Inst. Sci. India*, 2, 49-55.

Bose, R. C. (1936a), On the exact distribution and moment coefficients of the D^2-statistic, *Sankhyā*, 2, 143-154.

Bose, R. C. (1936b), A note on the distribution of differences in mean values of two samples drawn from two multivariate normally distributed populations, and the definition of the D^2-statistic, *Sankhyā*, 2, 379-384.

Neyman, J. and Pearson, E. S. (1936), Contributions to the theory of testing statistical hypotheses, I, *Stat. Res. Memoir*, 1, 1-37.

Bose, S. N. (1936), On the complete moment coefficients of the D^2-statistic, *Sankhyā*, 2, 385-396.

Martin, E. S. (1936), A study of the Egyptian series of mandibles with special reference to mathematical methods of sexing, *Biometrika*, 28, 149-178.

Fisher, R. A. (1936), The use of multiple measurements in taxonomic problems, *Ann. Eug.*, 7, 179-188.

Fisher, R. A. (1938), The statistical utilization of multiple measurements, *Ann. Eug.*, 8, 376-386.

Bose, R. C. and Roy, S. N. (1938), The distribution of the studentized D^2-statistic, *Sankhyā*, 4, 19-38.

Welch, B. L. (1939), Note on discriminant functions, *Biometrika*, 31, 218-220.

Goodwin, C. N. and Morant, G. M. (1940), The human remains of Iron Age and other periods from Maiden Castle, Dorset, *Biometrika*, 31, 295- .

Bhattacharya, D. P. and Narayan, R. D. (1941), Moments of the D^2-statistic for populations with unequal dispersions, *Sankhyā*, 5, 401-412.

Day, B. B. and Sandomire, M. (1942), Use of the discriminant function for more than two groups, *Jour. Amer. Statist. Assoc.*, 37, 461-472.

Wald, A. (1944), On a statistical problem arising in the classification of an individual into one of two groups, *Ann. Math. Statist.*, 15, 145-162.

von Mises, R. (1945), On the classification of observation data into distinct groups, *Ann. Math. Statist.*, 16, 68-73.

Cavalli, L. L. (1945), Alumni problemi della analisi biometrica di popolazioni naturali, *Mem. Ist. Idrobiol.*, 2, 301-323.

Rao, C. R. (1946), Tests with discriminant functions in multivariate analysis, *Sankhyā*, 7, 407-413.

Rao, C. R. (1947a), The problem of classification and distance between two populations, *Nature*, 159, 30-31.

Rao, C. R. (1947b), A statistical criterion to determine the group to which an individual belongs, *Nature*, 160, 835-836.

Brown, G. W. (1947), Discriminant functions, *Ann. Math. Statist.*, 18, 514-528.

Penrose, L. S. (1947), Some notes on discrimination, *Ann. Eug.*, 13, 228-237.

Smith, C. A. B. (1947), Some examples of discrimination, *Ann. Eug.*, 13, 272-282.

Rao, C. R. (1948), The utilization of multiple measurements in problems of biological classification, *Jour. Roy. Stat. Soc.* (B), 10, 159-203.

Rao, C. R. (1949a), On the distance between two populations, *Sankhyā*, 9, 246-248.

Rao, C. R. (1949b), On some problems arising out of discrimination with multiple characters, *Sankhyā*, 9, 343-366.

Aoyama, H. (1950), A note on the classification of data, *Ann. Inst. Statist. Math.*, 2, 17-20.

Wald, A. (1950), *Statistical Decision Functions*, Wiley: New York.

Rao, C. R. (1950), Statistical inference applied to classificatory problems, *Sankhyā*, 10, 229-256.

Hodges, J. L. (1950), Survey of discriminatory analysis, *USAF School of Aviation Medicine, Rep. No.1*, Randolph Field, Texas.

Wald, A. and Wolfowitz, J. (1950), Characterization of the minimal complete class of decision functions when the number of decisions and distributions is finite, *Proc. Second Berkeley Symp. on Stat. and Probability*.

Books

Rao, C. R. (1952), *Advanced Statistical Methods in Biometric Research* (Chapters 8 and 9), Wiley: New York.

Anderson, T. W. (1958), *An Introduction to Multivariate Statistical Analysis* (Chapter 6), Wiley: New York.
Sebestyn, G. S. (1962), *Decision making processes in pattern recognition*, Macmillan Co.: New York.
Uhr, L. M. (1966), *Pattern recognition theory: theory, simulations and dynamic models of form perception and discovery*, Wiley: New York.
Fu, K. S. (1968), *Sequential Methods in Pattern Recognition and Machine Learning*, Academic Press: New York York.
Helstorm, C. W. (1968), *Statistical Theory for Signal Detection*, Pergamon Press: New York.
Watanabe, M. S. (ed.) (1972), *Frontiers of Pattern Recognition*, Academic Press: New York.
Patrick, E. A. (1972), *Fundamentals of pattern recognition*, Prentice-Hall: Englewood Cliffs, New Jersey.
Anderson, T. W., Das Gupta, S. and Styan, G. P. H. (1972), *A Bibliography of Multivariate Statistical Analysis*, Oliver & Boyd: Edinburgh.

2. *Problems in classification; different situations*

The following are the main formulations of the classification problems along with some important variations.

Problem 1. Let $\pi_1,\pi_2,\ldots,\pi_m$ be m distinct populations (of experimental units). A random sample (of size $n\geq 1$, or a sequence) of experimental units is available from a population π_0 which is known to be exactly one of $\pi_1,\ldots,\pi_m$; the problem is to decide which one. In order to distinguish the populations, a vector (of p components, $p\geq 1$, or a sequence of components) X of real-valued characteristics of a unit is considered. The distribution of X in π_i is denoted by p_i (the corresponding c.d.f. of the p-dimensional distribution being F_i), $i=0, i,\ldots,m$. Since $P_0 = P_I$ for some $I = 1,2,\ldots,m$, the problem is to choose among the m decisions $I = i$ $(i=1,\ldots,m)$. A decision rule is based on the observations on the units to be classified and the available information on $P_1,\ldots,P_m$. If these distributions are not completely known, supplementary information on them is obtained through available samples from the populations $\pi_1,\ldots,\pi_m$. In this case, it is assumed that $F_1 \times F_2 \times \ldots \times F_m$ belongs to a certain set Ω of distributions and for each point $F_1 \times F_2 \times \ldots \times F_m$, the F_i's are taken to be different. The samples from the populations $\pi_1,\ldots,\pi_m$ will be called "training" samples (this term is used generally by engineers), and the P_i (or F_i) will be called "class-distributions."

Problem 2. Here π_0 is considered to be a mixture of the populations $\pi_1,\ldots,\pi_m$. Corresponding to each unit we define X as before and consider a number I which denotes the serial number $(1,2,\ldots,m)$ of the population to which the unit belongs. For the units to be classified, I is unobservable and the problem is to decide on the value of I from the knowledge of X. The distribution of X, given $I = i$, is P_i and the distribution of I is over the set $\{1,2,\ldots,m\}$. The problem will be termed as "known mixture" or "unknown mitture" according as the distribution of I is known or unknown. If the distribution of (X,I) is unknown, a (training) sample from π_0 (of size $N \geq 1$, or a sequence) is available to get information on it. A training sample may be of two types: (i) "Supervised" or "identified"--For each unit in the training sample both X and I are observable. (ii) "Unsupervised" or "unidentified"--For each unit in the training sample only X is observable.

Sometimes the units to be classified occur in a sequence and after the ith unit is classified its exact I-value becomes available. Thus for classifying the nth unit, the previous n-1 units form a supervised training sample. We shall call this case as (iii) post-supervised or post-identified.

It may also happen that the units to be classified do not come from the sample population, but, in a given sample, the number of units from each population is known.

Problem 2G. In the above problem, I is taken as a classificatory variable and it is artificial in nature. More generally, one may consider I as a continuous or discrete variable with physical meaning and the population π_i corresponds to $I \in S_i$, where $S_1,\ldots,S_m$ is a partition of the I-space. Marshall and Olkin (1968) incorporated the decision of observing I along with the m decisions in their formulation.

Instead of considering only the m decisions, one may also incorporate the possibility of reserving judgments, preferential decision, as well as consider a more general decision space as $P_0 \varepsilon \{P_{i_1},\ldots,P_{i_k}\}$ where $(i_1,\ldots,i_k)$ in a subset of $(1,2,\ldots,m)$ and $k = 1,2,\ldots,m-1$.

The populations $\pi_1,\pi_2,\ldots,\pi_m$ may represent m different "states" or points of time. In that case, one may get a training sample such that on each of its units X-observations are available at these m points of time. This would lead to "dependent" training sample.

Also it may not be possible to observe every component of X on each sampled unit. This would give rise to "incomplete" data. It may be mentioned that one may consider a general stochastic process instead of a finite-dimensional vector X.

The possibility of treating m as unknown is excluded in this review.

3. *Classification into known distributions; general theory*

Suppose that the distributions of X are $P_1,\ldots,P_m$ in $\pi_1,\ldots,\pi_m$, respectively, and the p.d.f. of X in π_i is given by f_i with respect to a σ-finite measure μ. We shall first consider the problem of classifying one unit from π_0 into one of $\pi_1,\ldots,\pi_m$ from the decision-theoretic viewpoint. The problem can be stated as a zero-sum two-person game where each person has m possible actions. Let $\ell(i,j) \geq 0$ be the loss for classifying the unit into π_i when it really belongs to π_j; assume $\ell(i,i) = 0$ for all $i = 1,\ldots,m$. A decision rule is given by $\delta = (\delta_1,\ldots,\delta_m)$ where $\delta_i(x)$ is the conditional probability of classifying into π_i, given the observation. The risk-vector of a rule δ is given by $\gamma(\delta) = (\gamma_1(\delta),\ldots,\gamma_m(\delta))$, where $\gamma_j(\delta) = \sum_{i=1}^{m} \ell_{ij} \int \delta_i(x)dP_j(x)$. When $\ell_{ij} = \ell_j$ for all $i \neq j$, $\gamma_j(\delta) = \ell_j \int[1-\delta_j(x)]dP_j(x) = \ell_j[1-\alpha_j(\delta)]$, where $\alpha_j(\delta)$ is the probability of correct classification (PCC) for the rule δ when π_j is the correct population. Correspondingly, the probabilities of misclassification (PMC) are given by $\int \delta_i(x)dP_j(x)$, $i \neq j$.

A prior distribution is given by $\xi = (\xi_1,\ldots,\xi_m)$, where ξ_j is the probability that π_j is the true population. (In case of mixed population, the distribution of X can be expressed as $\sum_j \xi_j P_j$). The ξ-Bayes risk of a rule δ is given by $R(\xi,\delta) = \sum_{j=1}^{m} \xi_j \gamma_j(\delta)$. The main results are as follows.

(i) A necessary and sufficient condition for a rule δ to be ξ-Bayes is that for any j (j=1,...,m), $\delta_j(x) = 0$ for all x (except possibly for a set of μ-measure 0) for which $L_j(x) > \min_{1\le i\le m} L_i(x)$, where $L_i(x) = \sum_{k=1}^{m} \ell_{ik}\xi_k f_k(x)$.

In particular, when $\ell_{ij} = 1$ for all $i \neq j$, the above inequality can be expressed as $\xi_j f_j(x) < \max_{1\le i\le m} \xi_i f_i(x)$.

(ii) The class of all admissible rules is complete (and hence minimal complete).

(iii) Every admissible rule is Bayes.

(iv) For every prior distribution ξ, there exists an admissible ξ-Bayes rule.

(v) There exists a least favorable distribution ξ^o and a minimax rule δ^M which is admissible and ξ^o-Bayes. For every minimax rule δ,

$$\gamma_i(\delta) \le R(\xi^o,\delta) = \max_{1\le i\le m} \gamma_i(\delta) \quad , \quad i=1,\ldots,m.$$

(va) If m = 2, there exists a unique minimax rule (and hence admissible Bayes) for which $\gamma_1(\delta^M) = \gamma_2(\delta^M)$.

(vb) Suppose $\ell_{ij} = \ell > 0$ for $i \neq j$, and the distribution $P_1,\ldots,P_m$ are mutually absolutely continuous. Then there exists a unique minimax rule δ^M for which

$$\gamma_1(\delta^M) = \ldots = \gamma_m(\delta^M) .$$

It can be shown that if either of the above two conditions is violated, the equality of the risk components of δ^M may not hold.

For proofs of the above results one may see Wald (1950, Section 5.1.1) and Ferguson (1967), although there are many papers and books (including Rao, 1952, Anderson, 1958) which deal with this problem and present results weaker than the above. For earlier work, see Welch (1939) and von Mises (1945). Raiffa (1961) considered comparisons among experiments along standard lines dealing with risk functions.

A Bayes rule may lead to large PMC's and there have been several attempts to overcome this difficulty. Anderson (1969) posed the classification problem with m+1 actions, the additional action being termed as a "deferred judgment" or "query". In that case, a decision rule δ is given by $(\delta_0, \delta_1, \ldots, \delta_m)$, where $\delta_0(x)$ is the conditional probability of suspending judgment. He considered the problem of maximizing $\sum_{i=1}^{m} \xi_i \alpha_i(\delta)$ subject to constraints given by

$$\int \delta_i(x) dP_j(x) \leq C_{ij} \ , \quad (i,j = 1,\ldots,m;\ i \neq j)$$

where C_{ij}'s are given constants, and obtained results on the existence, necessity, sufficiency and uniqueness of such solutions. Neyman and Pearson (1936) dealt with the maximization of the PCC to one population subject to the PMC's of m other populations being equal to specified levels. See also Lehmann (1959). Rao (1952) also considered the problem posed by Anderson but gave only sufficient conditions. There is a heuristic discussion in Rao (1952) on introducing doubtful decisions (or regions) or preferential decisions besides the m actions. Quesenberry and Gessaman (1968) considered the classification problem as a (2^m-1)-decision problem as follows:

$$\delta_{i_1,\ldots,i_s} : \text{means decide that } P_0 \in \{P_{i_1},\ldots,P_{i_s}\} \quad \text{for } s = 1,\ldots,m-1$$

$$\delta_0 : \text{means reserve judgment.}$$

where $(i_1,\ldots,i_s)$ is a subset of $(1,\ldots,m)$. They posed the problem of finding a rule which minimizes the probabilities of reserving judgment when the probabilities of wrong decisions (i.e. P_j (decide $P_0 \neq P_j$)) are controlled; they gave the solution for $m = 2$.

Marshall and Olkin (1968) (see Section 2 for their problem) gave some characterizations of minimum risk procedures. The possibility of observing the components of X sequentially was also considered. See also Cochran (1951) for a related problem.

Heuristic rules: A likelihood-ratio (LR) rule is defined by δ, where $\delta_j(x) > 0$ if (for some positive constants $C_1,\ldots,C_m$) $C_j f_j(x) < \max_{1\le i\le m} [C_i f_i(x)]$; see the result (a). In particular, if the C_i's are all equal, the rule will be called a maximum-likelihood (ML) rule.

For a distance function d defined for pairs of distributions, a minimum distance (MD) rule classifies X into F_i if $d(\hat{F}_0,F_i) = \min_{1\le j\le m} d(\hat{F}_0,F_j)$; ties may be resolved in some manner. In the above, $\hat{F}_0$ is an estimate of F_0 obtained from the sample (from π_0) to be classified, so that $d(\hat{F}_0,F_j)$ are defined; when $F_0 = F(\cdot;\theta)$, one may consider $\hat{F}_0 = F(\cdot;\hat{\theta})$.

Suppose we restrict our attention to rules belonging to a given class. Then one may find an optimum rule in that class (if it exists) by maximizing a weighted average of the PCC's or minimaxing PMC's. See Aoyama (1950) in this connection.

For $m = 2$, the classification problem is essentially the problem of testing a simple hypothesis against a simple alternative. In this case, there are well-known results on the asymptotic behavior of the error probabilities. See Kullback (1958), Chernoff (1952). For $m > 2$, see Hellman and Raviv (1970). Suppose $\varepsilon(F_1,F_2,t)$ is the average error probability for a rule $X \gtrless t$, when $m = 2$. Chernoff (1970) showed that

$$\sup_{F_i\varepsilon \mathcal{F}_i} \inf_t \varepsilon(F_1,F_2,t) = [2(1+\Delta^2)]^{-1} ,$$

where $\mathcal{F}_i$ is the class of all univariate distributions with means μ_i and variance σ_i^2, and $\Delta = |\mu_1-\mu_2|/(\sigma_1+\sigma_2)$. The same result is obtained if one considers only the LR tests. For other studies on error probabilities, see Bahadur (1971) and references therein.

The problem of distinguishing (i.e. finding sequential or non-sequential rules so that the PMC's can be controlled arbitrarily) between two sets of distributions is posed by Hoeffding and Wolfowitz (1958) and some necessary and sufficient conditions were obtained by them. In this framework, one considers a sequence of independent variables where the common distribution is known to belong to either of two given sets. Freedman (1967) extended some of these results when the possible distributions are countably many. Yarborough (1971) studied this problem with likelihood-ratio rules.

The papers dealing with discrimination between stochastic processes are mainly concerned with finding conditions for which two (or more) processes (i.e. the induced measures) are equivalent or singular. In case of equivalence, the next problem is to obtain the likelihood-ratio and study some rules based on it. For Gaussian processes, see Feldman (1958), Hájek (1958), Rao and Varadarajan (1963) and the references therein. Brown (1971) dealt with Poisson processes; Shepp (1965) and Kantor's (1967) results are concerned with distinguishing between a process and its translate. For general work in this area, see Kakutani (1948), Gikhman and Skorohod (1966), and Kraft (1955), and Adhikari (1957).

When we have a sample $(X_1,\ldots,X_m)$ from π_0 and the problem is to make decisions on their common distribution (which is known to be one of $F_1,\ldots,F_m$), one may treat the problem from the standard decision-theoretic view-point. It is also possible to use the compound-decision approach of Robbins (1951) in order to get some asymptotically good rules. The empirical Bayes approach may be used when the observations are from a mixed population.

When one has the possibility of getting observations from π_0 sequentially, it may be appropriate to consider the sequential m-decision problem. Out of a considerable literature, the following may be mentioned: Wald (1947), Wald (1950), Armitage (1950), Mallows (1953), Hoeffding and Wolfowitz (1958), Phatarford (1965), Simons (1967), Fu (1968, and the references therein), Meilijson (1969), Roberts and Mullis (1970), and Kinderman (1972).

Dvoretzky, Kiefer and Wolfowitz (1953) pointed out that most of the results of Wald (1947) extend to the case of stochastic processes in continuous time provided that the last observation is a sufficient statistic for the entire past and the log(LR) of this statistic at various points of time form a process with stationary and independent increments; e.g., sequentially testing the drift of a Brownian motion or the intensity of a Poisson process. Bhattacharya and Smith (1972) defined sequential probability ratio tests for testing a simple hypothesis against a simple alternative for the mean value function of a real Gaussian process with known covariance kernel; exact formulas are obtained for the error probabilities and the OC function.

References (3)

Neyman, J. and Pearson, E. S. (1936), see Ref. 1.
Welch, B. L. (1939), see Ref. 1.
von Mises, R. (1945), see Ref. 1.
Wald, A. (1947), *Sequential Analysis*, Wiley: New York.
Kakutani, S. (1948), On equivalence of infinite product measures, *Ann. Math.*, 49, 214-224.
Wald, A. (1950), *Statistical Decision Functions*, Wiley: New York.
Armitage, P. (1950), Sequential analysis with more than two alternative hypotheses and its relation to discriminant function analysis, *J. Roy. Statist. Soc.*, B, 12, 137-144.
Aoyama, H. (1950), see Ref. 1.
Cochran, W. G. (1951), Improvement by means of selection, *Proc. II Berkeley Symp. Math. Statist. Prob.*, Univ. of Calif. Press, 449-470.
Chernoff, H. (1952), A measure of asymptotic efficiency for tests of a hypothesis based on sum of observations, *Ann. Math. Statist.*, 23, 493-507.
Rao, C. R. (1952), see Ref. 1 (books).
Mallows, C. L. (1953), Sequential discrimination, *Sankhyā*, 12, 321-338.
Dvoretzky, A., Kiefer, J. and Wolfowitz, J. (1953), Sequential decision problems for processes with continuous time parameter: Testing hypotheses, *Ann. Math. Statist.*, 24, 254-264.
Kraft, C. (1955), Some conditions for consistency and uniform consistency of statistical procedures, *Univ. of Calif. Publ. Statist.*, 2, 125-142.
Adhikari, B. P. (1957), Analyse discriminante des measures de probabilite sur un espace abstrait, *C. R. Acad.*

Sci., 244, 845-846 (MR-18).
Anderson, T. W. (1958), see Ref. 1 (books).
Feldman, J. (1958), Equivalence and perpendiculariy of Gaussian processes, *Pacific Journ. Math.*, 8, 699-708.
Hájek, J. (1958), On a property of the normal distribution of any stochastic process, *Czech. Math. J.*, 8, 610-618.
Hoeffding, W. and Wolfowitz, J. (1958), Distinguishibility of sets of distributions, *Ann. Math. Statist.*, 29, 700-718.
Kullback, S. (1959), *Information Theory and Statistics*, Wiley: New York (Dover, 1968).
Lehmann, E. L. (1959), *Testing Statistical Hypotheses*, Wiley: New York.
Raiffa, H. (1961), see Ref. 6.
Rao, C. R. and Varadarajan, V. S. (1963), Discrimination of Gaussian processes, *Sankhyā*, A, 25, 303-330.
Shepp, L. A. (1965), Distinguishing a sequence of random variables from a translate of itself, *Ann. Math. Statist.*, 36, 1107-1112.
Phatarford, R. M. (1965), Sequential analysis of dependent observations, *Biometrika*, 52, 157-165.
Cacoullos, T. (1966), On a class of admissible partitions, *Ann. Math. Statist.*, 37, 189-195.
Gikhman, I. I. and Skorohod, A. V. (1966), On the densities of probability measures in function spaces, *Russian Math. Surveys*, 21, 83-156.
Simons, G. (1967), Lower bounds for average sample number of sequential multihypothesis tests, *Ann. Math. Statist.*, 38, 1343-1365.
Ferguson, T. S. (1967), *Mathematical Statistics: A decision-theoretic approach*, Academic Press: New York.
Freedman, D. A. (1967), A remark on sequential discrimination, *Ann. Math. Statist.*, 38, 1666-1676.
Marshall, A. W. and Olkin, I. (1968), A general approach to some screening and classification problems, *J. Roy. Statist. Soc.*, B, 30, 407-435.
Quesenberry, C. P. and Gessaman, M. P. (1968), see Ref. 7.
Fu, K. S. (1968), see Ref. 1 (books).
Anderson, J. A. (1969), Discrimination between k populations with constraints on the probabilities of misclassification, *J. Roy. Statist. Soc.*, B, 31, 123-139.

Kantor, M. (1969), On distinguishing translates of measures, *Ann. Math. Statist.*, 40, 1773-1777.

Meilijson, I. (1969), A note on sequential multiple decision procedures, *Ann. Math. Statist.*, 40, 653-657.

Chernoff, H. (1970), A bound on the classification error for discriminating between populations with specified means and variances, *Dept. Statist. Tech. Rep. #66*, Stanford Univ.

Robert, R. A. and Mullis, C. T. (1970), A Bayes sequential test of m hypotheses, *IEEE Trans. on Information Science*, IT-16, 91-94.

Hellman, M. and Raviv, J. (1970), Probability of error equivocation and the Chernoff bound, *IEEE Trans. Information Theory*, IT-16, 368-372.

Bahadur, R. R. (1971), Some limit theorems in statistics, *SIAM*, Bristol: England.

Brown, M. (1971), Discrimination of Poisson processes, *Ann. Math. Statist.*, 42, 773-776.

Yarborough, D. A. (1971), Sequential discrimination with likelihood-ratios, *Ann. Math. Statist.*, 42, 1339-1347.

Bhattacharya, P. R. and Smith, R. P. (1972), SPRT for the mean value function of a Gaussian process (to be published in *Ann. Math. Statist.*).

Kinderman, A. (1972), see Ref. 4.

4. *General Theory of Classification When the Information About the Distribution is Based on Samples*

When the class-distributions $F_1, F_2, \ldots, F_m$ (and the mixture probabilities $\xi_1, \ldots, \xi_m$, in case of a mixed population) are not completely known, information on them is available through a training sample (TS). As described in Section 2, a training sample may be obtained separately from each π_i $(i=1,\ldots,m)$, or from a mixture of these populations and, in that case, the sample may be supervised, unsupervised or post-supervised. Let n be the total size of the training sample and n_i be the size of the sample from π_i. Let n_0 be the size of the sample from π_0 which has to be classified and we shall denote such a sample by "CS".

Classification rules are generally devised using the following methods:

(a) *Plug-in rules:* Under complete knowledge of F_i's (and ξ_i's, in case of mixed population) a good rule (e.g., Bayes rule, minimax rule, LR rule, MD rule, etc.) δ is chosen. A plug-in rule $\hat{\delta}$ is obtained by replacing the F_i's (and ξ_i's) in δ by the corresponding estimates obtained from TS. When δ involves only the class densities or the parameters (in the parametric case) the corresponding estimates of the densities or the parameters are used in $\hat{\delta}$. In case of a MD rule using a distance function d, estimates of the distributions have to be chosen appropriately so that d is defined for these estimates.

(b) *LR rules and ML rule:* Suppose the class-densities are known except for some parameters. Let L(TS) denote the likelihood of the training sample and L_i(CS) denote the likelihood of CS under the hypothesis $\pi_0 = \pi_i$. Let $\lambda_i = \sup[L_i(CS)/L(TS)]$, the supremum being taken over the parametric space. A LR rule classifies CS into π_i, iff

$$k_i\lambda_i = \max_{1\le j\le m} [k_j\lambda_j] ,$$

where k_i's are non-negative constants; ties may be resolved in some manner. A ML rule is a LR rule with equal k_i's. The concept of a LR rule for the classification problem is due to Anderson (1951).

(c) *Best-of-class or constructive rules:* Such a rule is given by the one which optimizes certain specified criteria in a given class. One may consider Bayes rules, admissible rules, minimax rules or characterize a (reasonable) complete class following the general theory of statistical decision functions; the class of rules may be restricted by some invariance requirement. Note that in this case the action space of the statistician is finite. See Wald (1950), Kudo (1959, 1960). Another possibility is to consider some criteria depending on PMC's and use Neyman-Pearson theory and its extensions. See Rao (1954).

The criteria may be defined "empirically" as follows. Suppose the criterion for evaluating the performance of a rule is given by a real-valued function (e.g., PCC).

For each rule, an estimate of the value of the function corresponding to the rule is defined in terms of TS. Then an empirical best-of-class rule is the one for which such an estimate is the maximum in a given class of rules. See Glick (1969).

The main problem in obtaining plug-in rules is to get reasonable estimates of the distributions (or the densities, or the parameters). Generally, maximum-likelihood or some other consistent estimates are used. For estimation, especially in the non-supervised case, there is a huge literature and it is not possible to discuss these papers in this review. For an early work on non-supervised estimation, see Pearson (1894). For estimation by potential function method and stochastic approximation method, especially in the non-supervised case, see Fu (1968) and Patrick (1972).

General results on asymptotic properties of plug-in rules are given in Hoel and Peterson (1949), Fix and Hodges (1951), Das Gupta (1964), Van Ryzin (1966), Bunke (1967), Glick (1969, 1972).

See Sections 5, 6, 7 and Glick (1969, 1972), in particular, for studies on estimation of PMC's of a rule.

There is no systematic work on sequential rules. See Fu (1968), Patrick (1972) and Kurz and Woinsky (1969) in this connection. Kinderman (1972) studied some sequential rules based on distance functions and suggested some rules based on the idea of "tests with power 1" of H. Robbins.

Classifiability

Following the work of Hoeffding and Wolfowitz (*Ann. Math. Statist.*, 1958) on distinguishability of distributions, Das Gupta and Kinderman (1972) (see also Kinderman, 1972) introduced an important notion termed as "classifiability". Suppose $F_1 \times F_2 \dots \times F_m$ belongs to a certain set Ω of distributions. The set Ω is said to be classifiable finitely or sequentially if the PMC's can be controlled (arbitrarily) by some fixed sample-size rule or sequential rule, respectively, based on observations from $\pi_0, \pi_1, \dots, \pi_m$. Different conditions are obtained for Ω to be classifiable. The structure of Ω is studied for resolving the problem whether observations from all the populations or some (specific) of them are required (or sufficient) so as to get a rule when the PMC's are controlled (arbitrarily).

Compound-Decision and Empirical Bayes Approaches

Let $X_1, X_2, \ldots, X_n$ be independent random variables. The distribution of X_i is given by $F(\cdot, \theta_i)$ where $\theta_i \in [1, 2, \ldots, m]$ and for each i the problem is to choose one of the decisions $\theta_i = j$ $(j=1,\ldots,m)$; this will be called the ith component problem. The main bulk of the literature concerns $m = 2$ and we shall only describe this case. The general theory is given in Robbins (1951). Let $T_n = (t_1, \ldots, t_n)$ be a decision rule, where t_i and $1-t_i$ are the conditional probabilities of deciding $\theta_i = 1$ and 2, respectively, given the observations. Let R_n be the minimum value of the average of risks for the n component problem when one considers only fixed rules given by $t_i = t(x_i)$. It is shown in Robbins (1951), Hannan and Robbins (1955) that there exists $\{T_n\}$ with $T_i = t_i(x_1, \ldots, x_n)$ such that for large n the risk of T_n is uniformly (in the θ_i's) close to R_n; it is assumed that $F(\cdot, j)$ are known for all j. Hannan and Van Ryzin (1965) studied the rate of convergence of the risks of the above rules. Assuming that $X_1, X_2, \ldots, X_n$ occur in a sequence, Samuel considered "sequential" rules T_n with $t_i = t_i(x_1, \ldots, x_i)$. She (1963a) first characterized the minimal complete class under complete knowledge and in (1963b) proved a result similar to the previous one restricting to "sequential" rules. For a nonparametric method, see Johns (1961).

The empirical Bayes method, suggested by Robbins (1964), was used by Hudimoto (1968) in devising a rule for classifying observations from the mixed population π_0. When the class-distributions are unknown, they are estimated (by nonparametric methods) from a supervised TS. The risk of such a rule was also studied. Based on n_0 observations from a mixed distribution given by $\sum_{j=1}^{m} \xi_j F(x;\theta_j)$, Choi (1969) suggested to estimate the ξ_j's and the θ_j's by minimizing

$$\int [\Sigma \xi_j F(x;\theta_j) - \hat{F}(x)]^2 d\hat{F}$$

where $\hat{F}$ is the empirical c.d.f. These estimates are used to obtain a plug-in rule $\hat{\delta}$ from the Bayes rule δ. The asymptotic behavior of the conditional risk of $\hat{\delta}$ given the observations was studied.

Tanaka (1970) considered a method of approximating the difference of two posterior probabilities at the nth stage for classifying a sequence of observations, each into one of two distributions.

References (4)

Pearson, K. (1894), Contributions to the mathematical theory of evolution, *Phil. Trans. Roy. Soc., London*, 185, 71-110.

Hoel, P. G. and Peterson, R. P. (1949), A solution to the problem of optimum classification, *Ann. Math. Statist.*, 20, 433-438.

Wald, (1950), see Ref. 1.

Robbins, H. (1951), Asymptotically subminimax solutions of compound statistical decision functions, *Proc. II Berkeley Symposium on Probability and Statistics*, Univ. of California Press, 131-148.

Anderson, T. W. (1951), see Ref. 5.

Fix, E. and Hodges, J. L. (1951), Nonparametric discrimination: consistency properties, U.S. Air Force School of Aviation Medicine, *Report No.4*, Randolph Field, Texas.

Rao, C. R. (1954), A general theory of discrimination when the information about alternative population is based on samples, *Ann. Math. Statist.*, 25, 651-670.

Kudo, A. (1959, 1960), see Ref. 5.

Johns, M. V. (1961), see Ref. 7.

Samuel, E. (1963a), Note on a sequential classification problem, *Ann. Math. Statist.*, 34, 1095-1097.

Samuel, E. (1963b), Asymptotic solutions of the sequential compound decision problem, *Ann. Math. Statist.*, 34, 1079-1094.

Robbins, H. (1964), The empirical Bayes approach to statistical decision functions, *Ann. Math. Statist.*, 35, 1-20.

Das Gupta, S. (1964), see Ref. 7.

Hannan, J. F. and Van Ryzin, J. (1965), Rate of convergence in the compound decision problems for two completely specified distributions, *Ann. Math.*

Statist., 36, 1743-1752.

Van Ryzin, J. (1966), see Ref. 7.

Bunke, O. (1966), Über optimale verfahren der diskriminanzanalyse, *Abhandlg. Deutsch. Akad. Wiss., Klasse Math., Phys. Tech.*, 5, 35-41 (MR 32-6624).

Hudimoto, H. (1968), On the empirical Bayes procedure (1), *Ann. Inst. Statist. Math.*, 20, 169-185.

Fu, K. S. (1968), see Ref. 1 (books).

Kurz, L. and Woinsky, M. M. (1969), see Ref. 7.

Choi, K. (1969), Empirical Bayes procedure for (pattern) classification with stochastic learning, *Ann. Inst. Statist. Math.*, 21, 117-125.

Glick, N. (1969), Estimating unconditional probabilities of correct classification, Stanford Univ. Department of Statistics, *Tech. Report No.3*.

Tanaka, K. (1970), On the pattern classification problems by learning I, II, *Bull. Math. Statist.* (Japan), 14, 31-50, 61-74.

Kinderman, A. (1972), On some properties of classification: classifiability, asymptotic relative efficiency, and a complete class theorem, Univ. of Minnesota, Department of Statistics, *Tech. Report No. 178*.

Das Gupta, S. and Kinderman, A. (1972), On classifiability and requirements of different training samples, unpublished.

Patrick, E. A. (1972), see Ref. 1 (books).

5. *Classification Into Multivariate Normal Populations--Nonsequential Methods*

A. *Classification Into Two Multivariate Normal Distributions With the Same Covariance Matrix*

The distribution of X in π_i is $N_p(\mu_i,\Sigma)$, i=1,2. Suppose all the parameters μ_1, μ_2, and Σ are known and $X \sim N_p(\mu,\Sigma)$, where $\mu = \theta\mu_1 + (1-\theta)\mu_2$ for the classification problem $\theta = 0$ or 1. It is easily seen that $(\mu_1-\mu_2)'\Sigma^{-1}X$ is a sufficient statistic for θ. The class of Bayes rules is the same as the class of LR rules. Typically, a LR rule δ_c classifies X into $N_p(\mu_1,\Sigma)$, iff

$$T(x) \equiv T(x;\ \mu_1,\ \mu_2,\ \Sigma) \equiv ||x-\mu_1||^2_\Sigma - ||x-\mu_2||^2_\Sigma \leq c\ ,$$

where $||a-b||^2_\Sigma = (a-b)'\Sigma^{-1}(a-b)$. The minimax Bayes rule (0-1 loss) is given by δ_0; it is also called the minimum distance (MD) rule (for Mahalanobis distance). The PMC of δ_c is given by

$$\alpha_1(\delta_c) = \Phi(-\frac{c+\Delta^2}{2\Delta})\ ,\quad \alpha_2(\delta_c) = \Phi(\frac{c-\Delta^2}{2\Delta})\ ,$$

where $\Delta^2 = ||\mu_1-\mu_2||^2$, and Φ is the c.d.f. of $N(0,1)$. This classical case is treated in many papers and books, of which Welch (1939), Wald (1944), Rao (1952) and Anderson (1958) are worth mentioning. Recall that Fisher's LDF (in the population) is given by $(\mu_1-\mu_2)'\Sigma^{-1}x$ which maximizes $[a'(\mu_1-\mu_2)]^2/a'\Sigma a$ among all vectors a. Penrose (1947) suggested to consider the best LDF in terms of two linear functions of X given by the sum and a linear contrast of the components of X expressed in terms of their standard deviations; he called them the "size" and the "shape" respectively. He discussed the case when all the correlations are equal.

If the unknown parameters are structured in a special way, reasonable rules based on X can be found. For instance, Rao (1966) considered the following structure, relevant to growth models: $\mu_i = \nu_i + \beta\theta_i$ $(i=1,2)$, where ν_i, β are known but the vectors θ_i are unknown. By restricting to similar divisions of the sample space or by considering ancillary statistics, the problem is reduced to finding the usual LDF in terms of the projection of X on a space orthogonal to the column-space of β. This problem was originally posed by Burnaby (1966). Rao also treated the case when the covariance matrices are different.

Cochran (1962, 1964) studied the effects of the different components of X on Δ^2 (which determines the PMC of a LR rule), especially when all the correlations are equal.

When all the parameters are not known, random samples of sizes n_1 and n_2 from $N_p(\mu_1,\Sigma)$ and $N_p(\mu_2,\Sigma)$ are used to get information on the parameters. (Sampling is different in the mixed-population case.) The literature in this area consists of (i) suggestions of some heuristic rules, especially the plug-in LR rules, (ii) distributions of classification statistics and expressions for PMC, (iii) estimation of the PMC of a given rule, and (iv) derivation of constructive rules.

The rules considered in the literature are usually of the type involving a classification statistic Z and a cut-off point c (i.e., classifies into π_1, iff $Z<c$), where Z is a function of X, $\overline{X}_1$, $\overline{X}_2$, and S, where $\overline{X}_1$ and $\overline{X}_2$ are the sample mean vectors and S is the sample pooled covariance matrix (the divisor being $n_1+n_2-2 \equiv r$). The plug-in version of δ_c, denoted by $\hat{\delta}_c$, is based on the statistic $W \equiv T(X; \overline{X}_1, \overline{X}_2, S)$, when all the parameters are unknown. This statistic was proposed by Anderson (1951). More generally, one may consider a plug-in LR rule by replacing the unknown parameters in T by their respective estimates. Fisher (1936) and Wald (1943) suggested the plug-in LDF as the classification statistic, which is given by $U \equiv (\overline{X}_2-\overline{X}_1)'S^{-1}X$. Anderson (1951, 1958) proposed the LR rules which have the following classification statistic:

$$[r+(1+1/n_1)^{-1}||X-\overline{X}_1||_S^2]/[r+(1+1/n_2)^{-1}||X-\overline{X}_2||_S^2] \ .$$

When the cut-off point c is 1, this rule reduces to

$$V \equiv (1+1/n_1)^{-1}||X-\overline{X}_1||_S^2 - (1+1/n_2)^{-1}||X-\overline{X}_2||_S^2 \lessgtr 0 \ .$$

This is the same as $\hat{\delta}_0$, when $n_1 = n_2$. For known Σ, the LR rules involves the statistic V with S replaced by Σ. In the sequel, the rule $V \lessgtr 0$ will be called the ML rule and $\hat{\delta}_0$ will be called the MD rule; we shall use the same terminology when some of the known parameters are used instead of their estimates. Rao (1954) derived some rules restricting to invariance and local optimal conditions; the classification statistic for his rule (Σ unknown) will be called R. Matusita's (1967) minimum distance rules (for his distance function, see Section 5B) reduces to MD rules in this case; Matusita also considered the case when there are n observations to be classified and obtained some lower bounds for the PCC of the MD rule.

Rao (1946) suggested to test the hypothesis $\mu = (\mu_1+\mu_2)/2$ by Hotelling's T^2-test and use the MD rule when this test is significant. Brown (1947) considered a problem where $\mu_i = \alpha + \beta w_i$ $(i=1,2)$, w_i being the classificatory variable (e.g., age). α and β can be estimated from a training sample and using these estimates w is estimated for the observation to be classified; Brown extended this to more than 2 populations. Cochran (1964) posed the problem when the last q components of X have the same means in π_1 and π_2 and suggested a statistic W^* (similar in form to W) in terms of the residuals in the first p-q components of X after eliminating their (linear) regression on the last q components. Each of the statistics, U, V, W and R can be expressed as a linear function of the elements of a 2×2 random matrix

$$M = [m_{ij}] = (Y_1, Y_2)'A^{-1}(Y_1, Y_2) ,$$

where Y_1, Y_2 are independent $N_p(\cdot, I_p)$ vectors, and $A \sim W_p(n_1+n_2-2, I_p)$ independently of Y_1 and Y_2. In particular, for V, W and R the means of Y_1 and Y_2 are proportional, and, moreover, V is a constant multiple of m_{12}. Wald (1943) gave a canonical representation of U, and Harter (1951) derived its distribution when $p = 1$. Sitgreaves (1952) derived the distribution of M when the means of Y_1 and Y_2 are proportional; Kabe (1963) derived it without this restriction. Bowker (1960) showed that W can be represented as a (rational) function of two independent 2×2 Wishart matrices one of which is noncentral. Bowker and Sitgreaves (1961) used this representation to find an asymptotic expansion of the c.d.f. of W in terms of n_1^{-1} and Hermite polynomials, when $n_1 = n_2$. Sitgreaves (1961) derived the distribution of m_{12} and explicitly obtained the PMC of the MD rule when $n_1 = n_2$. Elfving (1961) obtained an approximation to the c.d.f. of W for large $n_1 = n_2$ and $p = 1$. In the univariate case, Linhart (1961) gave an asymptotic expansion for the average PMC of the MD rule in powers of $(n_1+n_2)/n_1n_2$ and Hermite polynomials in Δ^2. Teichroew and Sitgreaves (1961) used an empirical sampling plan to obtain an estimate of the c.d.f. of W. Okamoto (1963) considered the statistic W where the degrees of freedom r of S is not necessarily $n_1 + n_2 - 2$, and gave asymptotic expansions of

$$P[(W-\Delta^2/2)/\Delta < k/\pi_1] \quad \text{and} \quad P[(W+\Delta^2/2)/\Delta < k/\pi_2]$$

in terms of n_1^{-1}, n_2^{-1} and r^{-1} as n_1, n_2, and r tend to ∞ and n_1/n_2 tend to a finite positive constant. Anderson (1972) obtained asymptotic expansions of the above probabilities with Δ^2 replaced by $D^2 \equiv ||\overline{X}_1-\overline{X}_2||^2_S$. Memon and Okamoto (1971) obtained an asymptotic expression for the c.d.f. of $(V+\Delta^2)/2\Delta$, when $\mu = \mu_1$.

Cochran (1964) numerically compared the PMC's (computed from Okamoto-expansion) of the rules $W^* \lessgtr 0$ with those of $W \lessgtr 0$ when $n_1 = n_2$ is large. Memon and Okamoto (1970) derived an asymptotic expansion for the distribution of W^* and the PMC of the W^*-rule in terms of n_1^{-1}, n_2^{-1} and r^{-1}.

John (1959, 1960) derived the distributions of the statistics U, V, W and Rao's statistic (when Σ is known), S being replaced by Σ and obtained explicitly the PMC when the cut-off point is 0. Some bounds for the PCC were also given by John. When Σ is unknown, and S is used for Σ, some approximations are given for the distributions and the PCC's.

For $p = 1$, $\mu_1 < \mu_2$, Friedman (1965) considered a rule: $X \lessgtr \frac{\hat{\mu}_1+\hat{\mu}_2}{2}$ and compared its PMC with that of the rule $X \leq \frac{\mu_1+\mu_2}{2}$, with approximations for large sample size of the training sample.

Recently, Das Gupta (1972) proved that for a large class of rules (including the MD and the ML rules--Σ known and Σ unknown) the PCC's are monotonic increasing functions of Δ^2.

Let $\delta \equiv \delta(\cdot;\ \mu_1,\ \mu_2,\ \Sigma)$ be a decision rule when all the parameters are known. We shall denote the plug-in version of δ by $\hat{\delta}$ by replacing the unknown parameters by their respective (standard) estimates. The conditional error probabilities of $\hat{\delta}$, given $\overline{X}_1$, $\overline{X}_2$ and S, are given by

$$e_i(\hat{\delta}) = P[\hat{\delta} \text{ classifies X into } \pi_j|\overline{X}_1,\overline{X}_2,S;\ \mu = \mu_i]\ ,$$

$i \neq j$; $i,j = 1,2$. The unconditional error-probabilities of $\hat{\delta}$ are $\alpha_i(\hat{\delta}) = E(e_i(\hat{\delta}))$. An estimate of $e_i(\hat{\delta})$ is given by $\hat{e}_i(\hat{\delta})$ which is obtained by replacing the unknown parameters in $e_i(\hat{\delta})$ by their standard estimates. Similarly $\hat{\alpha}_i(\delta)$ and $\hat{\alpha}_i(\hat{\delta})$ are defined.

In the literature, the error-probabilities of the minimax rule δ_0 (parameters known) and its plug-in version $\hat{\delta}_0$ (the MD rule) are mostly considered. When Σ is known, John (1961) derived the distributions of $e_i(\hat{\delta}_0)$ and obtained their means; similar results are obtained when the cut-off point is not 0 and only approximations are given when Σ is unknown and n_1 and n_2 are large. In (1964) John considered the similar problem except that μ may be different from μ_1 or μ_2. John (1963) studied the conditional PMC's of the rules defined by the classification statistics $(1 + \frac{1}{n_1})^{-1}||X-\overline{X}_1||^2_\Sigma - \eta(1 + \frac{1}{n_2})^{-1}||X-\overline{X}_2||^2_\Sigma$ and Rao's statistic R, when Σ is known. Dunn and Varady (1966) empirically studied (Monte Carlo methods) $1 - \hat{\alpha}_i(\delta_0)$, $1 - e_i(\hat{\delta}_0)$ and $1 - \hat{e}_i(\hat{\delta}_0)$ and derived a confidence interval for the conditional error probabilities of $\hat{\delta}_0$. Geisser (1967) considered a prior measure for the parameters whose (improper) density is proportional to $|\Sigma|^{(p+1)/2}$. Using the posterior distribution of the parameters (given $\overline{X}_1$, $\overline{X}_2$ and S) he obtained confidence bounds for $e_i(\hat{\delta}_0)$; for large n_1, n_2 he used normal approximations. Several estimates of $e_1(\hat{\delta}_0)$, $\alpha_1(\hat{\delta}_0)$, $\alpha_1(\delta_0)$ are suggested in the literature of which the following are of main types: (i) Smith's (1947) reallocation or counting estimates, (ii) Lachenbruch's (1967) deletion-counting estimate, (iii) Fisher's estimate $\hat{e}_i(\hat{\delta}_0) = \Phi(-D/2)$ or the estimate obtained by replacing Δ in $\Phi(-\Delta/2)$ by some other estimate, (iv) the leading term in the Okamoto-expansion and replacing Δ^2 by its estimate, (v) estimates obtained from additional training sample. It follows from Hills (1966) that $\alpha_i(\hat{\delta}_0) > \alpha_i(\delta_0)$ when $n_1 = n_2$. For $p = 1$, Hills (1966) obtained the distribution of $\hat{e}_1(\hat{\delta}_0)$ and compared the expectations of $e_1(\hat{\delta}_0)$, $\hat{e}_1(\hat{\delta}_0)$ and those of the counting estimate by exact expressions and numerical computations. In 1967, Lachenbruch proposed the deletion-counting method for estimation. Lachenbruch and Mickey (1968) suggested some estimates of Δ^2 and studied empirically the behavior of the estimates (i)-(iv). Brofitt (1969) derived the uniformly minimum variance estimates of the mean values of Smith's and Lachenbruch's estimates and suggested some other estimators with smaller mean-square errors. Sorum (1971) obtained some estimates based on additional observations. For known Σ, she derived the means, the variances and approximation to the mean-square errors of most of the estimates and studied these estimates numerically when Σ is unknown (1972a, 1972b). Dunn (1971) studied the average PCC of $\hat{\delta}_0$ and Lachenbruch's estimates (using his estimate of Δ^2) for $n_1 = n_2$ by Monte Carlo methods. For $p = 1$, Sedransk and Okamoto (1971) obtained asymptotic expansions for the mean-square errors of several estimates. Recently, Das Gupta (1972) obtained some results on Fisher's and Smith's estimates which generalize Hill's (1966) results.

Chan and Dunn (1972) studied the effect of missing data on the PMC of $\hat{\delta}_0$ by Monte Carlo methods using several standard techniques of handling missing data. Srivastava and Zaatar (1972) derived the ML rule when Σ is known and the samples from the two populations are incomplete (all the p components are not available on each unit sampled) and showed that this rule is admissible Bayes. Lachenbruch (1966) posed the problem when the parent populations of the observations in the training sample are incorrectly identified. Mclachan (1972) derived asymptotic expressions for the mean and the variance of $e_i(\hat{\delta}_0)$ incorporating the possibility of incorrect identification of the training sample.

Following Glick (1972) it can be shown that as $n_1, n_2 \to \infty$, $\hat{\alpha}_i(\delta) \to \alpha_i(\delta)$ a.s. uniformly in the class of all rules (not based on training data). Furthermore, if δ is a LR rule, then $\alpha_i(\hat{\delta}) \to \alpha_i(\delta)$ a.s. and $\hat{\alpha}_i(\hat{\delta}) \to \alpha_i(\delta)$. For related results, see Glick (1969, 1972) and for slightly weaker results see Fix and Hodges (1950), Bunke (1964). Kinderman (1972) suggested a measure of the relative asymptotic efficiencies of two rules by the limit of the ratio of minimum total sample sizes required by the two rules to achieve a maximum probability of error α, as $\alpha \to 0$. In particular, he illustrated this concept by comparing a two-sample rule based on samples from π_0 and π_1 and a three-sample rule using Anderson's statistic when the populations are univariate normal with variance 1 and $\Delta = |\mu_1 - \mu_2| > 0$.

There are many ad hoc methods for choosing "good" components of the vector X. Cochran (1961, 1964) studied the effect of the different components of X on Δ^2, especially when all the correlations are equal. Urbakh (1971) made a similar study on Δ^2, as well as on Lachenbruch's estimate of Δ^2. Linhart (1961) made a numerical comparison of the effectiveness of selecting components by $\Phi(-\Delta/2)$ and the average PMC of $\hat{\delta}_0$. Weiner and Dunn (1966) also studied empirically three methods for selecting components.

In the normal case, Glick (1969) obtained some interesting results for the 'best-of-class' rules. Let C_{LD} be the class of all rules based on linear (discriminant) functions of X (i.e., partitioning the sample space into two half spaces). Let δ^* be a rule in C_{LD} which maximizes (in C_{LD}) the average (over some known prior or the standard estimates of the proportions in the mixture) of the proportions of the training sample correctly classified. Then this maximum value converges (a.s.) to the PCC of the best (Bayes) rule and the risk of δ^* converges a.s. to the Bayes risk as the sample size in the training sample increases to ∞.

When the training sample comes from a mixed population, different methods are available to estimate the parameters and the proportions in the mixture, if they are unknown. For the supervised case, there is not much change in the theory and methods from the usual case discussed before. For some asymptotic results see Glick (1969, 1972). In the non-supervised case, there is a good deal of literature; for this and relevant references, see Fu (1968), Patrick (1972); for an earlier work see Pearson (1894).

Rao (1954) derived an optimal rule in the class of rules for which the probabilities of error depend only on Δ^* using the following criteria: (i) to minimize a linear combination of the derivatives of the error-probabilities with respect to Δ at $\Delta = 0$ subject to the condition that the error-probabilities at $\Delta = 0$ leave a given ratio. (ii) The above criterion with the additional restriction that the derivatives of the error-probabilities at $\Delta = 0$ bear a given ratio. Rao separately treated the problem according as Σ is known or unknown. When Σ is known, Kudo (1959, 1960) showed that the ML rule has the maximum PCC among all translation-invariant rules δ for which the error-probabilities depend on Δ^2, and

$$\alpha_1(\delta;\ \Delta^2=\Delta_1^2) = \alpha_2(\delta;\ \Delta^2=\Delta_2^2)$$

for all Δ_1 and Δ_2 such that $(1+1/n_1)^{-1}\Delta_1^2 = (1+1/n_2)^{-1}\Delta_2^2$. He also showed that this rule is most stringent in the above class without the requirement of translation-invariance. When Σ is known, Ellison (1962) obtained a class of admissible Bayes rules which includes the MD and ML rules. In this case, Das Gupta (1962, 1965) showed that the ML rule is admissible Bayes (with a different prior and a general loss function) and minimax (unique minimax under some mild conditions). When Σ is unknown, similar results were obtained by Das Gupta (1962, 1965), restricting to the class of rules invariant under translation and the full linear group. For $p = 1$, $n_1 = n_2$, Bhattacharya and Das Gupta (1964) obtained a class of Bayes rules and showed that the MD rule is minimax Bayes. Srivastava (1964) also obtained a class of Bayes rules when Σ is unknown. Geisser (1964) used a prior (improper) density which is proportional to $|\Sigma|^{v/2}$, $v \leq n_1+n_2$ and $v = 0$ when Σ is known; he derived the (improper) Bayes rules for these priors which are the likelihood-ratio rules in respective cases. For similar analysis, see Geisser (1966). Kiefer and Schwartz (1965) indicated a method to obtain a broad class of Bayes rules which are admissible; in particular, they showed that the LR rules are admissible Bayes when Σ is unknown and $r+1 > p$. Marshall and Olkin (1968) derived Bayes rules for normal distributions in their special set-up. When $p = 1$, $n_1 = n_2$ and the number of observations to be classified is n (≥ 1), Kinderman (1972) characterized an essentially complete class of rules, invariant under translation and change of signs.

B. *Classification Into Two Multivariate Normal Populations With Different Covariance Matrices*

The distribution of X in π_i is taken as $N_p(\mu_i,\Sigma_i)$, $i = 0,1,2$; furthermore, it is known that Σ_1 and Σ_2 are different. Generally, three cases are considered: (i) $(\mu_0, \Sigma_i) = (\mu_i,\Sigma_i)$ for some $i=1,2$, (ii) $\mu_0 = \mu_i$, for some $i=1, 2$, (iii) $\Sigma_0 = \Sigma_i$ for some $i=1,2$.

When the parameters are known, the LR statistic was studied by Cavalli (1945) ($p=1$), Smith (1947), Okamoto (1963) ($\mu_0=\mu_1=\mu_2$), Cooper (1963, 1965), Bartlett and Please (1963) ($\mu_0=\mu_1=\mu_2=0$, $\Sigma_i = (1-\rho_i)I_p+\rho_i J_p$, $i=1,2$), Bunke (1964), Han (1968) ($\Sigma_i = (1-\rho_i)I_p+\rho_i J_p$, $i=1,2$), Hann (1969) ($\Sigma_1 = d\Sigma_2$, $d>1$), Han (1970) (Σ_i's are of circular type).

Kullback (1952, 1958) suggested a rule based on the linear statistic which maximizes the divergence J(1,2) between $N_p(\mu_1,\Sigma_1)$ and $N_p(\mu_2,\Sigma_2)$. He also obtained some partial results on deriving the optimum class of rules based on linear functions of X from the Neyman-Pearson viewpoint (i.e., minimizing one PMC by controlling the other). Clunies-Ross and Riffenburgh (1960) studied this problem geometrically. Anderson and Bahadur (1962) derived the minimax rule and characterized the minimal complete class after restricting to the class of rules based on linear functions of X. Banerjee and Marcus (1965) studied the form of this minimax rule.

Gilbert (1969) derived the PMC of a LR rule when the parameters are known and compared it with the PMC of the corresponding LR rule when $\Sigma_1 = \Sigma_2$. For the latter he obtained the optimum cut-off point for which the total PMC is minimized.

Lbov (1964) studied the PMC when p is large and the parameters are known. Grenander (1972) considered a similar problem.

Anderson (1964) studied the problem of choosing components by minimizing Bayes risk when the distributions are univariate normal.

When μ_0 equals either μ_1 or μ_2, and the covariance matrices are known, a class of admissible Bayes rules was obtained by Ellison (1962); in particular he showed that the MD and the ML rules are admissible Bayes.

Okamoto (1963) derived the minimax rule and the form of a Bayes rule when the parameters are known; he studied some properties of the Bayes' risk function, and suggested a method for choosing components. He also treated the case when Σ_i's are unknown, and the common value of μ_i's may be known or unknown. The asymptotic distribution of the plug-in log(LR) statistic was also obtained by Okamoto. Bunke (1964) derived the minimax rule and the form of a Bayes rule and proved that the plug-in minimax rule is consistent. Following the method of Kiefer and Schwartz (1965), Nishida (1971) obtained a class of admissible Bayes rules when the parameters are unknown.

Matusita (1967) considered a minimum distance rule and suggested its plug-in version by replacing the unknown parameters by their respective estimates; the distance between two distributions with p.d.f.'s p_1 and p_2 with respect to a σ-finite measure m was taken as

$$[\int(\sqrt{p_1(x)} - \sqrt{p_2(x)})^2 dm]^{1/2} .$$

He separately treated the different cases according as the μ_i's and Σ_i's are known or unknown, and obtained some bounds for the PCC.

When $\Sigma_1 = d\Sigma_2$ $(d>1)$, the distributions of the log(LR) statistic and its plug-in version (by replacing the mean vectors by their estimates) were derived by Han (1969). Similar results were obtained by Han (1970) when the Σ_i's are of "circular" type.

Chaadha and Marcus (1968) studied (mainly by simulation) the behavior of some estimates of a measure of divergence defined as $2(\mu_1-\mu_2)'(\Sigma_1+\Sigma_2)^{-1}(\mu_1-\mu_2)$.

Aoyoma (1950) considered rules of the form $X \lessgtr x_0$ and found the optimum value of x_0 which minimizes the PCC when X is a mixture of two univariate normal distributions; the mixture ratio may be known or unknown.

C. *Classification Into More Than Two Multivariate Normal Populations*

The distribution of X in π_i is taken as $N_p(\mu_i,\Sigma_i)$, $i = 0,1,\ldots,m$. In most of these cases the results for m=2 are extended in a straightforward way, and discussions on this case may be found in many papers cited in Sections 5A and 5B. In particular, see Fisher (1938), Day and Sandomire (1942), Brown (1947), Rao (1952), Anderson (1958), Rao (1963). In generalizing Fisher's LDF, one considers the eigenvectors of the "between means" matrix in the metric of "within error" matrix. For other criteria, see Uematu (1964).

Das Gupta (1962) considered the problem when $\mu_1,\ldots,\mu_m$ are linearly restricted (as in the linear model in MANOVA) and showed that the ML rule is admissible Bayes when the common covariance is known. Following Kiefer and Schwarz (1965), Srivastava (1967) obtained similar results when the common covariance matrix is unknown.

Cacoullos (1965a) considered the case when the distribution of X in π_i is $N_p(\mu_i,\Sigma)$, $i = 0,1,\ldots,m$, and μ_0 is not necessarily equal to one of $\mu_1,\ldots,\mu_m$; the problem is to choose a π_i which is nearest to π_0 (in the sense of

Mahalanobis-distance). When $\mu_1,\ldots,\mu_m$ and Σ are known, he obtained a unique invariant minimax rule allowing for indifference regions; for the m-decision problem he obtained a class of admissible invariant Bayes rules including the minimax rule. The limiting case of no indifference region was also studied in Cacoullos (1966). In his paper (1965b) he obtained a class of invariant Bayes rules when Σ and $\mu_1,\ldots,\mu_m$ are unknown (μ_0 known).

References (5)

Fisher, R. A. (1936), *Ann. Eugen.*, see Ref. 1.
Fisher, R. A. (1938), *Ann. Eugen.*, see Ref. 1.
Welch, B. L. (1939), *Biometrika*, see Ref. 1.
Day, B. B. and Sandomire, M. M. (1942), *J. Amer. Statist. Assoc.*, see Ref. 1.
Wald, A. (1944), *Ann. Math. Statist.*, see Ref. 1.
Cavalli, L. L. (1945), *Mem. Ist. Ital. Idrobiol.*, see Ref. 1.
Rao, C. R. (1946), *Sankhyā*, see Ref. 1.
Penrose, L. S. (1947), *Ann. Eugen.*, see Ref. 1.
Smith, C. A. B. (1947), *Ann. Eugen.*, see Ref. 1.
Brown, G. W. (1947), *Ann. Math. Statist.*, see Ref. 1.
Rao, C. R. (1948), *J. Roy. Statist. Soc.*, B, see Ref. 1.
Rao, C. R. (1949b), *Sankhyā*, see Ref. 1.
Rao, C. R. (1950), *Sankhyā*, see Ref. 1.
Aoyoma, H. (1950), *Ann. Inst. Statist. Math.*, see Ref. 1.
Anderson, T. W. (1951), Classification by multivariate analysis, *Psychometrika*, 16, 631-650.
Harter, H. L. (1951), On the distribution of Wald's classification statistic, *Ann. Math. Statist.*, 22, 58-67.
Rao, C. R. (1952), see Ref. 1 (books).
Sitgreaves, R. (1952), On the distribution of two random matrices used in classification procedures, *Ann. Math. Statist.*, 23, 263-270.
Kullback, S. (1952), An application of information theory to multivariate analysis, *Ann. Math. Statist.*, 23, 88-102.
Rao, C. R. (1954), A general theory of discrimination when the information on alternative hypotheses is based on samples, *Ann. Math. Statist.*, 25, 651-670.
Kudo, A. (1959), The classification problem viewed as a

two-decision problem I, *Mem. Fac. Sci. Kyushu Univ. Ser. A.*, 13, 96-125.

Kullback, S. (1959), *Information Theory and Statistics*, Wiley: New York (1968, Dover: New York).

John, S. (1959), The distribution of Wald's classification statistic when the dispersion matrix is known, *Sankhyā*, 21, 371-376.

Kudo, A. (1960), The classification problem viewed as a two-decision problem II, *Mem. Fac. Sci. Kyushu Univ. Ser. A.*, 14, 63-83.

John, S. (1960), On some classification statistics I, II, *Sankhyā*, 22, 301-308, 309-316. (Correction: *Sankhyā*, 23 (1961), 308.)

Bowker, A. H. (1960), A representation of Hotelling's T^2 and Anderson's classification statistic, *Contrib. Probability and Statistics* (Hotelling Vol.), 142-149.

Clunies-Ross, C. W. and Riffenburgh, R. H. (1960a), Linear discriminant analysis, *Pacif. Sci.*, 14, 251-256.

John, S. (1961), Errors in discrimination, *Ann. Math. Statist.*, 32, 1125-1144.

Sitgreaves, R. (1961), Some results on the distribution of the W-classification statistic, *Stud. Item. Anal. Pred.* (H. Solomon, Ed.), Stanford Univ. Press, Stanford, California, 241-251.

Bowker, A. H. and Sitgreaves, R. (1961), An asymptotic expansion for the distribution function of the W-classification statistic, *Ibid*, 293-310.

Bowker, A. H. (1961), A representation of Hotelling's T^2 and Anderson's classification statistic W in terms of simple statistics, *Ibid*, 285-292. (Reprint of Bowker (1960.)

Teichroew, D. and Sitgreaves, R. (1961), Computation of an empirical sampling distribution for the W-classification statistic, *Ibid*, 252-275.

Elfving, G. (1961), An expansion principle for distribution functions with applications to Student's t-statistic and the one-dimensional classification statistic, *Ibid*, 276-284.

Okamoto, M. (1961), Discrimination for variance matrices, *Osaka Math. Jour.*, 13, 1-39.

Linhart, H. (1961), Für wahl von variablen in der Tremanalyse, *Metrika*, 4, 126-139.

Cochran, W. G. (1961), On the performance of the linear discriminant function, *Bull. Intl. Statist. Inst.*,

34, 436-446. (Reprint in *Technometrics*, 6 (1964), 179-190.)

Anderson, T. W. and Bahadur, R. (1962), Classification into two multivariate normal distributions with different covariance matrices, *Ann. Math. Statist.*, 33, 420-431.

Ellison, B. E. (1962), A classification problem in which information about alternative distributions is based on samples, *Ann. Math. Statist.*, 33, 213-223.

Das Gupta, S. (1962), On the optimum properties of some classification rules, *Inst. Statist. Mimeo*, No.333, Univ. of N.C., Chapel Hill, (see abstract: *Ann. Math. Statist.*, 33, 1504).

Cooper, P. W. (1962), The hyperplane in pattern recognition, *Cybernetica*, 5, 215-238.

Cooper, P. W. (1962), The hypersphere in pattern recognition, *Inform. Control*, 5, 324-346.

John, S. (1963), On classification by the statistics R and Z, *Ann. Inst. Statist. Math.*, 14, 237-246. (Correction: *Ann. Inst. Statist. Math.*, 17 (1965), 113.)

Rao, M. M. (1963), Discriminant Analysis, *Ann. Inst. Statist. Math.*, 15, 11-24.

Kabe, D. G. (1963), Some results on the distribution of two random matrices used in classification procedures, *Ann. Math. Statist.*, 34, 181-185.

Okamoto, M. (1963), An asymptotic expansion for the distribution of linear discriminant function, *Ann. Math. Statist.*, 34, 1286-1301. (Correction: *Ann. Math. Statist.*, 39 (1968), 1358-1359.)

Bartlett, M. S. and Please, N. W. (1963), Discrimination in the case of zero mean differences, *Biometrika*, 50, 17-21.

Cooper, P. W. (1963), Statistical classification with quadratic forms, *Biometrika*, 50, 439-448.

John, S. (1964), Further results on classification by W, *Sankhyā* A, 26, 39-46.

Bhattacharya, P. K. and Das Gupta, S. (1964), Classification into exponential populations, *Sankhyā* A, 26, 17-24.

Bunke, O. (1964), Über optimale verfahren der diskriminazanalyse, *Abl. Deutsch. Akad. Wiss. Klasse Math. Phys. Tech.*, 4, 35-41 (MR 32-6624).

Uematu, T. (1964), On a multidimensional linear discriminant function, *Ann. Inst. Statist. Math.*, 16, 431-

437.

Anderson, T. W. (1964), On Bayes procedures for a problem with choice of observations, *Ann. Math. Statist.*, 35, 1128-1135.

Lbov, G. S. (1964), Errors in the classification of patterns for unequal covariance matrices, *Akad. Nauk. SSSR Sibirisk Otdel Inst. Mat. Vyc. Sistemy*, 14, 31-38 (MR 31-5724).

Geisser, S. (1964), Posterior odds for multivariate normal classification, *J. Roy. Statist. Soc. Ser. B*, 26, 69-76.

Srivastava, M. S. (1964), Optimum procedures for classification and related topics, Stanford University Department of Statistics *Tech. Report* No. 11.

Cochran, W. G. (1964), Comparison of two methods of handling covariates in discriminatory analysis, *Ann. Inst. Statist. Math.*, 16, 43-53.

Das Gupta, S. (1965), Optimum classification rules for classification into two multivariate normal populations, *Ann. Math. Statist.*, 36, 1174-1184.

Kiefer, J. and Schwartz, R. (1965), Admissible Bayes character of T^2-, R^2- and other full invariant tests for classical multivariate normal problem, *Ann. Math. Statist.*, 36, 747-770.

Ellison, B. E. (1965), Multivariate normal classification with covariance known, *Ann. Math. Statist.*, 36, 1787-1793.

Banerjee, K. S. and Marcus, L. F. (1965), Bounds in a minimax classification procedure, *Biometrika*, 52, 653-654.

Cooper, P. W. (1965), Quadratic discriminant functions in pattern recognition, *IEEE Trans. Inform. Theory IT*-11, 313-315.

Friedman, H. D. (1965), On the expected error in the probability of misclassification, *Proc. IEEE*, 53, 658-659.

Cacoullos, T. (1965a), Comparing Mahalanobis distances, I: Comparing distances between k normal populations and another unknown, *Sankhyā* A, 27, 1-22.

Cacoullos, T. (1965b), Comparing Mahalanobis distances, II: Bayes procedures when the mean vectors are unknown, *Sankhyā* A, 27, 23-32.

Cacoullos, T. (1966), On a class of admissible partitions, *Ann. Math. Statist.*, 37, 189-195.

Weiner, J. M. and Dunn, O. J. (1966), Elimination of variates in linear discriminant analysis, *Biometrics*, 22, 268-275.

Geisser, S. (1966), Predictive discrimination, *Proc. Internat. Symp. Multiv. Analysis* (P. R. Krishnaiah, Ed.), Academic Press: New York.

Cochran, W. G. (1966), Analyse des classifications d'ordre, *Rev. Statist. Appl.*, 14, 5-17.

Hills, M. (1966), Allocation rules and their error rates, *J. Roy. Statist. Soc. Ser. B*, 28, 1-31.

Burnaby, T. P. (1966), Growth invariant discriminant functions and generalized distances, *Biometrics*, 22, 96-110.

Rao, C. R. (1966), Discriminant function between composite hypothesis and related problems, *Biometrika*, 53, 339-345.

Lachenbruch, P. A. (1966), Discriminant analysis when the initial samples are misclassified, *Technometrics*, 8, 657-662.

Dunn, O. J. and Varady, P. V. (1966), Probabilities of correct classification in discrimination analysis, *Biometrics*, 22, 908-924.

Srivastava, M. S. (1967), Classification into multivariate normal populations when the population means are restricted, *Ann. Inst. Statist. Math.*, 19, 473-478.

Bunke, O. (1967), *Z. Wahr Theor. und Verwundte Gebiete* (MR 35-6267), see Ref. 4.

Geisser, S. (1967), Estimation associated with linear discriminants, *Ann. Math. Statist.*, 38, 807-817.

Day, N. E. and Kerridge, D. F. (1967), A general maximum likelihood discriminant, *Biometrics*, 23, 313-323.

Matusita, K. (1967), Classification based on distance in multivariate Gaussian case, *Proc. 5th Berk. Symp. Math. Stat. Prob.*, 1, 299-304, Univ. of California Press, Berkeley.

Lachenbruch, P. A. (1967), An almost unbiased method of obtaining confidence intervals for the probability of misclassification in discriminant analysis, *Biometrics*, 23, 639-645.

Fu, K. S. (1968), see Ref. 1 (books).

Chaddha, R. L. and Marcus, L. F. (1968), An empirical comparison of distance statistics for populations with unequal covariance matrices, *Biometrics*, 24, 683- .

Han, Chien Pai (1968), A note on discrimination in the

case of unequal covariance matrices, *Biometrika*, 55, 586-587.

Lachenbruch, P. A. and Mickey, M. R. (1968), Estimation of error rates in discriminant analysis, *Technometrics*, 10, 1-11.

Marshall, A. W. and Olkin, I. (1968), *J. Roy. Statist. Soc. Ser. B*, see Ref. 3.

Han, Chien Pai (1969), Distribution of discriminant function when covariance matrices are propositional, *Ann. Math. Statist.*, 40, 979-985.

Gilbert, E. S. (1969), The effect of unequal variance-covariance matrices on Fisher's linear discriminant function, *Biometrics*, 25, 505-516.

Glick, N. (1969), Stanford Univ. Department of Statistics *Tech. Report*, see Ref. 4.

Brofitt, J. D. (1969), Estimating the probability of misclassification based on discriminant function techniques, Ph.D. dissertation, Colorado State Univ., Fort Collins.

Han, Chien Pai (1970), Distribution of discriminant function in circular models, *Ann. Inst. Statist. Math.*, 22, 117-125.

Memon, A. Z. and Okamoto, M. (1970), The classification statistic W^* in covariate discriminant analysis, *Ann. Math. Statist.*, 41, 1491-1499.

Memon, A. Z. and Okamoto, M. (1971), Asymptotic expansion of the distribution of the Z-statistic in discriminant analysis, *J. Multiv. Anal.*, 1, 294-307.

Nishida, N. (1971), A note on the admissible tests and classifications in multivariate analysis, *Hiroshima Math. J.*, 1, 427-434.

Dunn, O. J. (1971), Some expected values for probabilities of correct classification in discriminant analysis, *Technometrics*, 13, 345-353.

Urbakh, V. Y. (1971), Linear discriminant analysis: loss of discriminating power when a variate is omitted, *Biometrics*, 27, 531-534.

Sorum, M. J. (1971), Estimating the conditional probability of misclassification, *Technometrics*, 13, 333-343.

Sedransk, N. and Okamoto, M. (1971), Estimation of the probabilities of misclassification for a linear discriminant function in the univariate normal case, *Ann. Inst. Statist. Math.*, 23, 419-436.

Sorum, M. J. (1972a), Three probabilities of misclassification, *Technometrics*, 14, 309-316.

Sorum, M. J. (1972b), Estimating the expected and the optimal probabilities of misclassification, *Technometrics*, 14, 935-943.

Glick, N. (1972), see Ref. 4.

Mclachan, G. J. (1972), Asymptotic results for discriminant analysis when the initial samples are misclassified, *Technometrics*, 14, 415-422.

Chan, L. S. and Dunn, O. J. (1972), The treatment of missing values in discriminant analysis I, *J. Amer. Statist. Assoc.*, 67, 433-477.

Kinderman, A. (1972), Univ. of Minnesota, Department of Statistics *Tech. Rep.* No.178, see Ref. 4.

Anderson, T. W. (1972), An asymptotic sxpansion of the distribution of the "Studentized" classification statistic W, Stanford Univ., Department of Statistics *Tech. Rep.* No.9.

Anderson, T. W. (1972), Asymptotic evaluation of the probabilities of misclassification by linear discriminant functions, Stanford Univ., Department of Statistics *Tech. Rep.* No.10.

Das Gupta, S. (1972), Probability inequalities and error in classification, Univ. of Minnesota, School of Statistics *Tech. Rep.* No.190.

Srivastava, J. N. and Zaatar, M. K. (1972), On the maximum likelihood classification rule for incomplete multivariate samples and its admissibility, *J. Multiv. Anal.*, 2, 115-126.

Grenander, U. (1972), Asymptotic distribution of quadratic forms: large deviations (unpublished).

Govindarajulu, Z. and Gupta, A. K. (1972), Some new classification rules for c univariate normal populations, Univ. of Michigan, Department of Statistics *Tech. Rep.* No.14.

6. *Discrete and Other Non-normal Distributions*

The papers are group according to the type of distribution and the nature of the problem considered. A short review is given for each paper.

(a) *Multinomial distribution.*

The random variable X is distributed as a multinomial distribution with k cells in each of the populations.

Matusita (1956):

A minimum distance rule is proposed based on samples of sizes n, n_1 and n_2 from the populations π, π_1 and π_2, respectively. The distance is computed for the sample c.d.f.'s and the distance function is taken as the square root of

$$||F - G||^2 = \sum_{i=1}^{k} (\sqrt{p_i} - \sqrt{q_i})^2 ,$$

where $(p_1,\ldots,p_k)$ and $(q_1,\ldots,q_k)$ are cell-probabilities corresponding to the distributions F and G, respectively. He obtained lower bounds for PCC and approximate value of the PCC when the sample sizes are large; the case $n = 1$ is also discussed.

Chernoff (1956):

The distribution of X in π_1 is the multinomial with equal cell-probabilities and a multinomial with unknown cell probabilities in π_2. A sample of size n_2 is available from π_2 and the problem considered is to classify a sample of size n_0 from π_0 into π_1 or π_2. Results are directed towards applications when the number of cells is large, n_0 and n_2 are large, and the ratio of error probabilities is either very large or very small. In a certain class, an 'optimal' rule is obtained which classifies into π_2 if the sum of the frequencies of all the cells for which the sample from π_2 provides non-zero frequencies is too large.

Wesler (1959):

The distribution of X in π_i is taken as a multinomial with cell probabilities being any permutation of a given probability vector $p^{(i)}$, (i=1,2). The problem considered is to classify a sample of n_0 observations from π_0, by minimaxing one error probability when the maximum of the other error probability is held fixed. He obtained an approximate solution for large n_0 and considered the case $k = 2$.

Cochran and Hopkins (1961):

They obtained the form of the Bayes rules and considered, in particular, the 'maximum likelihood' rule. For this rule they discussed the effect of 'plug-in' on the PMC and suggested a correction for bias.

Raiffa (1961):

See Section 3. Multinomial distributions and, in general, discrete distributions are included in the development of theories.

Hills (1966):

This paper contains some theoretical developments on the errors of misclassification for the 'ML' rule in the two-population case. In particular, it is shown that for $k = 2$ the PMC for the 'ML' rule is greater than the corresponding PMC for the ML rule obtained under complete knowledge of the distributions and Smith's reallocation estimate of the PMC underestimates the PMC of the ML rule. He obtained normal approximations for the expected value of the reallocation estimate, plug-in estimate of the PMC of the ML rule and its expected value. The effectiveness of these estimates are compared through a numerical study.

Bunke (1966):

For multinomial distributions, a property of the estimated (with empirical c.d.f.'s) minimax rule is studied from the asymptotic viewpoint.

Glick (1969):

The development is for general discrete distributions but also specialized for multinomial distributions. This paper generalizes some of the results of Cochran and Hopkins (1961) and Hills (1966) and furnishes rigorous proofs. The sample space of the random variable X is taken as $\{(x_1,x_2,\ldots,x_k,\ldots)\}$. The population π_0 is considered as a mixture of the populations $\pi_1,\ldots,\pi_m$. The rule δ (Bayes) which maximizes the PCC is dealt with throughout. Let $\hat{\delta}$ be the plug-in version of δ using the "supervised training" data. Let γ = PCC for δ, $c(\hat{\delta})$ = the conditional (given the training data) PCC for $\hat{\delta}$, $\hat{c}(\hat{\delta})$ = plug-in version of $c(\hat{\delta})$. The following results are obtained.

(i) $E\hat{c}(\hat{\delta}) \geq \gamma \geq c(\hat{\delta})$.

(ii) $\hat{c}(\hat{\delta}) \to \gamma$ a.s. and in quadratic mean when the sample size in the training data increases to ∞.

(iii) When $m = 2$, the bias of $\hat{c}(\hat{\delta})$ for estimating γ is at worst of order $1/\sqrt{n}$, where n is the size of the training data.

(iv) When $m = 2$, and the distributions are multinomial, $P(c(\hat{\delta}) = \gamma) \to 1$ as $n \to \infty$.

(v) Smith's reallocation estimate for PCC using $\hat{\delta}$ is equal to $\hat{c}(\hat{\delta})$.

(vi) Suppose $\{x_1,\ldots,x_k\}$ is enclosed in a finite interval. Consider partitions of this interval into k disjoint subintervals with x_i in exactly one subinterval. Let C_M be the collection of all m-partitions $(B_1,\ldots,B_m)$ such that B_i is a union (of at most k) subintervals containing x_i. The rule in C_M which maximizes the proportion of training data correctly classified is the same as $\hat{\delta}$. Moreover, $c(\hat{\delta}) \to \gamma$ a.s.

(vii) When $m = 2$, and the distributions are multinomials, he discussed some shortcomings of Lachenbruch's estimate and suggested some other estimates and studied their performances numerically.

(b) *Multivariate Bernoulli distributions.*

The random variable X is a p×1 vector and each component of X takes values 0 or 1.

Bahadur (1961):

$m = 2$. This paper gives some approximations to the log likelihood-ratio, e.g., normal approximation and approximations using various truncations of Bahadur's series representation for the probability functions. Some approximations to Kullback-Leibler symmetric information measure J are also obtained. These approximations are useful when J is small, p is large, and the interdependence among the components of X is not appreciable.

Solomon (1960, 1961):

$m = 2$. This is a numerical study of the effectiveness (PMC) and relative comparisons among rules based on the sum of the components, Fisher's LDF, LR statistic, and some truncated functions obtained from Bahadur's series representation for the probability functions.

Hills (1967):

$m = 2$. This is concerned with the problem of estimating log(LR) at a given point $X = x_0$. The following estimates are suggested.

(i) $(r_1/n_1)/(r_2/n_2)$, where r_i is the number of observations in a sample of size n_i from π_i which equal x_0.

(ii) 'Near neighbor' estimate of order 1,

$$\left(\frac{r_1 + r_1'}{n_1}\right) / \left(\frac{r_2 + r_2'}{n_2}\right)$$

where r_i' is the number of 'near neighbors' in a sample of size n_i from π_i whose x-value differ from x_0 in only *one* component.

(iii) 'Near neighbor' estimates of order > 1.

The distributions of these estimates are studied numerically. A step-wise method for selecting components using the Kullback-Leibler information measure J is suggested.

Elashoff et al. (1967):

$m = 2$. Fisher's LDF, two functions based on a logistic model, and a function based on the assumption of mutual independence of the components are considered as possible classification statistics. The effectiveness of these statistics is studied numerically.

Martin and Bradley (1972):

The probability function of X in π_i is taken as

$$p_i(x) = f(x)[1 + h_s(a_i, x)] ,$$

where h_s is a linear function of the orthogonal polynomials on the sample space of X. This paper deals with the estimation of a_i and f subject to some constraints.

(c) *Parametric non-normal continuous-type distributions.*

Cooper (1962, 1963):

The distribution of X in π_i is taken as a known multivariate distribution of Pearson type II or type VII. The LR statistic is studied.

Bhattacharya and Das Gupta (1964):

m = 2. The distribution of X in π_i is taken as a member of the one-parameter exponential family. A class of admissible Bayes rules is obtained.

Cooper (1965):

The p.d.f. of X in π_i is taken as

$$p_i(x) = A_i|\Sigma_i|^{-1/2} f_i[(Q_i(x))^{1/2}] ,$$

where Q_i is a positive definite quadratic form and $f_i(u)$ decreases as u increases from 0. The LR statistic is studied.

Day and Kerridge (1967):

The p.d.f. of X in π_i is taken as

$$p_i(x) = d_i \exp[-\frac{1}{2}(x-\mu_i)'\Sigma^{-1}(x-\mu_i)]f(x) .$$

Two cases are considered, namely, (i) $f(x) \equiv 1$, (ii) $\Sigma = I$ and $f(x) = 1$ if every component of x is either 0 or 1 and $f(x) = 0$, otherwise. The posterior probability of the hypothesis H_i: $\pi = \pi_i$, given X = x, is expressed as $\exp(x'b+c)/[1+\exp(x'b+c)]$. This paper mainly deals with the maximum likelihood estimates of b and c. For classification, it incorporates the idea of 'doubtful' decision.

Anderson (1972):

For the m-population, the posterior probability of H_i: $\pi = \pi_i$, given X = x, is taken as

$$p(H_i|x) = \exp(\alpha_{i0} + x'\alpha_i)p(H_m|x) ,$$

$$p(H_m|x) = 1/[1 + \sum_{i=1}^{m-1} \exp(\alpha_{i0} + x'\alpha_i)] .$$

This paper deals with the estimation of α's by the maximum-likelihood method.

(d) *Other cases.*

Kendall (1966):

Some heuristic rules are suggested based on categorization of data.

Marshall and Olkin (1968):

For their formulation (see Section 3) of the problem, X is considered as a binomial random variable with probability of success Y which is distributed as the uniform distribution on (0,1). The form of a Bayes rule is obtained.

References (6)

Johnson, P. O. (1950), The quantification of qualitative data in discriminant anayysis, *J. Amer. Statist. Assoc.*, 45, 65-76.

Matusita, K. (1956), Decision rule, based on the distance, for the classification problem, *Ann. Inst. Statist. Math.*, 8, 67-77.

Chernoff, H. (1956), A classification problem, Stanford University, Department of Statistics *Technical Report* No.33.

Wesler, O. (1959), A classification problem involving multinomials, *Ann. Math. Statist.*, 30, 128-133.

Linhart, H. (1959), Techniques for discriminant analysis with discrete variables, *Metrika*, 2, 138-149 (MR 21-6067).

Solomon, H. (1960), Classification procedures based on dichotomous response vectors, *Contrib. Probability and Statistics* (Hotelling Vol.), 414-423.

Solomon, H. (1961), Classification procedures based on dichotomous response vectors, *Stud, Item Anal. Pred.* (H. Solomon, Ed.), Stanford Univ. Press, Stanford, California, 177-186.

Bahadur, R. R. (1961), On classification based on responses to N dichotomous items, *Ibid*, 169-176.

Raiffa, H. (1961), *Ibid*, see Ref. 3.

Cochran, W. G. and Hopkins, C. E. (1961), Some classification problems with multivariate qualitative data, *Biometrics*, 17, 10-32.

Takakura, S. (1962), Some statistical methods of classification by the theory of quantification, *Proc. Inst. Statist. Math. Tokyo*, 9, 81-105 (MR 27-3063).

Cooper, P. W. (1962), see Ref. 5.

Cooper, P. W. (1963), Statistical classification with quadratic forms, *Biometrika*, 50, 439-448.

Bhattacharya, P. K. and Das Gupta, S. (1964), see Ref. 5.

Cooper, P. W. (1965), see Ref. 5.

Hills, M. (1966), see Ref. 5.

Bunke, O. (1966), Nichparametrische Klassifikations verfahren für qualitative und quantitative Beobachtungen, *Wiss Z. Humboldt Univ. Berlin Math. Naturwiss. Reihe*, 15, 15-18 (MR 36-1031).

Kendall, M. G. (1966), Discrimination and classification, *Proc. Internat. Symp. Multiv. Anal.* (P. R. Krishnaiah), 165-185, Academic Press: New York.

Hills, M. (1967), Discrimination and allocation with discrete data, *Applied Statist.*, 16, 237-250.

Elashoff, J. D., Elashoff, R. M. and Goldman, G. E. (1967), On the choice of variables in classification problems with dichotomous variables, *Biometrika*, 54, 668-670.

Day, N. E. and Kerridge, D. F. (1967), A general maximum likelihood discriminant, *Biometrics*, 23, 313-323.

Gilbert, E. (1968), On discrimination using qualitative variables, *J. Amer. Statist. Assoc.*, 63, 1399-1412.

Marshall, A. W. and Olkin, I. (1968), see Ref. 3.

Glick, N. (1969), see Ref. 4.

Martin, D. C. and Bradley, R. A. (1972), Probability models, estimation and classification for multivariate dichotomous populations, *Biometrics*, 28, 203-222.

Anderson, J. A. (1972), Separate sample logistic discrimination, *Biometrika*, 59, 19-36.

7. *Nonparametric or "Distribution-free" Methods*

The so-called nonparametric or distribution-free methods are used in statistical inference when one is concerned with a wide class of distributions which usually cannot be expressed as a parametric family with a finite number of parameters. When a statement regarding the probability of a certain statistical inference remains valid for every member in a given family of distributions, we call that a distribution-free inference with respect to that family; in particular, if the distribution of a statistic (used for inference) is the same for every member of a family of underlying distributions of the random variables involved, we say that the statistic is distribution-free with respect to that family. In the classification problem sometimes we face a similar situation when we devise rules for a broad class of underlying distributions whose structures cannot be expressed in simple parametric forms. However, unlike the problems of testing hypothesis or estimation, a "classification problem cannot be distribution-free" (Anderson, 1966) in the broad sense.

The available work in this area can be classified broadly into three main categories:

1) Consider a "good" rule (generally taken as a Bayes and/or an admissible minimax) assuming that the distributions are known. In this rule, replace the c.d.f.'s or the p.d.f.'s by their respective sample estimates. The rule thus obtained will be called a "plug-in" rule.

2) Use the statistics involved in devising some well-known tests for the nonparametric two-sample or k-sample problems.

3) Some ad-hoc methods which are typical for the classification problems, e.g., "minimum distance" rule.

In the literature, the main emphasis is (a) to study the asymptotic behavior (e.g., consistency, efficiency in some sense) of the rules, (b) to obtain some bounds for the PCC of a given rule, and (c) to study the small-sample performance.

Rules with Density Estimates

There are several papers in the literature describing different methods for estimating a p.d.f. and the properties of different estimates. The following papers are mentioned in this connection; these references may be found in Van Ryzin (1966), Fu (1968, book), Patrick (1972, book), and Glick (1972).

Rosenblatt (1956, *Ann. Math. Statist.*)
Parzen (1962, *Ann. Math. Statist.*)
Cencov (1962, *Soviet Math.*)
Watson and Leadbetter (1963, *Ann. Math. Statist.*)
Aizerman, Braverman and Rozonoer (1964, *Autom. Rem. Control*)-Potential function method.
Nadarya (1965, *Theory of Prob. and Appl.*)
Loftsgarden and Quesenberry (1965, *Ann. Math. Statist.*)
Van Ryzin (1965, see Ref. 7)
Cacoullos (1966, *Ann. Inst. Statist. Math.*)
Murthy (1966, *1st. Internat. Symp. Multiv. Anal.*)
Tsypkin (1966, *Autom. Telemekhanika*)-Stochastic approximation method.
Kashyap and Blaydon (1968, *IEEE Trans. Inform. Theory*)
Moore and Henrichon (1969, *Ann. Math. Statist.*)

As mentioned earlier, estimates of p.d.f.'s are used to obtain a plug-in rule for a given rule which involves density functions. Suppose δ^* is a Bayes rule with respect to a prior distribution ξ, assuming that the densities in the m populations are known. Let $R(\xi,\delta)$ be the Bayes risk of a rule δ and $\hat{\delta}^*$ be the plug-in rule obtained from δ^* by replacing the densities by their respective estimates (based on a training sample). Van Ryzin (1966) defined the notion of "Bayes-risk consistent" by the following:

$$P[R(\xi,\hat{\delta}^*] - R(\xi,\delta^*) \geq \varepsilon] \rightarrow 0$$

as the sample sizes in the training sample tend to ∞. Van Ryzin also defined the Bayes risk consistency of order α_N by the following:

$$P[q_N\{R(\xi,\hat{\delta}^*) - R(\xi,\delta^*)\} \geq \varepsilon\alpha_N] \rightarrow 0$$

as N = minimum of the sample sizes $\rightarrow \infty$ and q_N is any sequence $\rightarrow 0$ as $N \rightarrow \infty$. With respect to these notions, he studied some plug-in rules with different density estimates. For related results, see Van Ryzin (1965).

Glick (1969, 1972) obtained some properties of non-randomized plug-in rules assuming that the training data come from a mixed population (with unknown mixture ratios). Let $\gamma(\delta)$ be the PCC of a rule δ and δ^* be the rule which maximizes $\gamma(\delta)$ assuming that the class-densities and the mixture rations are known. Let $\hat{\gamma}(\delta)$ be a plug-in estimate of $\gamma(\delta)$ by replacing the densities by their respective estimates. Glick's results are as follows:

i) If $\hat{f}_i \to f_i$ (density in π_i) a.s. $(i=1,\ldots,m)$ as the sample sizes in the training data increase to ∞, then

$$\hat{\gamma}(\delta) \to \gamma(\delta) \quad \text{a.s.} ,$$

uniformly in the class of all rules (not based on training data).

ii) If $\hat{f}_i \to f_i$ a.s. (in probability)

$$\gamma(\hat{\delta}^*) \to \gamma(\delta^*)$$
$$\hat{\gamma}(\hat{\delta}^*) \to \gamma(\delta^*)$$

a.s. (in probability).

iii) If the density estimates are pointwise unbiased, then

$$E[\hat{\gamma}(\hat{\delta}^*)] \geq \gamma(\delta^*) \geq \gamma(\hat{\delta}^*) .$$

For other results, see the books by Fu (1968) and Patrick (1972).

Fix and Hodges (1951) also considered the density-plug-in rules (of which the nearest neighbor rules have drawn much attention) and studied the consistency of such rules.

Bunke (1966) considered the plug-in rule $\hat{\delta}$ obtained from a restricted minimax rule δ by replacing the distributions involved by the respective empirical c.d.f.'s. He showed that asymptotically the rule $\hat{\delta}$ has the same Bayes-minimax property.

Nearest Neighbor (NN) Rules

In 1951, Fix and Hodges propored a classification rule for the two-population problem based on nonparametric estimates of the p.d.f.'s. Their method of estimating a density f can be described as follows: Let $X_1,\ldots,X_n$ be i.i.d. r.v.'s with the common p.d.f. f which is continuous at x. Let $\{S_n\}$ be a sequence of sets in the sample space with corresponding volumes $\{V_n\}$, such that

i) $\lim_{n\to\infty} \sup_{y\in S_n} ||x-y|| = 0,$

ii) $\lim_{n\to\infty} nV_n = \infty$.

Let k_n be the number of observations that lie in S_n. Then

$$\hat{f}(x) \equiv \frac{k_n}{nV_n} \overset{p}{\to} f(x)$$

when $k_n \to \infty$, $n \to \infty$. Rosenblatt (1956) used this approach for

$$S_n = \{y: ||x-y|| \leq h_n\}, \quad \lim h_n = 0 .$$

Parzen (1962) replaced this set S_n by kernels $K_n(y,x)$. More generally,

$$\hat{f}(x) \equiv \frac{k_n}{nV_n} ,$$

where $V_n = \int K_n(x,y)dy$, $k_n = n\int K_n(x,y)dF_n(y)$, and F_n is the empirical c.d.f. based on $X_1,\ldots,X_n$. Watson and Leadbetter (1963) determined the best kernel which minimizes the integral square error for some specific f. Fix and Hodges (1951) also considered the sets S_n which depend on the sample $X_1,\ldots,X_n$; they suggested that S_n be defined as a "ball" with respect to some distance function d, centered at x, just large enough to contain k observations. For the m-population problem, one may also consider m different sequences of such sets. These estimates were studied by Loftsgarden and Quesenberry (1965).

The K-NN rule, as proposed by Fix and Hodges (1951) is described as follows. Let $\{X_{ij};\ j=1,\ldots,n_i\}$ be a random sample from the ith population, $i = 1,\ldots,m$. Let X be the observation to be classified. Consider a distance function d and order all the values $d(X_{ij},X)$, $j=1,\ldots,n_i$; $i=1,\ldots,m$. The K-NN rule assigns X to the population π_i, if $K_i/n_i = \max_j (K_j/n_j)$, where K_i is the number of observations from π_i in the K observations "nearest" to x; ties may be resolved in some manner. For $m = 2$, $n_1 = n_2 = n$, they showed that the PCC's of the K-NN rule (with d as the Euclidean distance) tend to the respective PCC's of the "maximum likelihood" rule when $n \to \infty$, $K \equiv K_n \to \infty$, $K_n/n \to 0$. Fix and Hodges (1953) obtained the exact and asymptotic expression for the PMC's of the NN rule when $p = 1$, $K = 1, 3$ and the parent distributions are normal with the same covariance matrix. For this normal case, they (numerically) compared the NN-rule with the ML rule for $p = 1,2$; $k = 1,3$.

Cover and Hart (1967) considered the mixed-population case and proposed a K-NN rule which assigns X to the population π_i, if $K_i = \max_j K_j$. They showed, under mild regulariy conditions, that when the sample space is a separable metric space, and the distributions admit densities with respect to a measure, the limiting Bayes risk (0-1 loss function) of their 1-NN rule is bounded below by R^* and bounded above by $R^*(2-R^*m/(m-1))$, where R^* is the minimum Bayes risk (assuming that the distributions are known). Another result of Cover and Hart is as follows: Let $X, X_1, X_2, \ldots$ be a sequence of i.i.d. r.v.'s in a separable metric space. Then X_n' = nearest neighbor to X among $X_1, \ldots, X_n$, tends to X with probability 1 as $n \to \infty$. In a later paper, Cover (1968) studied the rate of convergence of the Bayes risk of their 1-NN rule. In the above notation, let $\gamma(X, X_n')$ be the conditional Bayes risk of the 1-NN rule, given X and X_n', and let $\gamma^*(X)$ be the conditional Bayes risk, given X, under complete knowledge of the distributions. Peterson (1970) studied different modes of convergence of

$$\gamma(X, X_n') - 2\gamma^*(X)[1-\gamma^*(X)]$$

under appropriate conditions. In a recent paper, Goldstein (1972) has studied some asymptotic properties of the K_n-NN rules and obtained a consistent upper bound for its PMC.

In 1966, Whitney and Dwyer considered the K-NN rule (of Cover and Hart) when the observations in the training sample are correctly identified with probability $\beta > 1/2$. Hellman (1970) modified the K-NN rule of Cover and Hart such that if at least K' of the K nearest neighbors to X come from the same population, then X is assigned to that population; otherwise, decision is withheld. Specht (1966) noted that if the densities (p-variate) in the Bayes rule (mixed-population case) are replaced by the corresponding Parzen's estimate with

$$K_n(x,y) = \frac{1}{(2\pi\sigma^2)^{p/2}} \exp(-\frac{1}{2\sigma^2}||x-y||^2)$$

then the plug-in rule is the same as the 1-NN rule of Cover and Hart for σ sufficiently small.

In 1966, Patrick proposed another NN-rule in a more general framework. He considered different distance functions d_i such that

$$\lim_{\varepsilon \to 0} [\max_y\{||x-y||: d_i(x,y) < \varepsilon\}] = 0 ,$$

and the set $\{y: d_i(y,x) = \varepsilon\}$ has zero volume for all $\varepsilon>0$ and all x. He suggested the following estimate of $f_i(x)$:

$$\hat{f}_i(x) = \frac{K_i(x)}{(n_i+1)V_i} ,$$

where $K_i(x)$ is a positive-integer and V_i is the volume of a d_i-neighborhood S_{in} of X depending on the training sample. Using these estimates he proposed the plug-in rule, obtained from the Bayes rule. For the special case, Patrick's NN rule assigns X to π_i if the Kth nearest neighbor to X in the sample from π_i is closest to X than that for a sample from any other population. An excellent account of these NN rules is given in Patrick's book (1972); see also the paper by Patrick and Fisher (1970). Pelto (1969) studied some estimates of the PMC of a NN-rule.

Rules Based on Distances Between Empirical c.d.f.'s

For classification into two discrete distributions Matusita (1956) proposed the minimum distance rule based on Matusita distance between the empirical c.d.f.'s and obtained some lower bounds for the PCC's. (See also Section 6.) Das Gupta (1964) considered the minimum distance rule (with arbitrary distance) for the m-population problem and showed the consistency of such rules under appropriate conditions. He also obtained a lower bound for the PCC of such rules and specialized this to the minimum Kolmogorov-distance rule.

Best-of-Class Rules

The systematic development of this concept is due to Glick (1969). Suppose that the observation X to be classified comes from a mixture of m distributions. Consider a collection ϕ of ordered m-partitions of the sample space; for any such ordered partition $S = (S_1,\ldots,S_m)$, $X \in S_i$ leads to the decision that X comes from the ith population π_i. Let $\gamma(S)$ be the PCC of the rule S. Define

$$\gamma(\phi) = \sup_{S\in\phi} \gamma(S) .$$

Let $X_1,\ldots,X_N$ be a supervised training sample. Then the "reallocation estimate" of $\gamma(S)$ is given by

$$\tilde{\gamma}(S) = \sum_{i=1}^{m} \frac{n_i}{N} \int_{S_i} d\hat{F}_i(x)$$

where n_i is the number of observations from π_i and $\hat{F}_i$ is the corresponding empirical c.d.f. Define

$$\tilde{\gamma}(\phi) = \sup_{S\in\phi} \tilde{\gamma}(S) .$$

If a rule $\tilde{S} \in \phi$ exists such that $\tilde{\gamma}(\phi) = \tilde{\gamma}(\tilde{S})$ then $\tilde{S}$ is called a "best-of-class" rule in ϕ. The results obtained by Glick (1969) are stated below:

i) $E(\tilde{\gamma}(\phi)) \geq \gamma(\phi)$.

ii) $\sup_{S\in\phi} |\tilde{\gamma}(S)-\gamma(S)| \to 0$ a.s. as $N \to \infty$.

iii) Let H_ν be the collection of all subsets of the sample space which are intersections of at most ν open half spaces. Let $\phi(\nu_1,\nu_2)$ be the collection of all ordered m partitions $S = (S_1,\ldots,S_m)$ such that for each i, either S_i or its complement is a union of at most ν_2 sets, each of which is a member of H_{ν_1} or the complement of a member.

Let $\phi \subset \phi(\nu_1,\nu_2)$ be a collection of ordered m-partitions. Then, as $N \to \infty$

(a) $\tilde{\gamma}(\phi) \to \gamma(\phi)$ a.s.

(b) $|\tilde{\gamma}(\phi)-\gamma(\tilde{S})| \to 0$ a.s.

(c) $\gamma(\tilde{S}) \to \gamma(\phi)$ a.s.

It is to be noted that these results tacitly assume the existence of $\tilde{S}$. For $m = 2$, the collection of all hyperplane partitions coincide with $\phi(1,1)$. The collection of all "interval" m-partitions, denoted by ϕ_I, is a subset of $\phi(2,2)$. When $\phi = \phi_I$, $\gamma(\phi) \geq \gamma(\tilde{S})$.

Historically, Aoyama (1950) first considered this approach for the two-population univariate case. Restricting to rules of the form $X \lessgtr x_0$, he studied the cut-off point x_0 which maximizes the PCC. Stoller (1954) assumed $m = 2$ and the two distributions are such that an interval partition is the best one. Restricting to the class of all interval partitions (with known order) he proved the results (i), (iii)(a), (iii)(c) of Glick only "in probability" instead of "a.s.". Hudimoto (1956) also considered the special case treated by Stoller and obtained an upper bound for the c.d.f. of $|\tilde{\gamma}(S)-\gamma(S)|$, where S is a rule with a given cut-off point ξ. Furthermore, he showed that the cut-off point $\hat{\xi}$ corresponding to the best-in-class rule $\tilde{S}$ is a consistent estimate of ξ. In a later paper, Hudimoto (1957) gave better bounds for the distribution of $\tilde{\gamma}(S)$ and obtained lower bounds for the c.d.f.'s of $\tilde{\gamma}(S) - \gamma(S)$, $\tilde{\gamma}(\phi) - \gamma(\phi)$, where ϕ is the class of all (known) ordered interval partitions and $m = 2$.

Rules Based on Tolerance Regions

The idea of using tolerance regions for classification was first suggested by Anderson (1966), although it is implicit in the work of Fix and Hodges (1951). For the univariate case, Anderson suggested some variations of NN rules; vector observations may be "ranked" (using them to define blocks) and then a univariate method can be applied. Other heuristic methods proposed by Anderson are as follows. Use the pooled training sample to construct "blocks". An observation X is classified into π_i if the block to which X belongs is defined by majority of observations from π_i. For the two-population problem, construct two sets of blocks separately based on the observations from π_1 and π_2. Let B_1 and B_2 be the blocks in the two sets which contain X. Consider the number of observations from π_2 in B_1 and the number of observations from π_1 in B_2 and classify X according to the larger number.

Quesenberry and Gessaman (1968) also suggested to use tolerance regions for the m-population classification problem with $2^m - 1$ decisions (instead of m decisions) described below:

$$\delta_{i_1,\ldots,i_s} : \text{decide } P \in \{P_{i_1},\ldots,P_{i_s}\},\ s = 1,\ldots,m-1$$

$$\delta_0 : \text{reverse judgment}$$

where $(i_1,\ldots,i_s)$ is a subset of $(1,2,\ldots,m)$. For each j, sample observations from π_j are used to construct a tolerance region for P_j. They suggested a decision rule obtained by partitioning the sample space using the standard union-intersection method with the A_j's. The PMC's may be controlled by appropriately choosing the number of blocks used for A_j $(j=1,\ldots,m)$. When the underlying distributions have some appropriate structure, the tolerance regions A_j can be so chosen that the resulting rule δ is consistent with the rule δ^* (i.e., $P_j(\delta=\delta^*) \to 1$, for each j) which minimizes the probabilities of reserving judgment subject to the size restrictions for the PMC's under complete knowledge of $P_1,\ldots,P_m$. However, in practice, the information concerning the distributions may not be sufficient enough so as to construct the above rule δ. Anderson and Benning (1970) partially resolved this difficulty by using clustering techniques to get information on the likelihood-ratios. Patrick and Fisher (1970) used tolerance regions for estimating p.d.f.'s and plug-in rules. (See the discussion on NN rules.) Gessaman and Genaman (1972) suggested some procedures based on statistically equivalent blocks and studied them by Monte Carlo methods.

Rules Based on Ranks--Analogy With Rank Tests

The idea of using the statistics in the standard nonparametric rank-tests for devising classification rules was suggested by Das Gupta (1962, 1964). Das Gupta considered a rule which decides $P_0 = P_i$ if $|W_i|$ is the smaller of $|W_1|$ and $|W_2|$, where W_i is the Wilcoxon statistic based on samples from π_0 and π_i; he proved that this rule is consistent. Hudimoto (1964) modified this rule by taking W_i instead of $|W_i|$ when $F_1(x) \geq F_2(x)$ for all x; he derived a bound for the PCC of this rule and in a later paper (1965) studied it when ties may be present. Kinderman (1972) proposed a class of rules based on linear rank statistics as follows: Suppose n observations are available from each of three populations π_0, π_1, π_2. Define $N = 3n$,

$$T_{nj} = n^{-1} \sum_{i=1}^{N} E_{Ni} L_{ji} , \; j = 0,1,2 ,$$

where E_{Ni} is a sequence of scores and L_{ji} is 1 if the ith ordered observation in the pooled sample is from π_j, and 0 otherwise. Kinderman's rule classifies the observations from π_0 into π_1, iff $2T_{n0} - T_{n1} - T_{n2} > 0$; he assumed that the distribution in π_2 differs from that in π_1 by a positive shift in translation. He computed the relative asymptotic efficiency (in Pitman's sense) of this rule with the rule obtained by replacing the T_{nj} by the corresponding sample mean of the observations (from π_j) and specialized his results to "Wilcoxon's rank-sum" scores and "normal" scores. Govindarajulu and Gupta (1972) considered similar linear rank statistics for the m-population problem when the sample sizes may be different and obtained a rule based on them which asymptotically controls the average (with respect to a known prior) PCC.

For the two-population problem, a sequential rule based on Mann-Whitney statistics was proposed and studied by Woinsky and Kurz (1969). (See Fu's book (1968), for some other nonparametric sequential rules.)

An Empirical Bayes Approach

Johns (1961) considered the two-category classification problem when I is considered a random variable and the two categories are defined by a partition of the I-space. (See Section 2.) Following the empirical Bayes approach, he proposed a rule δ_N based on a training sample of size N and showed that the Bayes risk of δ_N tends to the minimum Bayes risk under complete knowledge of the distribution of (X,I). He treated the following three cases: (i) X is discrete-valued (supervised training sample); (ii) X is of continuous type (supervised training sample); (iii) X is discrete-valued (post-supervised training sample). It may be noted that when I is treated as a classificatory varible and the loss function is 0-1, his rules reduce to NN rules.

Selection of Variables

On the basis of random samples from two p-variate distributions, Patrick and Fisher (1969) devised a method for obtaining a q-dimensional ($q<p$) linear subspace of R^p such that the two induced q-variate marginal distributions are most "separated". Their method is based on nonparametric estimates (Murthy's extension of Parzen's estimate) of the p.d.f.'s and a 'separation' or distance criterion. For related work, see Patrick's book (1972) and Meisel's paper (1971). A nonparametric sequential method for including additional variates for classification is given in Smith and Yau (1972). For other methods, see Fu's book (1968), Wu (1970), Davisson et al. (1970).

Other Results

Suppose that the c.d.f. of X is F_i in π_i ($i=1,2$), where F_i has the mean μ_i and the covariance matrix Σ. Recall that the maximum likelihood rule classifies X into π_1, iff

$$X'\Sigma^{-1}(\mu_1-\mu_2) > \frac{1}{2}(\mu_1+\mu_2)'\Sigma^{-1}(\mu_1-\mu_2)$$

when the distributions F_i are N_p. Using the well-known one-sided Chebyshev-inequality, Zhezhel (1968) showed that the maximum PMC of such a rule is $(1 + \Delta^2/4)^{-1}$, where $\Delta^2 = ||\mu_1-\mu_2||^2_\Sigma$, for all possible such F_i's.

Albert (1963) considered the classification problem where the supports of X are S_1 and S_2 in π_1 and π_2, respectively, where S_i's are *unknown* disjoint subsets of a Hilbert space such that the convex hulls of S_i's are at a positive distance apart. Samples are drawn from $S_0 \cup S_1$ sequentially and at the nth stage a decision rule is given based on post-supervised training sample such that the PMC's tend to 0 as $n \to \infty$.

References (7)

Aoyama, H. (1950), *Ann. Inst. Statist. Math.*, see Ref. 1.

Fix, E. and Hodges, J. L. (1951), Nonparametric discrimination: Consistency properties, *U.S. Air Force School of Aviation Medicine, Report No.4*, Randolph Field, Texas.

Fix, E. and Hodges, J. L. (1953), Nonparametric discrimination: Small sample properties, *Ibid, Report No. 11*.

Stoller, D. C. (1954), Univariate two-population distribution-free discrimination, *Jour. Amer. Statist. Assoc.*, 49, 770-777.

Matusita, K. (1956), *Ann. Inst. Statist. Math.*, See Ref. 6.

Hudimoto, H. (1956), On the distribution-free classification of an individual into one of two groups, *Ann. Inst. Statist. Math.*, 8, 105-112.

Hudimoto, H. (1957), A note on the probability of the correct classification when the distributions are not specified, *Ann. Inst. Statist. Math.*, 9, 31-36.

Johns, M. V. (1961), An empirical Bayes approach to nonparametric two-way classification, *Studies in Item Analysis and Prediction* (H. Solomon, Ed.), Stanford University Press, Stanford, California.

Das Gupta, S. (1962), Univ. of N.C., Chapel Hill *Mimeo No. 333*, see Ref. 5.

Albert, A. (1963), A mathematical theory of patter recognition, *Ann. Math. Statist.*, 34, 284-299.

Das Gupta, S. (1964), Nonparametric classification rules, *Sankhyā A*, 26, 25-30.

Hudimoto, H. (1964), On a distribution-free two-way classification, *Ann. Inst. Statist. Math.*, 16, 247-253.

Hudimoto, H. (1964), On the classification I. The case of two populations, *Proc. Inst. Statist. Math. Tokyo*, 11, 31-38 (MR-29).

Hudimoto, H. (1965), On the classification II, *Proc. Inst. Statist. Math. Tokyo*, 12, 273-276 (MR-32).

Van Ryzin, J. (1965), Nonparametric Bayesian decision procedures for (pattern) classification with stochastic learning, *Proc. IV Prague Conf. on Information Theory, Statistical Decision Functions, and Random Processes*.

Van Ryzin, J. (1966), Bayes risk consistency of classification procedures using density estimation, *Sankhyā A*, 26, 25-30.

Bunke, O. (1966), see Ref. 6.

Anderson, T. W. (1966), Some nonparametric multivariate procedures based on statistically equivalent blocks, *Proc. 1st Internat. Symp. Multiv. Anal.* (P. R. Krishnaiah, Ed.), Academic Press: New York, 5-27.

Whitney, A. W. and Dwyer, S. J., III (1966), Performance and implementation of the K-nearest neighbor decision rule with incorrectly identified training samples, *Proc. IV Annual Allerton Conf. on Circuit Theory and System Theory*, Champaign, Illinois.

Specht, D. F. (1966), Generation of polynomial discriminant functions for pattern recognition, presented at IEEE Pattern Recognition Workshop, Puerto Rico.

Patrick, E. A. (1966), Distribution-free, minimum conditional risk learning systems, Purdue Univ. School of Elec. Engin. *Tech. Report EE66-18*, Lafayette, Indiana.

Cover, T. M. and Hart, P. E. (1967), Nearest neighbor pattern classification, *IEEE Trans. Inform. Theory*, IT-16, 26-31.

Fu, K. S. (1968), Rates of convergence for nearest neighbor procedures, *Proc. Hawaii Internat. Conf. on System Sciences*, 413-415.

Quesenberry, C. P. and Gessaman, M. P. (1968), Nonparametric discrimination using tolerance regions, *Ann. Math. Statist.*, 39, 664-673.

Zhezhel, Y. N. (1968), The efficiency of a linear discriminant function for arbitrary distributions, *Engineering Cybernetics*, 6, 107-111.

Pelto, C. R. (1969), Adaptive nonparametric classification, *Technometrics*, 11, 775-792.

Glick, N. (1969), *Stanford Univ. Tech. Report*, see Ref. 4.

Kurz, L. and Woinsky, M. M. (1969), Sequential nonparametric two-way classification with prescribed maximum asymptotic error, *Ann. Math. Statist.*, 40, 445-455.

Patrick, E. A. and Fisher, F. P. (1969), Nonparametric feature selection, *IEEE Trans. Inform. Theory*, IT-15, 577-584.

Patrick, E. A. and Fisher, F. P. (1970), Generalized K nearest decision rule, *Jour. Information and Control*, 16, 128-152.

Peterson, D. W. (1970), Some convergence properties of a nearest neighbor rule, *IEEE Trans. Inform. Theory*, IT-16, 26-31.
Anderson, M. W. and Benning, R. D. (1970), A distribution-free discrimination procedure based on clustering, *IEEE Trans. Inform. Theory*, IT-16, 541-548.
Davisson, L. D., Feustel, E. A. and Modestino, J. W. (1970), The effects of dependence on nonparametric detection, *IEEE Trans. Inform. Theory*, IT-16, 32-41.
Wee, W. G. (1970), On feature selection in a class of distribution-free pattern classifiers, *IEEE Trans. Inform. Theory*, IT-16, 47-55.
Hellman, M. E. (1970), The nearest neighbor classification rule with a reject option, presented at the IEEE Internat. Convention on Information Theory, Holland.
Meisel, W. S. (1971), On nonparametric feature selection, *IEEE Trans. Inform. Theory*, IT-17, 105-106.
Glick, N. (1972), *Jour. Amer. Statist. Assoc.*, see Ref. 4.
Patrick, E. A. (1972), Prentice Hall, see Ref. 1 (books).
Goldstein, M. (1972), K_n-nearest neighbor classification, *IEEE Trans. Inform. Theory*, IT-18, 627-630.
Smith, S. E. and Yau, S. S. (1972), Linear sequential pattern classification, *IEEE Trans. Inform. Theory*, IT-18, 673-678.
Govindarajulu, Z. and Gupta, A. K. (1972), Certain nonparametric classification rules: Univariate case, Michigan Univ. Statist. Dept. *Tech. Report 17*.
Gessaman, M. P. and Genaman, P. H. (1972), A comparison of some multivariate discrimination procedures, *Jour. Amer. Statist. Assoc.*, 67, 468-472.

8. Miscellaneous References

(a) *On distance functions.*

Pearson, K. (1926), *Biometrika*, see Ref. 1.
Fréchet, M. (1929), Sur la distance de deux variable aléatoires, *C.R. Acad. Sci.*, Paris, 188, 368-370.
Mahalanobis, P. C. (1930), *J. Asiatic Soc.*, Bengal, see Ref. 1.

Mahalanobis, P. C. (1936), *Proc. Nat. Inst. Sci.*, India, see Ref. 1.

Hoel, P. G. (1944), On statistical coefficients of likeness, *Univ. Calif. Publ. Math.*, 2(1), 1-8, (MR-6).

Bhattacharya, A. (1946), On a measure of divergence between two multinomial populations, *Sankhyā*, 2, 401-406 (MR-8).

Rao, C. R. (1947), *Nature*, see Ref. 1.

Ivanović, B. V. (1954), Sur la discrimination des ensembles statistiques, *Publ. Inst. Statist.*, Univ. of Paris, 3, 207-269 (MR-16).

Adhikari, B. P. and Joshi, D. D. (1956), Distance, discrimination et résumé exhaustif, *Publ. Inst. Statist.*, Univ. of Paris, 5, 57-74 (MR-19).

Fréchet, M. (1957), Sur la distance de deux lois de probabilité, *C.R. Acad. Sci.*, Paris, 244, 689-692 (MR-18).

Fréchet, M. (1959), Les definitions de la Somme et du product Scalaires en terms de distance dans un space abstrait (avec supplement), *Cal. Math. Soc. (Golden Jubilee vol.)*, 1, 151-157, 159-160.

Kullback, S. (1959), Information theory and Statistics, Wiley (Dover-1968).

Samuel, E. and Bachi, R. (1964), Measure of distances of distribution functions and some applications, *Metron*, 23, 83-121.

Ali, S. M. and Silvey, S. D. (1966), A general class of coefficients of divergence of one distribution from another, *Jour. Roy. Statist. Soc.*, Series B, 28, 134-142.

Matusita, K. (1967), On the notion of affinity of several distributions and some of its applications, *Ann. Inst. Statist. Math.*, 19, 181-192.

(b) *Clustering.*

Macqueen, J. (1967), Some methods for classification and analysis of multivariate observations, *Proc. 5th Berkely Symposium on Probability and Statistics*, 2, Univ. of Calif.

(c) *Review.*

Hodges, J. L. (1950), see Ref. 1.

Miller, R. G. (1962), Statistical Prediction by discrimi-

nation analysis, *Amer. Meteor. Soc.*, Boston.
Nagy, G. (1967), State of the art in pattern recognition, *Proc. IEEE,* 56, 836-860.

METHODS AND APPLICATIONS OF EQUAL-MEAN DISCRIMINATION

M. M. Desu and S. Geisser
State University of New York at Buffalo
and University of Minnesota

1. *Introduction*

Until the beginning of the past decade the problem of discrimination, initiated by Fisher (1936), has been investigated mainly with respect to mean vectors of several multidimensional normal populations with a common covariance matrix. Previously there had been some investigations of linear discriminants for unequal covariance matrices by Smith (1947), Kossack (1945), and also Greenhouse and Kullback as given in Chapter 13 of Kullback (1959). Subsequently, other research workers considered this problem from various points of view, in particular Clunies-Ross and Riffenburgh (1960), Anderson and Bahadur (1962). In 1961 Okamoto studied the problem with respect to differing covariance matrices of two multidimensional normal populations with a common mean vector. Besides developing the theory, he presented a biometrical application.

In this paper we shall attempt to sketch an account of the various ways of looking at the problem focusing on the case where the equal means are the zero vector. In section 2 we shall give a formal setup of the problem, without specific assumptions about the distributions, and make some general remarks. In section 3 we discuss the problem with reference to normal distributions with common mean vector and differing covariance matrices from the classical viewpoint. The results of this section represents the work of Okamoto (1961) and Bartlett and Please (1963). Section 4 deals with Bayesian approaches to the problem summarizing work of Geisser (1964), Geisser and Desu (1967, 1968) and of Enis and Geisser (1970). The penultimate section presents an example illustrating the application of some of the methods presented in earlier sections. The last section discusses the problem of discriminating amongst multiple births based only on twin data, a natural application of the zero mean vector case.

2. *Setup of the Problem and Some Remarks*

We assume that there are two populations Π_1, Π_2 and the observations (which are p-dimensional vectors) from Π_i are random samples from the distribution defined by the continuous c.d.f. $F(\cdot|\theta_i)$ with the density function $f(\cdot|\theta_i)$ (i=1,2). Given an observation z, a priori we assume that it comes from either Π_1 or Π_2. The task is to devise a procedure that will determine, subject to uncertainty, from which population the observation z at hand emanates or a function which "best" separates the two populations. This is essentially the problem of discrimination.

The likelihood principle and the sufficiency principal imply that we need to consider only procedures based on the ratio of the likelihoods $\lambda(\theta) = f(z|\theta_1)/f(z|\theta_2)$ where $\theta = \theta_1 \cup \theta_2$ the set of distinct parameters. Intuition and the likelihood principle have led to the following discrimination procedure based on $U = \log \lambda$:

$$\text{Assign } z \text{ to } \Pi_1 \text{ if and only if } U > c, \quad \text{otherwise to } \Pi_2 \;; \tag{2.1}$$

where c is some specified constant. Of course there is no particular reason to choose U above any other monotonic increasing function of λ. It will be immaterial in what way we assign z when $U = c$, since we shall assume such an event occurs with probability zero. Actually the likelihood principle gives the rule R : "assign z to Π_1 if and only if $\lambda > c^*$." The rule (2.1) is equivalent to rule R. Instead of (2.1) one can use a rule of the form "assign z to Π_1 if and only if $h(\lambda) > h(c^*)$," where h is any convenient monotone increasing function.

The problem can also be presented as a two decision problem using the decision theory framework of Wald, e.g., Anderson (1958), Okamoto (1961). Let w_1 (w_2) be the loss incurred in deciding that z comes from $\Pi_2(\Pi_1)$ when actually it comes from $\Pi_1(\Pi_2)$. If we let q_i be the prior probability that z comes from Π_i (i=1,2), then it is clear that the (Bayes) rule, which minimizes the average risk with respect to $q = (q_1, q_2)$, is a non-randomized decision rule and it has the form (2.1) where $c = \log(q_2 w_2 / q_1 w_1)$. It can also be demonstrated that at least one minimax rule exists, which has the same form (2.1).

In most practical situations we do not know the densities $f(\cdot|\theta_i)$ completely, in the sense that we know only the functional form of f and perhaps some components of the parametric vector θ_i. As such one cannot use the rule (2.1). However there are at least three alternatives at one's disposal.

2.1 Classical Approaches

Estimates of the unknown parameters are obtained from the set of all observations $X = (X_1, X_2)$, where X_i is the sample from Π_i (i=1,2). In U the unknown parameters are replaced by their sample analogues to obtain V. Using this function V in (2.1) for U one obtains a discrimination rule, which could be used for assigning future observations. Thus, the rule is:

$$\text{Assign } z \text{ to } \Pi_1 \text{ if and only of } V > c, \quad \text{otherwise to } \Pi_2 . \tag{2.2}$$

Again we note that we could have chosen a function h other than U and specified some optimum estimation property to obtain the estimator of h. This property could have led to different rules for different h although intuitively we should have preferred to obtain the same rule irrespective of h. One way to solve this dilemma is to use an estimation procedure such as maximum likelihood. For this problem this is an invariant procedure since the m.l.e. of $h(\lambda)$, $\hat{h}(\lambda) = h(\hat{\lambda})$ where $\hat{\lambda}$ is the m.l.e. of λ. Hence we may assert that the rule is invariant with respect to this procedure. In practice it is customary however to apply the estimating procedures to θ rather than $\lambda(\theta)$, though the latter would be preferable. A second and more theoretically satisfying method is to obtain a rule which minimizes the error of classification with regard to future observations. Although h, when it is known, will satisfy this objective, it is extremely difficult, if not impossible, to obtain such a rule or to demonstrate whether any estimator of h or any other function of the observations will achieve this goal where θ is unknown. A third way of viewing the problem is to select a "simple" function of z to maximally separate the populations with regard to some distance criterion. Of course linearity in z has immediate advantages in terms of simplicity, recalling of course that when we are dealing with normal populations having the same dispersion that U is linear in z. Here the primary emphasis is on the form of the discriminatory function and secondarily on its optimal estimation with regard to a distance criterion. Although this seems to be the procedure of choice when assumptions about the functional form of the distributions are lacking, Fisher (1936), we note that Clunies-Ross and Riffenbourgh (1960) as well as Anderson and Bahadur (1962) apply this method even when the functional forms are assumed known. They derive procedures for the case of two Normal populations differing in means and dispersions with a view towards optimizing the discriminatory power with regard to linear functions. This approach will be of some value when the means are unequal but relatively useless when they are equal. In the later case a quadratic discriminant would appear to be closer to the mark, as this is indicated for Normal populations and may be robust for symmetric populations.

2.2 A Semi-Bayesian Approach

The first of a second set of alternatives is based on a Bayesian estimation technique as a replacement for classical estimation procedures. Since q_i is the prior probability that z emanates from Π_i $(i=1,2)$, one obtains the posterior probability that z belongs to Π_i as

$$\Pr(z \in \Pi_i|\theta) = q_i f(z|\theta_i)/[q_1 f(z|\theta_1) + q_2 f(z|\theta_2)], \qquad (2.3)$$

where θ is the set of all distinct parameters. The rule then, equivalent to (2.1), is to assign z to Π_1 if and only if $\Pr(z \in \Pi_1|\theta) > \Pr(z \in \Pi_2|\theta)$, or $U > \log q_2 q_1^{-1} = c$. In this approach, which is the counterpart of the first discussed of the classical procedures, the focus is also on estimating U a function of λ and consequently the unknown set θ. Here we can utilize the posterior distribution $P(\theta|X)$ assuming the injection of a prior for θ. Again we are faced with the problem of which monotone function $h(\lambda(\theta))$ to estimate and what estimating procedure to use. We noted before that maximum likelihood estimation yielded an invariant rule for the frequentist approach. We observe here that if we choose the posterior median of h, this will yield the same rule irrespective of h, i.e., an invariant rule, since the posterior median of $h(\lambda)$ is $h(\lambda_m)$ where λ_m is the posterior median of λ. However, in practice this is generally a very difficult computation to make. Hence we shall settle for the rule:

$$\text{Assign } z \text{ to } \Pi_1 \text{ if and only if } E_\theta[U] > c, \quad \text{otherwise to } \Pi_2 , \tag{2.4}$$

as this computation is manageable in most of the problems we shall discuss.

2.3 *Complete Bayesian Approach*

Instead of estimating $\log\{f(z|\theta_1)/f(z|\theta_2)\}$ in the previous approach, one can estimate $f(z|\theta_i)$ by $\int f(z|\theta_i)dp(\theta|X)$ where $P(\theta|X)$, as before, stands for the posterior distribution of θ. This average is called the *predictive density of z* in Π_i and is denoted by $f(z|X,\Pi_i)$. The ratio of these predictive densities may be used to define the following discrimination rule:

$$\text{Assign } z \text{ to } \Pi_1 \text{ whenever } W = \log\left\{\frac{f(z|X,\Pi_1)}{f(z|X,\Pi_2)}\right\} > c, \quad \text{otherwise to } \Pi_2 . \tag{2.5}$$

This rule is equivalent, when $e^c = q_2q_1^{-1}$, to assigning z to that population which has maximum posterior predictive probability and hence has the virtue of minimizing the total predictive error of misclassification similar to the second classical goal. A third Bayesian possibility, when there is reason for emphasis on a class of discriminating functions, is to choose that particular function of the class which minimizes the predictive error of classification with respect to $f(z|X,\Pi_i)$ and q_i $(i=1,2)$, see Enis and Geisser (1971). In what follows, the emphasis is on the approach 2.2 and the predictive density approach of 2.3 with discussion focused on the multivariate normal case with equal mean vectors and different covariance matrices. The case where the covariance matrices have uniform structure and the common mean vector is zero, receives special attention.

3. *Discrimination for Covariance Matrices of the Multivariate Normal Populations from the Classical Viewpoint*

In this section we shall examine the problem of discrimination on the assumption that $F(\cdot|\theta_i)$ is the cdf of p-variate normal distribution with mean vector μ and covariance matrix Σ_i. In this case the quantity U is given by

$$\begin{aligned} U &= \frac{1}{2}\log[|\Sigma_2|/|\Sigma_1|] - \frac{1}{2}(z-\mu)'(\Sigma_1^{-1}-\Sigma_2^{-1})(z-\mu) \\ &= \frac{1}{2}\log[|\Sigma_2|/|\Sigma_1|] - \frac{1}{2}Q \quad , \text{ say.} \end{aligned} \tag{3.1}$$

Q is called the quadratic discriminant function. It may be noted that the rule (2.1) can be expressed in terms of Q viz.,

$$\text{Assign } z \text{ to } \Pi_1 \text{ if and only if } \quad Q < \log\left[\frac{|\Sigma_2|}{|\Sigma_1|}\right] - 2c. \tag{3.2}$$

As pointed out in section 2, Q involves unknown parameters. We shall assume that we have random samples of size n_1 and n_2 from Π_1 and Π_2 respectively. One need distinguish the two cases corresponding to the common mean vector μ known or unknown. If μ is not known it is often estimated by the pooled sample mean vector, though this is not the most efficient estimator. We denote by $\hat{\mu}$ the estimator of μ or the known value of μ. If $\hat{\Sigma}_1$ and $\hat{\Sigma}_2$ are estimators of Σ_1 and Σ_2, then the rule assumes the form:

$$\text{Assign } z \text{ to } \Pi_1 \text{ if and only if} \quad \hat{Q} < \log_e[|\hat{\Sigma}_2|/|\hat{\Sigma}_1|] - 2c \;, \tag{3.3}$$

where $\hat{Q} = (z-\hat{\mu})'(\hat{\Sigma}_1^{-1} - \hat{\Sigma}_2^{-1})(z-\hat{\mu})$. For further details about the rule (3.3) one is referred to Okamoto (1961).

Bartlett and Please (1963) studied the rule (3.2) for the case when the common mean vector is the zero vector and the covariance matrices have uniform structure with the same correlation coefficient ρ, viz., $\Sigma_1 = (1-\rho)I + \rho E$ and $\Sigma_2 = \sigma^2[(1-\rho)I + \rho E]$. (I is the identity matrix and E is the matrix of the same dimension as I, with all elements being equal to 1.) In this case (3.2) reduces to:

$$\text{Assign } z \text{ to } \Pi_1 \text{ if and only if} \quad a - \frac{\rho}{1 + (p-1)\rho}(a+b) < \frac{(1-\rho)p \log \sigma^2}{1 - \sigma^{-2}} - 2c \;, \tag{3.4}$$

where $a = \text{tr } zz'$ and $a + b = \text{tr}(Ezz')$. For the purpose of discriminating between like-sexed monozygotic and dizygotic twins, where the zero mean vector arises naturally, they used the rule (3.4) with $c = 0$ and with ρ replaced by its estimate based on like-sexed twin data. We have included an example in section 5 illustrating their method of discrimination.

If we assume that the covariance matrices have general uniform structure, namely, $\Sigma_i = \sigma_i^2[(1-\rho_i)I + \rho_i E]$, the quantity U of (3.1) reduces to

$$2U_1 = p \log\left(\frac{\sigma_2^2}{\sigma_1^2}\right) + (p-1) \log\left[\frac{1-\rho_2}{1-\rho_1}\right] + \log\left[\frac{1+(p-1)\rho_2}{1+(p-1)\rho_1}\right]$$

$$- \Bigg[\left\{\frac{1}{\sigma_1^2(1-\rho_1)} - \frac{1}{\sigma_2^2(1-\rho_2)}\right\} a + \left\{\frac{\rho_1}{1-\rho_1}\,\frac{1}{1+(p-1)\rho_1}\,\frac{1}{\sigma_1^2} \tag{3.5}$$

$$- \frac{1}{\sigma_2^2} \frac{\rho_2}{1-\rho_2} \frac{1}{1+(p-1)\rho_2} \Bigg\} (a+b) \Bigg] .$$

Thus the rule (3.2) now reads as:

$$\text{Assign } z \text{ to } \Pi_1 \text{ if and only if } U_1 > c . \qquad (3.6)$$

If we take $\sigma_1^2 = 1$, $\sigma_2^2 = \sigma^2$ and $\rho_1 = \rho_2 = \rho$ the rule (3.6) reduces to (3.4).

4. *Bayesian Discrimination Methods for Covariance Matrices of Multivariate Normal Populations with Common Mean Vector*

In this section we shall present the discrimination rules obtained by using alternatives 2.2 and 2.3 of section 2 for the problem of discrimination outlined in the previous section. The results of this section are excerpted from Geisser (1964), Geisser and Desu (1967,1968), Enis and Geisser (1970).

Here we assume that the common mean vector is known and wihout loss of generality it is taken as the zero vector. The assumption of zero-mean vector is appropriate when our observation vector represents paired differences of a set of characteristics from either monozygotic or dizygotic like sexed twin populations.

First we shall present results on the posterior densities of the parameters. These are followed by the results on the two discrimination rules obtained under methods 2.2 and 2.3 of section 2.

Let $(X_{1i}, \ldots, X_{N_i i})$ denote the sample from Π_i and let

$$N_i S_i = \sum_{\alpha=1}^{N_1} X_{\alpha i} X'_{\alpha i} , \quad (i = 1, 2) . \qquad (4.1)$$

We assume that $N_i > p$ $(i=1,2)$. Let $d\Sigma_i^{-1}$ stand for

$\prod_{i\geq j} d\sigma^{ij}$ and let us use a convenient prior density for Σ_i^{-1},

$$g(\Sigma_i^{-1})d\Sigma_i^{-1} \propto |\Sigma_i|^{p+1/2}\, d\Sigma_i^{-1} . \tag{4.2}$$

Under these assumptions the posterior density of Σ_i^{-1} is

$$P(\Sigma_i^{-1}|S_i, \Pi_i) \propto |\Sigma_i|^{\frac{1}{2}(N_i-1-p)} \exp\{-\frac{1}{2}\operatorname{tr} \Sigma_i^{-1}N_iS_i\}. \tag{4.3}$$

As mentioned earlier in section 2, we can use this posterior density to obtain the predictive density of z or to obtain the $E[\log \lambda(\theta)]$.

4.1 Semi-Bayesian Approach

Even though it would be of some interest to obtain the posterior median of $\lambda(\theta)$, its evaluation presents a problem. Thus, we proceed to obtain $E[\log \lambda(\theta)]$. Enis and Geisser (1970) showed that

$$E[\log \lambda(\theta)] = E[U] = V + B(p,N_1,N_2) \tag{4.4}$$

where

$$V = U = \frac{1}{2}\log \frac{|S_2|}{|S_1|} - \frac{1}{2} z'(S_1^{-1}-S_2^{-1})z , \tag{4.5}$$

$$B(p,N_1,N_2) \tag{4.6}$$

$$= \frac{1}{2} \sum_{k=1}^{p} \psi\frac{N_1+1-k}{2} - \psi\frac{N_2+1-k}{2} - p \log \frac{N_2}{N_1} .$$

In (4.6), ψ stands for the Digamma function, defined as

$$\psi(\alpha) = \frac{d \log \Gamma(\alpha)}{d\alpha} . \tag{4.7}$$

It is interesting to note that B of (4.6) becomes zero when $N_1 = N_2$, so that $E(U) = V$. In that case we obtain the rule:

$$\text{Assign } z \text{ to } \Pi_1 \text{ if and only if } V > c , \tag{4.8}$$

which is the same as the rule obtained in the classical approach. If $N_1 \neq N_2$, the only change in (4.8) is that the constant c will be decreased by $B(p, N_1, N_2)$. Thus this semi-Bayesian approach and the classical approach give essentially the same rule, since $B(p, N_1, N_2)$ tends to zero as both N_1 and N_2 increase.

4.2 *Complete Bayesian Approach Based on Predictive Densities*

Geisser (1964) considered the general problem of discrimination of normal populations using this approach. We shall only mention the results pertinent to our problem. Using (4.3), Geisser obtains the predictive density of z as

$$f(z|S_i,\Pi_i) \propto \frac{\Gamma\{\frac{1}{2}(N_i+1)\}|N_iS_i|^{\frac{1}{2}N_i}}{\Gamma\{\frac{1}{2}(N_i-p+1)\}|N_iS_i+zz'|^{\frac{1}{2}(N_i+1)}} . \tag{4.9}$$

Hence, for $N_1 = N_2 = N$

$$\begin{aligned} W &= \log \frac{f(z|S_1, \Pi_1)}{f(z|S_2, \Pi_2)} \\ &= \frac{1}{2} N \log \frac{|NS_1|}{|NS_2|} + \frac{1}{2}(N+1) \log \frac{|NS_2+zz'|}{|NS_1+zz'|} . \end{aligned} \tag{4.10}$$

Thus this approach gives the rule:

$$\text{Assign } z \text{ to } \Pi_1 \text{ if and only if } W > c , \tag{4.11}$$

where W is given by (4.10). We may also compute the posterior odds ratio

$$R_1 = \frac{q_1\left(1 + \frac{z'T_1^{-1}z}{N+1}\right)^{-\frac{N+1}{2}}}{q_2\left(1 + \frac{z'T_2^{-1}z}{N+1}\right)^{-\frac{N+1}{2}}}$$

where $T_i = (N+1)^{-1}NS_i$. If we estimate the posterior odds ratio inserting the estimator T_i for Σ_i in the ratio of densities we get

$$R_2 = \frac{q_1}{q_2}\, e^{\frac{1}{2}(z'T_2^{-1}z - z'T_1^{-1}z)}$$

Further if $z'T_1^{-1}z \leq z'T_2^{-1}z$, then

$$1 \leq R_1 \leq \frac{q_1}{q_2}\left(1 + \frac{z'T_2^{-1}z - z'T_1^{-1}z}{N+1}\right)^{\frac{N+1}{2}} \leq R_2$$

and conversely if $z'T_1^{-1}z \geq z'T_2^{-1}z$ then $R_2 \leq R_1 \leq 1$. This implies that the predictive posterior probability that z belongs to Π_1 or Π_2 will be closer to 1/2 than the estimated posterior probability, if estimated in this particular way. At any rate this blurring effect which decreases with N seems to be not unreasonable.

We also note that the complete Bayesian and the Semi-Bayesian approaches involve some difficult integrals which are not easy to evaluate when the common mean is unknown. On the other hand, when the means are assumed different and unknown the results are easy to obtain (Geisser [1964], Enis and Geisser [1970]).

4.3 Results for Uniform Covariance Structure

Here we assume that

$$\Sigma_i = \sigma_i^2[(1-\rho_i)I + \rho_i E]$$

where $1 > \rho_i > -(p-1)^{-1}$, $i = 1,2$. Taking the prior density of σ_i, ρ_i as

$$g(\sigma_i, \rho_i) \propto [\sigma_i(1-\rho_i)\{1 + (p-1)\rho_i\}]^{-1}, \tag{4.13}$$

Geisser and Desu (1968) obtained the posterior density of σ_i, ρ_i. It is convenient to consider the posterior den-i-ty of $\alpha_i = \sigma_i^2\{1 + (p-1)\rho_i\}$ and $\theta_i = \sigma_i^2(1-\rho_i)$. Let

$$A_i = \operatorname{tr} \sum_{\alpha=1}^{N_i} X_{\alpha i} X'_{\alpha i} = \sum_{\alpha} \sum_{j=1}^{p} X_{j\alpha i}^2 = pN_i s_i^2 ,$$

$$A_i + B_i = \operatorname{tr}\{E \sum_{\alpha=1}^{N_i} X_{\alpha i} X'_{\alpha i}\} \tag{4.14}$$

$$= \sum_{\alpha} (\sum_{j=1}^{p} X_{j\alpha i})^2 = pN_i s_i^2[1 + (p-1)r_i] .$$

The joint posterior density of α_i and θ_i is

$$P(\alpha_i, \theta_i | A_i, B_i, \Pi_i)$$

$$= (A_i + B_i)^{\frac{1}{2}N_i} \{(p-1)A_i - B_i\}^{\frac{1}{2}(p-1)N_i} \tag{4.15}$$

$$\cdot [\Gamma(\tfrac{1}{2}N_i)\Gamma\{\tfrac{1}{2}(p-1)N_i\}]^{-1} \alpha_i^{-\frac{1}{2}(N_i+2)} \theta_i^{-\frac{1}{2}[(p-1)N_i+2]}$$

$$\cdot \exp[-\tfrac{1}{2}\{\frac{A_i + B_i}{\alpha_i} + \frac{(p-1)A_i - B_i}{\theta_i}\}] .$$

It can be shown that

$$E(\log \lambda) = E(U_1) = V_1 + \frac{p-1}{2}[\psi\{\frac{(p-1)N_2}{2}\} - \psi\{\frac{(p-1)N_1}{2}\}]$$

$$+ \frac{1}{2}\{\psi(\frac{N_2}{2}) - \psi(\frac{N_1}{2})\} , \qquad (4.16)$$

so that $E(\log \lambda) = V_1$ whenever $N_1 = N_2$. The function V_1 is given by

$$V_1 = \hat{U}_1 = \frac{(p-1)}{2} \log \frac{s_2^2(1-r_2)}{s_1^2(1-r_1)} + \frac{1}{2} \log \frac{\{1+(p-1)r_2\}s_2^2}{\{1+(p-1)r_1\}s_1^2}$$

$$+ \frac{\{(p-1)a - b\}}{2p} \left[\frac{1}{s_2^2(1-r_2)} - \frac{1}{s_1^2(1-r_1)}\right] \qquad (4.17)$$

$$+ \frac{a+b}{2p} \left[\frac{1}{s_2^2\{1+(p-1)r_2\}} - \frac{1}{s_1^2\{1+(p-1)r_1\}}\right]$$

Thus for $N_1 = N_2$, the Semi-Bayesian approach gives the same rule as the rule one would have used under the Classical Approach by substituting the usual estimates for the parameters. In particular the rule is:

$$\text{Assign } z \text{ to } \Pi_1 \text{ if and only if } V_1 > c , \qquad (4.18)$$

where V_1 is given by (4.17).

Geisser and Desu (1968) obtained the predictive density of z as

$$f(z|A_1, B_1, \Pi_1) \propto [\Gamma(\tfrac{1}{2}\nu_1)/\Gamma(\tfrac{1}{2}N_1)][\Gamma\{\tfrac{1}{2}(p-1)\nu_1\}/\Gamma\{\tfrac{1}{2}(p-1)N_1\}]$$

$$\cdot \frac{(A_1 + B_1)^{\frac{1}{2}N_1}\{(p-1)A_1 - B_1\}^{\frac{1}{2}(p-1)N_1}}{(\bar{A}_1 + \bar{B}_1)^{\frac{1}{2}\nu_1}\{(p-1)\bar{A}_1 - \bar{B}_1\}^{\frac{1}{2}(p-1)\nu_1}} \qquad (4.19)$$

where

$$\nu_1 = N_1 + 1, \quad \bar{A}_1 = A_1 + a \quad \text{and} \quad \bar{B}_1 = B_1 + b \;. \tag{4.20}$$

By changing the suffix 1 in (4.19) to 2 we obtain $f(z|A_2, B_2, \Pi_2)$. In the case when $N_1 = N_2 = N$, the logarithm of the ratio of the predictive densities is given by

$$\begin{aligned} W_1 = {} & N \log \frac{A_1 + B_1}{A_2 + B_2} + N(p-1) \log \frac{(p-1)A_1 - B_1}{(p-1)A_2 - B_2} \\ & + (N+1) \log \frac{\bar{A}_2 + \bar{B}_2}{\bar{A}_1 + \bar{B}_1} + (N+1)(p-1) \log \frac{(p-1)\bar{A}_2 - \bar{B}_2}{(p-1)\bar{A}_1 - \bar{B}_1} \;. \end{aligned} \tag{4.21}$$

Thus this complete Bayesian approach gives the rule:

$$\text{Assign } z \text{ to } \Pi_1 \text{ if and only if } W_1 > c \;, \tag{4.22}$$

where W_1 is given by (4.21).

5. *An Example*

An interesting application of the above techniques is when we are interested in the discrimination between monozygotic and dizygotic (with like sex) pairs of twins. If the observations are the differences between the observations on each pair, the zero-mean vector assumption is appropriate. Here we examine the extent to which physical measurements could be used to distinguish the twin pairs, even though better discrimination can generally be achieved by more sophisticated observations on blood groups, though even these will not always be conclusive.

In his paper, Stocks (1933) recorded measurements as well as age and sex on 832 children in Elementary and Central Schools of the London County Council during the period 1925 to 1927. He recorded the following measurements for most of the children.

1. Height
2. Weight
3. Head length

4. Head breadth
5. Head circumference
6. Interpupillary distance
7. Systolic blood pressure
8. Diastolic blood pressure
9. Pulse interval
10. Respiration interval
11. Strength of grip, right hand
12. Strength of grip, left hand
13. Visual acuity, right eye
14. Visual acuity, left eye

For the twin pairs he recorded their finger prints and facial resemblance and derived an empirical criterion, based on the finger prints, height and head measurements, for separating the like sexed twins into monozygotic and dizygotic pairs.

For illustrating the discrimination techniques discussed in the previous sections, 30 pairs of female twins were selected from Stock's data. 15 of these pairs were judged monozygotic by Stock's criterion and the other 15 pairs were judged dizygotic. Following Bartlett and Please (1963) only 10 of the above measurements were included in the analysis, numbers 8, 10, 13 and 14 being omitted. From each group only 10 observations were used in constructing the discriminant functions and these discriminant functions were used to classify all 30 observations.

For each variate the difference between the first and second recorded twin was taken as the observation. The data was analyzed under two different models-uniform covariance structure and an arbitrary covariance structure.

5.1 *Discrimination Under the Uniform Covariance Model*

Here we first standardized the data as in Bartlett and Please (1963). Using only the first 10 observation vectors of each group, the ratios of sums of squares of the dizygotic twins to those of the monozygotic twins for the 10 variates were obtained. The average of these ratios was found to be $\hat{\sigma}^2 = 4.5135$. The variance of each variate was estimated from the sum of squares of the mono-

and dizygotic pairs dividing by $\hat{\sigma}^2$. After standardizing, the average variance of each variate in monozygotes is 0.0994, whereas the corresponding value in dizygotes is 0.4505. From the standardized data the values of X (= a) and Y (= a+b) were obtained for each one of the 30 twin pairs. These were plotted in Figures 1 and 2. The points corresponding to observations not used in obtaining the rules were circled.

Figure 1 gives the 2 curves representing the discrimination rules obtained under the assumption $\rho_1 \neq \rho_2$. These are given by (4.18) and (4.21) with $c = 0$. The straight line is given by the equation

$$X - 0.0353\, Y = 1.7496 .$$

If $\phi(Y)$ is defined as follows

$$\phi(Y) = 140.3900 + 11 \log_e \frac{16.9910 + Y}{141.3350 + Y} ,$$

then the equation of the curve in Figure 1 is

$$X = 0.1\, Y - 19.5787 + 11.3418 \frac{e^{\phi(Y)/99} + 1}{e^{\phi(Y)/99} - 1} .$$

An examination of Figure 1 reveals that both rules classify the data at hand in the same manner. In other words, no difference in the discriminatory power of the two rules is apparent. In the monozygotic group of the 10 observations used for obtaining the discriminant functions and one of the 5 additional observations are misclassified. In the dizygotic group one of the 5 observations not used in obtaining the discriminant function is misclassified.

Figure 2 gives the plot of the data and the straight lines which correspond to the discrimination rules obtained under the assumption $\rho_1 = \rho_2$. One straight line corresponds to the discrimination rule obtained by using sample estimates in the rule given by (3.4). It is called L_1 and its equation is

$$X - 0.0652\, Y = 1.6229 .$$

This line is the one that one would use if he followed the technique of Bartlett and Please (1963). The other line corresponds to the discrimination rule obtained by adopting the complete Bayesian approach with the assumption that the common ρ is known. The sample value is taken as the value of the parameter. The theoretical results pertaining to this case are not included in the earlier sections, but they are available in Geisser and Desu (1968). It may also be pointed out that exact results under this approach, for the case that the common ρ is unknown, are not available. The equation of the second line L_2 is

$$X - 0.0652\, Y = 1.6665 .$$

Again, no difference in the way these two lines classify the data is apparent.

5.2 *Discrimination Under General Covariance Model*

The same data was used to derive the discrimination rule (4.11). It has been found that

$$W = 7.9308 + 5.5[\log\{1 + z'(NS_2^{-1})z\} - \log\{1+z'(NS_1)^{-1}z\}].$$

Using the rule (4.11) with $c = 0$, all the 20 observations (10 from each group) used in deriving W have been correctly classified. Among the 10 additional observations (five from each group) three from the monozygotic group are misclassified. None from the dizygotic group were misclassified. Although the arbitrary covariance and the uniform covariance structures misclassified the same number of individuals, the pattern of misclassification possibly indicates that the uniform structure is more appropriate for this data, or that the structure itself is relatively unimportant for discrimination as compared with the differences in the covariance matrices.

6. *Multiple Birth Discrimination*

As was shown in the previous example, an important application of discrimination based on unequal covariance matrices occurs most naturally with regard to like sexed twins. We shall extend this to like sexed triplets, incidentally demonstrating that we need only twin data in order to obtain a discriminatory procedure for triplets. For the univariate extensions see Richter and Geisser (1960). For ease in exposition we shall assume that we have a large enough data set of like sexed twins pairs so that the estimates of the covariance matrices are essentially the actual covariance matrices. Previously in the twin discrimination application we discussed assumptions concerning the structure of the covariance matrices and implicitly implied that the monozygotic covariance matrix was "smaller" than the dizygotic. We now make this more precise by introducing a component of covariance matrix model.

Let $R_1, R_2, \ldots$ be a sequence of p-dimensional independent random variables such that R_i is $N(\mu_i, \Sigma_N)$. Further let μ_i be an observation on a random variable M which is $N(\mu, \Sigma_B)$. Let x,y be p-dimensional observations on a pair of like sexed twins from the same mother. Then x,y are interpreted as observations on R_i, R_j respectively. If $i = j$, the pair are monozygotic twins; if $i \neq j$, the pair are dizygotic twins. Assume that Σ_W, the within-egg covariance matrix, is constant for all mothers. Hence $x - y$ is $N(0, 2\Sigma_W)$ if x,y are each observations on R_i. Now suppose x,y are observations on R_i and R_j respectively. Then $x - y$ is $N(\mu_i - \mu_j, 2\Sigma_W)$ given $\mu_i - \mu_j$. Since $\mu_i - \mu_j$ is distributed as $N(0, 2\Sigma_B)$ we have that $x - y$ is unconditionally distributed as $N(0, 2\Sigma_W + 2\Sigma_B)$.

Hence the posterior probability that a future twin pair $z = x - y$ based on the p characteristics is dizygotic is $\Pr(\pi_d | z) = \dfrac{\phi_d}{\phi_d + \gamma\phi_m}$. ϕ_d and ϕ_m represent the density of a $N(0, 2\Sigma_W + 2\Sigma_B)$, $N(0, 2\Sigma_W)$ variable respectively and γ the relative frequency of monozygotic twins to dizygotic like sexed twins in the population from which the future twin pair is to be drawn.

We now consider that we are presented with a set of like sexed triplets x, y, w and are to determine their zygotic status, i.e., whether they arose from 1, 2, or 3 eggs and if from two which is the non-identical triplet. Let $z_1 = x - y$, $z_2 = x - w$. Then we can calculate the various unconditional distributions of (z_1, z_2), which are presented below. All distributions of (z_1, z_2) are multivariate normal with zero vector mean and covariance matrix given in the table.

Case	Covariance Matrix of the Joint Distribution of (z_1, z_2)	Relative Frequency
1 egg (x, y, w)	$\begin{array}{cc} 2\Sigma_W & \Sigma_W \\ \hline \Sigma_W & \Sigma_W \end{array}$	1
2 egg (x), (y,w)	$\begin{array}{cc} 2\Sigma_W + 2\Sigma_B & \Sigma_W + 2\Sigma_B \\ \hline \Sigma_W + 2\Sigma_B & 2\Sigma_W + 2\Sigma_B \end{array}$	$\frac{2}{3}\gamma$
2 egg (x,y), (w)	$\begin{array}{cc} 2\Sigma_W & \Sigma_W \\ \hline \Sigma_W & 2\Sigma_W + 2\Sigma_B \end{array}$	$\frac{2}{3}\gamma$
2 egg (x,w), (y)	$\begin{array}{cc} 2\Sigma_W + 2\Sigma_B & \Sigma_W \\ \hline \Sigma_W & 2\Sigma_W \end{array}$	$\frac{2}{3}\gamma$
3 egg (x),(y),(w)	$\begin{array}{cc} 2\Sigma_W + 2\Sigma_B & \Sigma_W + \Sigma_B \\ \hline \Sigma_W + \Sigma_B & 2\Sigma_W + 2\Sigma_B \end{array}$	γ^2

Hence the posterior probability can easily be calculated for each case and depends only on Σ_W, Σ_B and γ, parameters obtainable from twin data. Here of course the rule is to classify (x,y,w) to that case (population) which maximizes the posterior probability when the parameters are known or the estimate of that quantity when the parameters are unknown. When the parameters are unknown, there are interesting estimation problems both in the classical and Bayesian approaches but these estimates can be wholly based on twin data. However, we shall not belabor these points at present. The procedure is extendable to higher order births, Geisser (1972), though the calculations tend to be tedious. Of course these variables may also be combined with other criteria of a discrete nature such as blood groups which by itself may be conclusive in discriminating between a twin pair. There is, however, a not insubstantial number of twins for which no single conclusive method exists and the combination of inconclusive but informative criteria would be particularly useful in the classification of these cases.

REFERENCES

Anderson, T. W. (1958), *An Introduction to Multivariate Statistical Analysis*, John Wiley: New York.

Anderson, T. W. and Bahadur, R. R. (1962), Classification into two multivariate normal distributions with different covariance matrices, *Ann. Math. Statist.* 33, 420-437.

Bartlett, M. S. and Please, N. W. (1963), Discrimination in the case of zero mean differences, *Biometrika*, 50, 17-21.

Clunies-Ross, C. W. and Riffenburgh, R. H. (1960), Geometry and linear discriminants, *Biometrika*, 47, 185-189.

Enis, P. and Geisser, S. (1970), Sample discriminants which minimize posterior squared error loss. *S. African Statist. J.*, 4, 85-93.

Enis, P. and Geisser, S. (1971), Optimal predictive linear discrimination, *Ann. Math. Statist.*, 42, Abstract p. 2179.

Fisher, R. A. (1936), Use of multiple measurements in taxonomic problems, *Ann. Eugen.*, 7, 179-188.

Geisser, S. (1964), Posterior odds for multivariate normal classification, *J. R. Statist. Soc.* B, 26, 69-76.

Geisser, S. (1972), Multiple birth discrimination, *Proceedings of the Dalhousie Symposium on Multivariate Statistical Inference*, Halifax, Canada.

Geisser, S. and Desu, M. M. (1967), Bayesian zero-mean uniform discrimination, *Research Report No. 10, Department of Statistics, State University of New York at Buffalo*, 1-19.

Geisser, S. and Desu, M. M. (1968), Predictive zero-mean uniform discrimination, *Biometrika*, 55, 519-524.

Kossack, C. F. (1945), On the mechanics of classification, *Ann. Math. Statist.*, 16, 95-98.

Kullback, S. (1959), *Information Theory and Statistics*, John Wiley: New York.

Okamoto, M. (1961), Discrimination for variance matrices, *Osaka Math. Journal*, 13, 1-39.
Richter, D. L. and Geisser, S. (1960), A statistical model for diagnosing zygosis by ridge-count, *Biometrics*, 16, 110-114.
Smith, C. A. B. (1947), Some examples of discrimination, *Ann, Eugen.*, 13, 272-282.
Stocks, P. (1933), A biometric investigation of twins, Part II, *Ann. Eugen.*, 5, 1-55.

COMPUTER GRAPHICAL ANALYSIS AND DISCRIMINATION

W. J. Dixon and R. I. Jennrich
University of California, Los Angeles

Summary

A data analytic approach to discrimination will be presented in which computer graphics is used as an aid in examining the data for possible initial misclassifications and unequal covariance strutures. This is achieved by noting the data (also summarized by ellipses if desired) projected on various planes defined by the centroids of selected triplets of groups. Weighted discriminations are compared for discrimination computed under both linear and non-linear assumptions.

Introduction

The practicing data analyst often finds the classical theories of mathematical statistics of limited usefulness unless he devotes great care and attention to the needs of his particular problem. This appears to be particularly true in multivariate analysis. For general discussions see Gnanadesikan and Wilk (1969) and Rao (1962).

We present here an example of an approach to data analysis where the goal is discrimination, and the approach taken is a mixture of traditional theory, some modifications of traditional theory, graphical capacity to monitor and to permit interaction with the data analysis and, of course, computer support to the whole process.

The program to be described has been written for three primary reasons:

(i) To investigate the advantages of carrying out a discriminant analysis of data on an interactive graphic console.

(ii) To provide the first BMD program designed specifically for the frequently encountered problem of unequal covariance matrices and to include a provision for specifying relative costs of misclassification as well as prior probabilities for each group.

(iii) To investigate the value of several techniques for graphically representing groups of data. The problem is to draw, in two dimensions, meaningful pictures of higher dimensional data.

Most of the program is classical in theory. The application of this theory, however, appears to have been limited, perhaps because of lack of implementation. The classical theory can be found in Anderson (1958). The last frame described is new and considerably more experimental and, perhaps, more interesting. It will be discussed in greater detail.

Linear discriminant analysis programs make the assumption, tacitly at least, that the populations considered have about the same covariance structure and differ primarily in mean shifts. These programs work poorly, of course, when applied to populations with nearly equal means and quite distinguishable covariance structures. Although classification can be accomplished quite straight straightforwardly, the familiar linear discriminant functions are not obtained. This results both from the introduction of unequal covariance matrices and the option to use unequal costs of misclassification. These are important options in a variety of problems including diagnosis. The failure to detect cancer, for example, is surely a more costly error than the false assumption that a patient has the disease when the latter can be easily resolved by a simple biopsy.

A pictorial introduction to the problem

Consider two elliptical data swarms (Figure 1). For purposes of illustration assume one wishes to display these in one dimension. Projection on the y-axis is clearly a poor idea. Projection on the x-axis does a better job of separating the swarms but one can do better still by transforming first. One can map Figure 1 into Figure 2 by means of a linear transformation. Projection on the x-axis then produces, in one dimension, essentially all the separation present in two.

These pictures raise the basic questions:

(i) What plane do we project on?
(ii) How do we project?

The second may be rephrased as follows:

(ii) How do we transform before projection?

To address (i), assume we have succeeded in transforming the data into spherical swarms. In the case of three or fewer groups one clearly wants to project onto a plane containing the group means. More than three groups can be looked at in subsets of three. In the case of 4 or more groups, one might, however, look at the plane defined by the first two canonical variables as is done in the discriminant analysis program BMD07M, Dixon (1970). In the case of 3 or fewer groups this produces nothing new. The advantages one might gain from doing it, however, are:

(i) It is probably the best single picture summary we can produce when many groups are involved.

(ii) It follows a suggestion of C. R. Rao and is certainly less arbitrary than using the first three means encountered.

We turn to the question of how to project or, specifically, how to transform. If the original data have the same sample covariance matrix, or matrices which are scalar multiples of each other, we may, by the appropriate linear transformation, produce new swarms whose sample covariance matrices are scalar multiples of the identity. As a consequence, the transformed swarms tend to be circular. When the original swarms do not have the same shape, a compromise is necessary. A natural compromise is to transform so that the pooled covariance matrix of the transformed swarms is the identity. This is the program's default setting. It produces swarms which are "on the average" circular. Another compromise is to transform so that the sample covariance matrix of a specified swarm is a scalar multiple of the identity. A generalization of these two is a transform based on any specified linear combination of the sample covariance matrices. This is a program option.

To aid the eyes in looking at the transformed projected swarms, each is summarized by an ellipse. This is the ellipse whose image is a circle of radius r under any transformation which maps the projected swarm into one whose sample covariance matrix is the identity. If the data are approximately normally distributed, the ellipse corresponding to r = 1, the default setting, contains about 50% of the data and density.

If the original swarms have about the same shape, the projected swarms will be roughly circular as will the summary ellipses. In general (under default settings at least) they will be circular "on the average".

The Program

The program will be described frame by frame with the description of each frame introduced by its name.

PROBLEM: This frame is the first to appear when the program is executed. It gives the initial number of variables and groups, and their input format. Like almost all parameters in the program, these can be changed by simply over-typing new values. Each frame contains a list of all frame names at the bottom. A transfer to any desired frame is executed by a light pen on the appropriate name. The data are read upon exit from the Problem Frame.

SELECT: This frame gives the investigator the opportunity to select or change the variables and gorups he wishes to use in his analysis. One may wish to remove a group with too few cases or to remove variables which are part of the definition of the groups themselves.

DATA: This frame displays the data in any specified group. This includes case numbers, all variable values, and the inclusion status. Initially the status of every case is "in", but detection on its status will change it from "in" to "out" and vice versa. This frame also contains means and standard deviations for all variables in group displayed. Detection on the group number causes a display of the data in the next group. Whenever a case, variable or group, is added or deleted during execution, all parameters in all frames are immediately updated. If a case is deleted the means and standard deviations displayed in this frame will also change appropriately.

PRIOR: This frame allows the user to specify prior probabilities for each group and a matrix specifying the costs of misclassification. The initial prior probabilities are equal as are the costs of misclassification. Over-typing can change these, if desired, to any other values.

EQ. COV: This frame displays likelihood ratio statistics for testing the equality of the covariance matrices simultaneously for all of the selected groups and for each possible pair within the selected groups. Under normality, these statistics are asymptotically χ^2-distributed. The appropriate degrees of freedom and corresponding probability values are given for each.

POST: Assuming normal distributions and using the usual sample estimates for population parameters, but not assuming equal covariance matrices, this frame displays for each case in any specified group, the posterior probability that it came from any of the groups used in the analysis and identifies the group with highest posterior probability. Also for each case it gives the posterior expected loss resulting from a classification into each group and identifies the group with smallest loss.

CLASS: This frame gives four matrices summarizing the classification of the cases in any selection of groups into those groups selected for analysis. The groups classified may include new groups as well as any of the groups used to derive the classification criteria. The matrices are based on (i) posterior probability and the assumption of unequal covariance matrices, (ii) posterior expected loss with unequal covariance matrices, (iii) posterior probability under the assumption of equal covariance matrices, (iv) posterior expected loss assuming equal covariance matrices. The latter two are included to provide a convenient check against the results for the equal covariance case.

PLOT: The purpose of this frame is to produce informative two-dimensional pictures of multidimensional data. These pictures can reveal interesting properties of a set of data which a classical analysis will fail to give or possibly only hint at. They often clarify results which are otherwise difficult to understand. For example, one might find pictorial evidence that a set of groups lie quite close together and might be considered as a single group, or that one group appears to be composed of several parts which should be viewed as separate groups. One may also spot outliers, misclassified data, and visual evidence of dissimilarities in covariance structure.

It is not at all clear how successful one can be in an effort to draw meaningful two-dimensional pictures of higher dimensional data. Experience with printer plots from BMD07M, a linear discriminant analysis program, is encouraging, however. The output of this frame is more experimental than the more or less classical outputs of the previous frames.

In essence, this frame simply plots a variety of pairs of linear functions of the original data. The aim is to produce a good two-dimensional picture of a higher dimensional data set. One way to start, and the way the program does start, is to select three group means. These define a two-dimensional plane. It remains only to project the data points onto this plane. A natural way is "straight down" or in mathematical terms "perpendicularly in the Euclidean norm." Unfortunately, unless the data in each group are rather spherically distributed, such a projection will lead to unnecessary overlap between groups. Moreover, for those who worry about such things, such a projection is not invariant under change of scale.

In the equal covariance case, the groups are transformed so that all groups are spherical, or nearly so, and then the straight down projection is used. For the unequal covariance case anticipated by this program, one can transform the data so that a specified group is spherical and then project straight down to view the shape and location of the other groups relative to the circularized one. What is actually done is a little more general. The user specifies a weighted average of the group covariance matrices. He can specify that all weight be assigned to a particular covariance matrix, that equal weights be given to those matrices whose means define the projection plane, or that weights be proportional to group sizes. These are all natural selections. The data are then transformed in such a way that if this user-specified composite covariance matrix were the covariance matrix for each group, they would all be circularized and a straight down projection would produce a set of circular groups. This, of course, does not happen in general, but the groups are more or less circular on the average and hopefully unnecessary overlap can be avoided with the appropriate choice of weights.

Actually it is difficult to look at superimposed scattergrams. To aid the eyes, each group is summarized by an ellipse whose shape and size corresponds to that of the data in the group. The ellipses are normal-theory constant-density ellipses for each group and, unless otherwise specified by the user, are scaled to include about 50% of the cases in the group.

In addition to plotting selected means, scattergrams and ellipses, the investigator may identify a point by means of a light pen detect. The point's group and case number appear. He may then remove the case if he wishes (with another light pen detect or he can return to the DATA or POST frames and examine the case in greater detail). Points deleted from the analysis may be plotted separately and they may be reentered into the same or a different group.

A variety of views of the data are obtained by varying the groups, cases, and variables included as well as the selection of projection planes and modes of projection.

Examples

It is not possible to adequately illustrate the use of this program short of an actual demonstration at the console. Its true power comes from the capacity to move from one portion of the program to another, to examine various portions, to add or delete variables, cases or groups from the analysis and to immediately observe the results of this modification by viewing the data as projected in various directions.

One can, however, get some feel for the use of the program from a motion picture. The film runs about ten or fifteen minutes and shows some detail for Fisher's iris data. Several portions of the analysis are shown for Mahalanobis' anthropological data and for the F-310

(Standard Oil) data on pollution from automobile exhaust.

Example 1

Fisher's iris data (Fisher, 1936). These data consist of 4 observations on 50 samples each of 3 groups of iris. These data appear in many places in the literature. The measurements are sepal length and width, and petal length and width. The groups are setosa, versicolor and virginica. The problem is not only classical. It is very simple. The data are well separated by group and there are no obvious outliers or other disturbing factors. It will be noted however that the covariance matrices show considerable differences.

Example 2

Mahalanobis' anthropological data (Mahalanobis et al, 1949). Only a portion of these data are used: 10 groups of 50 each but including all 13 variables. The ten groups are:

1. Ahir	4. Basti	7. Panika
2. Kurmi	5. Brahmin (others)	8. Majhi
3. Kahar	6. Chero	9. Bhatu
		10. Habru

The variables are age, frontal breadth, bizygomatic breadth, nasal length, head length, upper face length, sitting height, stature, bigonial breadth, head breadth, nasal breadth, nasal depth, total face length. The analysis presented by the original authors is very complete, 214 journal pages. The problems of measurement definition, data acquisition, computational problems, alternate methods of analysis, etc. are thoroughly and admirably done. It is difficult to estimate the size of the task. One wonders at the energy and determination to invert so many large matrices.

The most interesting point to present here is the ease with which we were able to reproduce the discrimination chart given as chart 8(b) in their article.

Example 3

The Chevron F-310 data (1971). Data are available

for a number of models and makes of cars. Exhaust emissions were measured before and after the cars had been driven about 2000 miles.

The groups are General Motors, Chrysler, Ford, Datsun, Toyota, Mercedes, Fiat, Rambler, Volkswagen and (Volvo, Triumph, Opel, English Ford).

The variables are model year, before or after 1966, number of cylinders, displacement, automatic or manual, number of carburetor valves, starting readings for odometer, hydrocarbons, carbon monoxide, nitrogen oxide, and final readings of these last four. In our analysis we took differences of these four readings. It is clear that no appreciable differences were observable except possibly for the Mercedes.

Example 4

Cell scan data*. These data are essentially lengths of cell boundaries for various steps of a shrink algorithm for 50 cells each of seven types: monocyte, basophil, lymphocyte, eosinophil, neutrophil, erythrocyte, reticulocyte. For the first discrimination, 32 variables are used: 16 measurements at each of two illuminating wave lengths. For the second discrimination 22 variables are used, five polynomial variables being used for ten ordered original measurements.

Good separation is obtained for three subgroups; the first two listed above, then the next three and finally the last two.

*Personal communication, Perkin-Elmer Corp.

REFERENCES

Chevron Research Company (1971), Report presented at the 64th Annual Meeting of the Air Pollution Control Association, Atlantic City, N. J.

Dixon, W. J. (Editor) (1970) BMD, Biomedical Computer Programs, University of California Press, Los Angeles.

Fisher, R. A. (1936), The use of multiple measurements in taxonomic problems, *Ann. Eugen.*, 7, part 2, 179-188.

Gnanadesikan, R. and Wilk, M. B. (1969), Data analytic

methods in multivariate statistical analysis, *Multivariate Analysis*, Vol.2, Academic Press: New York.
Mahalanobis, P. C., Majumdar, D. N. and Rao, C. R. (1949), Anthropometric survey of the United Provinces, 1941: A statistical study, *Sankhyā*, 9, 89-324.
Rao, C. R. (1962), *Advanced Statistical Methods in Biometric Research*, Ch. 8-9, Wiley: New York.

APPENDIX

Graphics Discriminant Analysis

1. Read specified variables and groups and form

$$\text{Group Means} = \hat{\mu}_i = (x_{i\cdot k})$$

$$\text{Group Cov Mat} = \hat{\Sigma}_i = \left(\frac{1}{n_i-1}\Sigma_j(x_{ijk}-x_{i\cdot k})(x_{ijl}-x_{i\cdot l})\right)$$

2. Test the equality of the group covariance matrices using

$$\chi^2 = \sum_{i=1}^{g}(n_i-1)\log\frac{|\hat{\Sigma}|}{|\hat{\Sigma}_i|} \to \chi^2((g-1)p(p+1)/2)$$

where

$$\hat{\Sigma} = \frac{1}{n-g}\sum_{i=1}^{g}(n_i-1)\hat{\Sigma}_i .$$

3. Read prior probabilities and compute posterior probabilities

$$p_i(x) = p_i f_i(x) / \sum_{i=1}^{g} p_i f_i(x)$$

$$f_i(x) = |\hat{\Sigma}_i|^{-1/2}\exp(-\tfrac{1}{2}(x-\hat{\mu}_i)(\hat{\Sigma}_i)^{-1}(x-\hat{\mu}_i)^T)$$

for each group i and specified cases x not necessarily those previously used.

4. Read cost matrix (c_{ij}) and compute the posterior expected loss

$$L_i(x) = \Sigma_j \, c_{ij} p_j(x)$$

for each group i and specified cases x.

5. Read metric weights α_i and compute

$$W = \sum_{i=1}^{g} \alpha_i \hat{\Sigma}_i$$

and its inverse W^{-1}.

6. Read three points p_1, p_2, p_3 (usually group means) and compute

$$\tilde{u}_1 = ((p_2 - p_1) W^{-1} (p_2 - p_1)^T)^{-\frac{1}{2}} (p_2 - p_1)$$

$$\tilde{\tilde{u}}_2 = p_3 - p_1 - (\tilde{u}_1 W^{-1} (p_3 - p_1)^T) \tilde{u}_1, \quad \tilde{u}_2 = (\tilde{\tilde{u}}_2 W^{-1} \tilde{\tilde{u}}_2)^{-\frac{1}{2}} \tilde{\tilde{u}}_2$$

$$u_i = \tilde{u}_i W^{-1} \; ; \quad i = 1,2 \; .$$

7. Compute the projections

$$v_i(x) = u_i (x - \overline{x})^T \; ; \quad i = 1,2$$

for specified cases x and plot the points

$$(v_1(x), \, v_2(x)) \; .$$

Included $x = p_1, p_2, p_3$, with special identification.

8. For selected groups k, compute the 2 by 2 covariance matrices

$$\hat{\Gamma}_k = (u_i \, \hat{\Sigma}_k \, u_j^T) \; ,$$

find their eigenvectors w_i and values λ_i, and plot the ellipses,

$$e(\theta) = \begin{pmatrix} v_1(\bar{x}_k) \\ v_2(\bar{x}_k) \end{pmatrix} t + (\lambda_1^{\frac{1}{2}} \cos\theta) w_1 + (\lambda_2^{\frac{1}{2}} \sin\theta) w_2 \; ;$$

$$0 \le \theta < 2\pi \; .$$

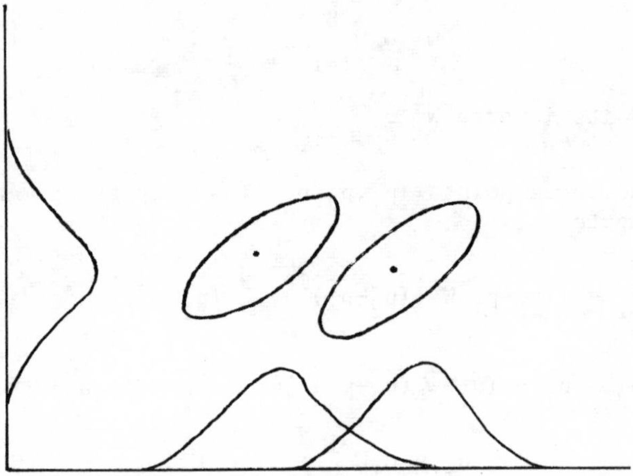

Figure 1

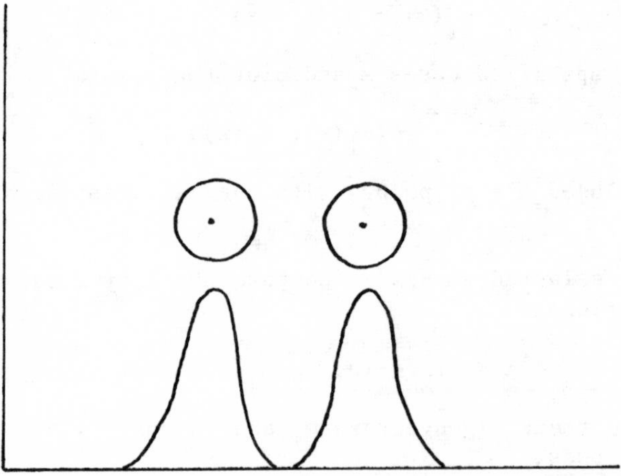

Figure 2

DISCRIMINANT PROBLEMS ABOUT GAUSSIAN PROCESSES

Daniel Dugué
University of Paris

Suppose we have $n + 1$ independent gaussian process $X_1(t),\ldots,X_n(t)$, $X(t)$ $(0 \leq t \leq 1)$. We assume that the covariance function is the same for the $n + 1$ processes, and the mean value $E(X_i(t)) = f(t)$ is independent of i for $i=1,\ldots,n$. We shall further suppose that the covariance and mean value functions belong to L^2. This short paper is intended to give a way of testing the null hypothesis $E(X_i(t)) = E(X(t))$ for $0 \leq t \leq 1$.

We shall give four ways of defining a squared distance of $X(t)$ to the set $\{X_i(t)\}$ and we shall give the characteristic functions of these four distances in the null hypothesis. Let

$$\Delta^{(1)} = \int_0^1 \left[\frac{\sum_{i=1}^{n} X_i(t)}{n} - X(t)\right]^2 dt = \int_0^1 (\overline{X}(t)-X(t))^2 dt \;;$$

$$\Delta^{(2)} = \int_0^1 (\overline{X}^2(t) - X^2(t))dt \;;$$

$$\Delta^{(3)} = \int_0^1 \frac{1}{n} \sum_{i=1}^{n} X_i^2(t)dt - \int_0^1 X^2(t)dt \;;$$

$$\Delta^{(4)} = \int_0^1 \frac{1}{n} \sum_{i=1}^{n} [X_i(t)-X(t)]^2 dt \;.$$

First, let us recall some results about the integral $\int_0^1 X^2(t)dt$, $X(t)$ being a gaussian process with covariance function $C(t_1,t_2)$ and $E[X(t)] = f(t)$. If the set of eigenvalues of $C(t_1,t_2)$ is $\{\lambda_k\}$ and the normalized eigenfunctions $\{\phi_k(t)\}$ with

$$\phi_k(t_1) = \lambda_k \int_0^1 C(t_1,t_2)\phi_k(t_2)dt_2 \quad \text{and} \quad \int_0^1 \phi_k^2(t)dt = 1 ,$$

then for a complete system $\{\phi_k\}$ we have

$$C(t_1,t_2) = \sum_{k=0}^{\infty} \frac{1}{\lambda_k} \phi_k(t_1)\phi_k(t_2) .$$

Setting

$$a_k = \int_0^1 f(t)\phi_k(t)dt ,$$

we have

$$E[\exp iu \int_0^1 X^2(t)dt]$$

$$= \prod_{k=1}^{\infty} (1 - \frac{2iu}{\lambda_k})^{-1/2} \exp\left[\frac{1}{2} \sum_{k=1}^{\infty} a_k^2\lambda_k\left\{\frac{1}{1-(2iu/\lambda_k)} - 1\right\}\right]$$

$$= \theta_1(u)\cdot\theta_2(u) ,$$

θ_1 being the infinite product and θ_2 the exponential function. It is easy to see that

$$E[\int_0^1 X^2(t)dt] = \sum_{k=1}^{\infty} \frac{1}{\lambda_k} + \sum_{k=1}^{\infty} a_k^2 = \int_0^1 C(t,t)dt + \int_0^1 f^2(t)dt$$

and

$$\mathrm{Var}(\int_0^1 X^2(t)dt) = 2 \sum_{k=1}^{\infty} \frac{1}{\lambda_k^2} + 4 \sum_{k=1}^{\infty} \frac{a_k^2}{\lambda_k}$$

$$= 2 \int_0^1\int_0^1 C^2(t_1,t_2)dt_1dt_2 + 4 \int_0^1\int_0^1 C(t_1,t_2)f(t_1)f(t_2)dt_1dt_2.$$

For full demonstration of these points, we refer to Dugué (1972). Now for $\Delta^{(1)}$, since $X_1,\ldots,X_n$, X are independent and have the same covariance function, the covariance function of $\overline{X}(t)-X(t)$ is $\frac{n+1}{n} C(t_1,t_2)$ and under the null hypothesis the mean value is zero. We have of course

$$\phi_k(t_1) = \frac{n}{n+1}\lambda_k \int_0^1 \frac{n+1}{n} C(t_1,t_2)\phi_k(t_2)dt_2 ,$$

and the eigenvalues of $\frac{n+1}{n} C(t_1,t_2)$ are $\frac{n}{n+1}\lambda_k$. Thus

$$E[\exp iu\,\Delta^{(1)}] = \theta_1(\frac{n+1}{n} u) .$$

We get

$$E[\Delta^{(1)}] = \frac{n+1}{n} \int_0^1 C(t,t)dt ;$$

$$Var(\Delta^{(1)}) = 2(\frac{n+1}{n})^2 \int_0^1\int_0^1 C^2(t_1,t_2)dt_1dt_2 .$$

For $\Delta^{(2)}$ we note that $\overline{X}(t)$ has mean value $f(t)$ and covariance function $\frac{1}{n} C(t_1,t_2)$. The eigenvalues are multiplied by n and

$$E\left[\exp iu \int_0^1\left[\frac{\sum_i X_i(t)}{n}\right]^2 dt\right] = \theta_1(\frac{u}{n})\left[\theta_2(\frac{u}{n})\right]^n .$$

Since X is independent of $\{X_i\}$, under the null hypothesis

$$E[\exp iu\,\Delta^{(2)}] = \theta_1(\frac{u}{n})[\theta_2(\frac{u}{n})]^n\theta_1(-u)\theta_2(-u) .$$

Thus we have

$$E[\Delta^{(2)}] = -\frac{n-1}{n}\sum\frac{1}{\lambda_k} = -\frac{n-1}{n}\int_0^1 C(t,t)dt ,$$

$$Var(\Delta^{(2)}) = \frac{n^2+1}{n^2}\, 2 \int_0^1\int_0^1 C^2(t_1,t_2)dt_1dt_2$$

$$+ \frac{n+1}{n}\, 4 \int_0^1\int_0^1 C(t_1,t_2)f(t_1)f(t_2)dt_1dt_2 .$$

The characteristic function of $\Delta^{(3)}$ under the null hypothesis will be

$$E[\exp iu\,\Delta^{(3)}] = [\theta_1(\tfrac{u}{n})\theta_2(\tfrac{u}{n})]^n\theta_1(-u)\theta_2(-u).$$

Hence

$$E[\Delta^{(3)}] = 0\ ,\qquad \mathrm{Var}(\Delta^{(3)}) = \frac{n+1}{n}\Big[2\int_0^1\int_0^1 C^2(t_1,t_2)dt_1dt_2 + 4\int_0^1\int_0^1 C(t_1,t_2)f(t_1)f(t_2)dt_1dt_2\Big].$$

The problem is more elaborate for $\Delta^{(4)}$. Putting $X_i - f = Y_i$, $X - f = Y$ we have $\sum_i(X_i-X)^2 = \sum_i(Y_i-Y)^2$ and the Y_i have a zero mean value and the same covariance as Y. The $(n+1)\times(n+1)$ matrix which is associated with the quadratic form $\sum_i(X_i-X)^2$ is

$$U = (u_{ij}) = \begin{bmatrix} n & -1 & -1 & \dots & -1 \\ -1 & 1 & & & \\ -1 & & 1 & & 0 \\ \vdots & & & & \\ \vdots & & 0 & & \\ -1 & & & & 1 \end{bmatrix}$$

that is

$$\begin{aligned} u_{ij} &= n \text{ if } i = j = 1 \\ u_{ij} &= +1 \text{ if } i = j \neq 1 \\ u_{ij} &= -1 \text{ if } i = 1,\ j \neq 1,\ i \neq 1,\ j = 1 \\ u_{ij} &= 0 \text{ otherwise} . \end{aligned}$$

The characteriestic equation is

$$\begin{aligned}(n-\lambda)(1-\lambda)^n - n(1-\lambda)^{n-1} &= (1-\lambda)^{n-1}[(n-\lambda)(1-\lambda)-n] \\ &= \lambda(1-\lambda)^{n-1}(\lambda-(n+1)) .\end{aligned}$$

The eigenvalues are $\lambda = n + 1$, $\lambda = 0$ and $\lambda = 1$ of multiplicity $n - 1$. Thus the matrix which is associated with the quadratic form

$$\sum_{i=1}^{n} \sum_{j=1}^{s} \left[X_i(\tfrac{j}{s}) - X(\tfrac{j}{s}) \right]^2$$

is $U \otimes I_s$, I_s being the s×s unit matrix. The covariance matrix of $X(\frac{j}{s}), X_1(\frac{j}{s}), \ldots, X_n(\frac{j}{s})$, j running from 1 to s, is $I_{n+1} \otimes C$, where C is the covariance matrix of $X(\frac{j}{s})$. It is now easy to write down

$$E\left(\exp[iu\,\Delta^{(4)}]\right) = \theta_1(\tfrac{n+1}{n} u) \left[\theta_1(\tfrac{u}{n})\right]^{n-1}$$

$$E[\Delta^{(4)}] = 2 \int_0^1 C(t,t)dt \; ;$$

$$\mathrm{Var}[\Delta^{(4)}] = \frac{n+3}{n} 2 \int_0^1\int_0^1 C^2(t_1,t_2)dt_1 dt_2 \; .$$

From the four characteristic functions, it is possible to find the distribution function in an easy way as $\Delta^{(1)}$, $\Delta^{(2)}$, $\Delta^{(3)}$ and $\Delta^{(4)}$ are sums of series of independent χ^2 variables.

REFERENCE

Dugué, D. (1972), Characteristic functions and Bernoulli numbers, *Journal of Multivariate Analysis*, 2, 230-235.

THE BASIC PROBLEMS OF CLUSTER ANALYSIS

M. G. Kendall
Chairman, Scientific Control Systems (Holdings) Ltd.

Summary

The paper discusses in non-technical terms the problems of cluster analysis: the definition of cluster and shape, the setting up of a suitable distance function, the treatment of polytomized variables, the routines for delimiting clusters, the assessment of reliability and the role of probability.

1. Cluster analysis as a separate subject is of comparative recent emergence; or perhaps it would be better to say, only recently have a number of topics in the analysis of classification become distinguished one from another. It may help if I begin with some demarcation of the constituent parts of the broader subject as I understand it.

(a) *Pattern Recognition*

The general object here is to devise some mechanical means of recognizing a *pre-determined* pattern; for example, to program a machine to distinguish among the letters of the alphabet, or to distinguish in context a specified shape such as a type of finger print. It is an essential part of the subject that the form of the pattern (with perhaps some degree of tolerance) is specified beforehand. If the expression be allowed, the machine is instructed what pattern to look for.

(b) *Classification Analysis*

In this situation there are no pre-determined patterns. Given a set of objects (languages, biological organisms, sciences) the problem is to classify them in some useful way. In general there are several ways in which any aggregate of individuals may be broken down into classes. For any one classificatory system the sub-classes and their sub- sub-classes are hierarchic; that is to say, the individuals are grouped into species, the species are then grouped into genera, and so on. The most familiar example, perhaps, is the classification of forms of life along Linnean lines.

(c) *Cluster Analysis*

In its statistical meaning, this subject relates to the classification of individuals into groups based on the observation on each individual of the values of a number of variables (which may be continuous, discontinuous or even, in the limit, dichotomous). Characteristically we have n individuals, on each of which is observed p variables. There are no predetermined patterns. In one way this type of analysis is of more restricted scope than classification, because we require every variable to be recorded for each individual, whereas in general classificatory distinctions we may subdivide classes into subclasses on the basis of criteria which vary from one class to another.

(d) *Discrimination Analysis*

The typical situation here is that we are given the existence of two or more classes and given for certain a sample of individuals from each, the object being to set up some rule under which an individual of unknown origin can be allotted to the correct class in some optimal sense such as the minimization of error. It is characteristic that the classes are given as separate and distinct, whereas in cluster analysis we do not know beforehand whether the set of individuals under scrutiny fall into separate classes or not.

2. In a way, the problem of clustering is anterior to that of discrimination. Not until we know, or can assume as a hypothesis, that the individuals fall into groups can we consider the problem of discriminating among the groups. Nevertheless, the two subjects have developed in the reverse order. Discrimination analysis has been fairly intensively studied by statisticians for nearly forty years. Cluster analysis, as I am using the term, has only recently come to statistical notice. The reason, I think, is that a serviceable routine for cluster analysis requires the aid of an electronic computer, and in fact can, in practice, call for substantial amounts of machine time.

Cluster and Shape

3. The fundamental problem in cluster analysis is to define what we mean by "cluster". One thinks of this word in terms of sets of points in two or three dimensions, but even in such relatively simple cases the word is ambiguous. In everyday speech, perhaps, we think of a cluster as a group in which all the individuals are close to some central point and as far as possible to one another, like a bunch of grapes. But what are we to say of the particles which compose one of Saturn's rings, which are certainly a grouping, but a hollow one; or the tracks of a particle in a Wilson cloud chamber, which is an organized series of droplets but a linear one? And if we allow a scatter of points inside an ellipse to constitute a cluster, what are we to say of two such shapes with common centre and major axes at right-angles - are they one cluster or two overlapping clusters?

4. It seems to me that we may have to contemplate several forms of cluster analysis, distinguished according to the kind of cluster in which we are interested. This would occasion no great difficulty if we always knew beforehand what sort of cluster would interest us. In many cases we do not, being concerned with the existence of some degree of organization (in the sense of contiguity) among members of the set but being unaware what form it might take, at least in the preliminary exploratory stage.

5. In one, two or three dimensions we can always exhibit the situation geometrically and use subjective judgements (which are not to be despised) in making at least a provisional clustering. Even this is a nuisance in three dimensions, or in any number of dimensions if there are a lot of points to be considered. In more than three dimensions - and cases occur where we may have as many as a hundred - it is impossible.

6. Even the definition of "shape" raises some interesting but difficult problems, especially in more than three dimensions. With some difficulty (by the use of linear programming methods) we can determine the convex hull of n points in p dimensions, but even if we can describe the shape of the hull this tells us nothing about the distribution of points inside it. Hitherto the determination of shape has not been considered by statisticians as particularly important. Nor, indeed, has it been. But if it becomes so there are a number of interesting lines of approach.

(a) It may be possible (but will not be easy) to compare an observed distribution with shapes which have been previously specified, e.g. as hyperspheres, quadrics, toroids, simplices and so on.

(b) Some idea of condensation may be obtained by calculating the distances between all pairs of points and exhibiting the totality as a frequency distribution.

(c) The eigenvalues of the dispersion matrix of a set of points give some idea of its relative extension in orthogonal directions.

(d) If the distances of the points from their centre of gravity are arranged as a frequency distribution, the shape of the resultant throws some light on the "shape" of the set. For instance, if they are uniformly dispersed inside a sphere the frequency function will be a backward J-shape; if they are distributed in an anchor ring the distribution will not attain zero values.

(e) With a powerful computer and a visual display unit, perhaps the most effective way will be to explore the set by many different trials; e.g. by dividing the space by a hyperplane through the centre of gravity of the points at various orientations and counting the points on either side.

7. Although matters such as these have some importance in cluster analysis, they also bear on other branches of multivariate analysis. Anyone familiar with elementary regression and the associated scatter diagram will know how useful the latter is in exposing outliers, suggesting the kind of curve to fit or suggesting some variate transformations. Over the past fifty years mathematics has tended to discount subjective impressions gained from visual inspection, but the practising statistician cannot afford to neglect any method of feeling his way in p dimensions, however intuitive and however empirical.

Scale

8. Our notion of cluster is derived from our experience of aggregates of points in ordinary two- or three-dimensional space. The units in which we measure length, breadth and height are identical and therefore of comparable scale. In general cluster analysis this often ceases to be true. We may, for example, have individual humans classified by height, age, weight and salary, each of which is measured in a different unit. The "closeness" of a pair of individuals clearly depends on how we scale these variables. We are, therefore, faced at the outset with the problem of constructing a distance function from a set of p variables which are, in a strict sense, non-comparable. There does not appear to be any simple solution to this problem. (It is well known that principal components, for example, are not invariant under change of scale.)

9. One method, which I personally prefer, is to standardise all the variables at the outset to have unit variance. If we have reason to suppose that some variables are more important than others in the clustering process, we can make allowance by decreasing the variance for the more important ones by some numerical factor, which has, of course, to be guessed at or determined on extraneous grounds. If the variables are correlated it may be better to transfer to principal components (which are uncorrelated) in order to avoid an element of double counting.

10. The suggestion has been made that we can find distance functions which are scale invariant. Such a one appears to be the Mahalanobis distance. Given a set of points x_{ij}, $i = 1,2,\ldots p$; $j = 1,2,\ldots,n_1$ and another set x_{ij}, $i = 1,2,\ldots,p$; $j = 1,2,\ldots,n_2$, let $\bar{x}_{i1}$ be the means of the first set and $\bar{x}_{i2}$ the means of the second set, and let (c_{ij}) be their pooled dispersion matrix with inverse c^{ij}. Then the Mahalanobis distance is defined as

$$D^2 = \sum_{j,k=1}^{p} c^{jk}(\bar{x}_{j1}-\bar{x}_{j2})(\bar{x}_{k1}-\bar{x}_{k2}) \ . \qquad (1)$$

This is indeed scale-invariant but it is not a suitable measure for our purposes.

(a) There seems no reason why we should pool the dispersion of the two groups.

(b) The measure does not exist as a distance between two points.

(c) The standardisation involved is not very different from the straightforward reduction to unit variance.

In regard to these last two points, consider the case $p = 2$ and let the dispersions be c_{11}, c_{12}, c_{22}. Then the distance becomes

$$\frac{c_{22}(\bar{x}_{11}-\bar{x}_{12})^2}{c_{11}c_{22} - c_{12}^2} + 2c_{12}\frac{(\bar{x}_{11}-\bar{x}_{12})(\bar{x}_{21}-\bar{x}_{22})}{c_{11}c_{22} - c_{12}^2} + c_{11}\frac{(\bar{x}_{21}-\bar{x}_{22})^2}{c_{11}c_{22}-c_{12}^2}$$

If each group comprises only one member $c_{11}c_{22} - c_{12}^2$ vanishes. If c_{12} is zero the distance reduces to a simple standardisation to unit variance.

Distance

11. There are probably many ways in which a "distance" between two "points" can be measured. In any case it seems impossible to set up a quantifiable system of cluster analysis without some measure or other. For most purposes I prefer the simple Euclidean distance.

$$d^2(A,B) = \sum_{j=1}^{n} (x_{iA} - x_{iB})^2 . \tag{2}$$

This at least has the advantage that it is invariant under rotation of the axes (and hence under transformation to principal components). It also has the advantage that the mean of the squared distances between pairs of n points is twice the mean of the square distances from their centre of gravity, so that if we are interested only in the average square-distance among a set of points the calculations can be reduced from order $\frac{1}{2}n(n-1)$ to order $2n$. But other methods may be more suitable in certain cases and it would be interesting to have experimental studies of alternatives. For example

(a) One might define a cluster as one for which the greatest distance between any pair of points in it falls below a certain value.

(b) We might similarly define it as a group for which the average distance (as distinct from distance squared) between individuals falls below some given value.

But in any case, separation into clusters involves, in some sense, not merely the distances within a cluster but the distances between clusters. For the latter one thinks naturally of the distance between centres of gravity. Other measures involve serious practical difficulties in mathematics or arithmetic.

12. Cases often arise, especially in market research or sociology, in which the "variables" are discontinuous classifications and may even be dichotomies. Thus, a set of respondents may be recorded according to sex, possession or non-possession of a university degree, or social class. It is not easy to bring these variables into a distance function. The best I have been able to suggest so far is to replace the classifications by ranks. It is known that, for some purposes at least, classification can be regarded as heavily tied rankings. In default of anything better I should be inclined to regard even a dichotomy as a tied ranking. But for clustering purposes there are special features which we may illustrate by the case where there are three dichotomies involved. If we represent them by (0,1) variables the possible cases may be considered as falling on the corners of a cube (see Figure 1). If we regard as a cluster only those members which all fall on one corner there is no difficulty. All we are then saying is that we cluster together those members which are identical in every respect (so far as these three variables are concerned). But if we relax this criterion and allow as a cluster those members which have two qualities in common, we see that there are many different ways of carrying out the clustering. Any two vertices of the cube which are joined by an edge have two common qualities. There is no unique solution, and indeed there are many equivalent ones.

13. This is one aspect of a problem which occurs in many multivariate contexts, for example in the analysis of multiple contingency tables and in discriminant analysis. The treatment of multiple classifications, especially those which are not orderable (e.g. grouping by religion or by type of crime) seem to be peculiarly intractable. It may, of course, be that in cluster analysis we are asking the wrong question in trying to cluster the data. For instance, in market research, wherein we might be trying to delimit classes of consumer, it would probably be better, given dichotomy by sex, to cluster men and women separately, and only attempt to match the two groups if they both exhibited the same kind of clustering, in which case the sex-variable could be ignored. But it remains an open question how we should deal with those cases where a great many dichotomous variables are involved and consideration in the separate cells leads to trivially small numbers.

Delimitation of Clusters

14. Suppose that we have settled to our own satisfaction a suitable distance matrix and can compute all the distances between pairs of points. The next problem is how we use these distances, $\frac{1}{2}n(n-1)$ in number, to classify our individuals. One can draw a useful distinction here between the clustering of variables and the clustering of individuals. In general p, the dimension number, is reasonably small and rarely exceeds 30. On the other hand, n, the sample number, may be very large, usually more than a hundred and sometimes several thousands. In the former case we can conveniently cluster variables directly from the correlation matrix. This will be a p×p matrix with units down the diagonal, about which it is symmetric, so that there are $\frac{1}{2}p(p-1)$ correlations to consider. The matrix can be printed out in a form which is not very difficult to analyse by hand. It will be remembered that the correlation between two variables is a fair measure of their "closeness", and indeed if we represent them by vectors in an n-dimensional space, the cosine of the angle between them is the correlation coefficient. One may then proceed by inspecting the matrix for the largest correlation. If this is greater than some acceptable value of closeness (to be determined on arbitrary grounds) the two variables, say i, j, may be taken to form the nucleus of a cluster. We then look at the correlations of the other variables with i and with j. If there is a variable k such that the average of the three correlations (i,j), (i,k), (j,k) attain an acceptable level, k is added to the cluster; and so on until no further variable can be added without reducing the average correlation below the acceptable level. We then remove these variables from the matrix and repeat the process on the remainder: and so on until all clusters are delimited. There may, of course, be residual "clusters" of one variable only.

15. The process is too tedious when n is large. It does, however, have the advantage of being scale-invariant and some rough idea can be obtained of the shape of the cluster from the correlations themselves. Incidentally, I prefer this method of grouping variables to the more familiar one of extracting non-orthogonal principal components or factors by factor-analysis and rotation.

16. A good many methods have been put forward by different authors for the clustering of individuals. A review and a large number of references are given in Cormack (1971). They do not by any means lead to identical or even to similar results. An account of the method I prefer is given in my paper (Kendall, 1971). In effect, it starts with a provisional division of the space and proceeds iteratively to define clusters on the basis of closeness of individuals to the centre of gravity of the cluster as compared to distances between clusters. Another method which has been proposed is to start by dividing the set of points into two groups, based on the same kind of principle, a kind of comparison of interclass and intraclass variance. The disadvantage, as it seems to me, is that if we then proceed to further clusters we have to subdivide the two groups each into two groups and hence produce a bifurcated hierarchy which may be far from optimal.

17. One thing at least is clear, that solution of the problem by any kind of exhaustive enumeration of all possibilities is out of the question. Even with two groups the number of possibilities to be examined is of the order of 2^n, and with more the number far exceeds that. No foreseeable computer could cope with such large numbers of possibilities. It would follow that, whatever methods stand the test of experience (and, as I have already said, different methods may be required for different types of cluster) they must be of an iterative trial-and-error character.

The Role of Probability

18. The traditional part of statistical theory which is based on probability has, in general, a very small role to play in cluster analysis. The individuals which we are trying to cluster may not be a sample at all - they may comprise the whole of the population under examination, as when we try to classify all the towns in a county or all the programmes on a television network. Even if they are a sample in the sense of being a sub-set of what is known to be a larger aggregate they are not necessarily, or even frequently in practice, a random sample. The visible stars are not a random sample of cosmic matter, nor are the known bacteria a random sample of all existent bacterial forms.

19. The consequence is that, in general, we cannot set standard errors, or confidence bands, or conduct exact significance tests, on any particular clustering. The question thus remains, how do we assess the reliability of our results? "Reliability" may have three meanings in this context. If the whole population of individuals is available for scrutiny, our results are only as reliable as we choose to regard them; that is to say, the definition of cluster, the metric employed, and the degree of tightness of bunching of the clusters which we are prepared to accept are matters of personal taste or relevance to the enquiry in hand, and in no way depend on probabilistic variation. The one possible exception to this statement might be that if the raw data are subject to observational error, the resultant clustering might be unreliable in the sense that more accurate observation might give different results.

20. Where samples are concerned the results may be unreliable in the sense that scrutiny of the whole population would give different answers. If the samples are not random this cannot be expressed in probabilistic terms. The best one can do, I think, is to split the sample into two or more sub-samples and perform an analysis on each. If the results are consistent and some reproduceability is present we have greater confidence in their common elements. Given a random sampling situation it might be possible to set up some statistic to measure the stability of the clustering under repeated sampling. The problem is a challenging one but is, I think, likely to prove formidable.

21. The third factor which may affect reliability is the very simple one that we may have omitted to measure an important variable. A biologist, working on criteria such as shape and habitat, might be inclined to classify a shark as much closer to a dolphin than either is to a rabbit. But if he takes account of blood temperature, skeletal structure and type of reproduction the conclusion is very different. There is, of course, no statistical remedy for leaving out important information. One can only hope that, if a clustering process is to be meaningful it has been preceded by a careful examination of all the possible contributory variables.

The Discarding of Redundant Variables

22. There can occur a kind of converse situation to that of missing variables, namely the one in which we have more variables than are necessary. Or to put the thing another way, if we arrive at a set of clusters, which of the variables are doing the work and which can be discarded as not relevant? This is not an easy question to answer, and it is frequently necessary to have resource to the primary data to decide it. One useful practical possibility, having delimited the cluster, is then to turn the problem into one of discrimination. Given two clusters we can find a discriminant in the usual way and then examine it to see which variables are important. This entails comparing the clusters pair by pair, and it may well be that variables which are important in separating one pair are not so for another. But all this would be useful information.

23. Notwithstanding what was said in previous paragraphs, there are cases where we have to have regard to the relative risk of occurrences of the individuals under examination. Consider, for example, a two-dimensional map on which we plot the occurrence of some rare disease such as leukemia. The points will cluster for the simple reason that there are more people at risk in the more densely populated areas. However, this is not the sort of clustering with which we should usually be concerned. Overlying the condensation due to population densities there may be clustering due, for example, to contagion or to local conditions or to inbreeding. One would like to dissect any clustering due to these causes from those due to the distribution of population. If we can get population figures for sufficiently small areas (e.g. polling districts in the United Kingdom) it may be possible to weight the distance metric to take account of differing liabilities to appearance in the sample. I have suggested one such method in the paper under reference (Kendall, 1971), but I think that further research on this important problem is desirable.

REFERENCES

Cormack, R. M. (1971), A review of classification, *Jour. Roy. Statist. Soc.*, 134, 321-367.
Kendall, M. G. (1971), Cluster Analysis, *Proceedings Hawaiian Conference on pattern recognition.*

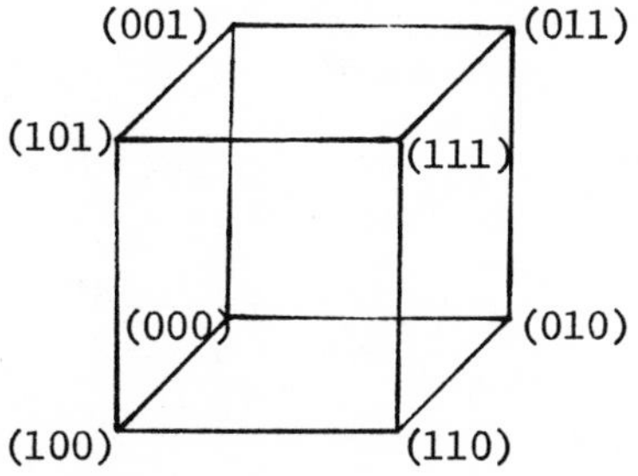

Fig. 1 Configuration of 3 dichotomies

SOME RESULTS ON THE MULTIPLE GROUP DISCRIMINANT PROBLEM

Peter A. Lachenbruch
University of North Carolina

1. *Introduction*

The problem of assigning an individual to one of two groups on the basis of a set of observations has been extensively studied since the introduction of the linear discriminant function (LDF) by Fisher (1936). The behavior of the LDF has been considered from the point of view of its distribution when samples are used to estimate the coefficients by Anderson (1951, 1973) and Okamoto (1963), estimation of error rates by Dunn (1971), Hills (1966), Lachenbruch and Mickey (1968), and robustness to various departures from the assumptions by Gilbert (1968, 1969), Lachenbruch (1966) and Lachenbruch, Sneeringer and Revo (1968). Recently, Glick (1972) has shown the asymptotic optimality of "plug in" estimators of correct classification rates. Relatively little has been done regarding the problem of assigning an observation to one of $k > 2$ groups. In this study, two different approaches are reviewed, some of their properties studied, and compared.

The first approach assumes that the populations π_i have density functions $f_i(\underset{\sim}{x})$ whose form is known, but whose parameters may not be known. Then, it can be shown that the rule which maximizes the mean probability of correct classification is to assign the unknown observation $\underset{\sim}{x}$ to π_i if $p_i f_i(\underset{\sim}{x}) = \max p_j f_j(\underset{\sim}{x})$ where p_i is the a priori probability of an individual belonging to π_i. In this study, it is assumed that the p_i are equal so the rule becomes "assign $\underset{\sim}{x}$ to π_i if $f_i(\underset{\sim}{x}) = \max_j f_j(\underset{\sim}{x})$." If the parameters of the f_i are unknown, they may be suitably estimated, e.g. by maximum likelihood, and the rule applied to the estimated functions. Computer programs can be written to perform the calculations for any given set of density functions, and many are already available for the special case of normal populations with the same covariance matrices. In this case, the rule is equivalent to "assign $\underset{\sim}{x}$ to π_i if

$$(\underset{\sim}{x} - \frac{1}{2}\underset{\sim}{\mu}_i)' \Sigma^{-1}\underset{\sim}{\mu}_i = \max_j(\underset{\sim}{x} - \frac{1}{2}\underset{\sim}{\mu}_j)' \Sigma^{-1}\underset{\sim}{\mu}_j ,$$

where $\underset{\sim}{\mu}_i$ is the mean of π_i and Σ is the common covariance matrix." In the sampling situation, $\underset{\sim}{\mu}_i$ is replaced by $\overline{\underset{\sim}{x}}_i$ and Σ is replaced by S, the usual maximum likelihood estimates. This method will be referred to as the Multiple Discriminant Function or MDF method.

The second approach to the multiple group problem is based on an extension of Fisher's original approach to the two group problem. Let B be the between groups mean square and cross-product matrix and W be the within groups mean square and cross-product matrix; then, for two groups Fisher suggested finding the linear combination $\underset{\sim}{\lambda}'\underset{\sim}{x}$, which maximized $\phi = \underset{\sim}{\lambda}'B\underset{\sim}{\lambda}/\underset{\sim}{\lambda}'W\underset{\sim}{\lambda}$ which is the ratio of the between groups mean square to the within groups mean square of the linear combination $\underset{\sim}{\lambda}'\underset{\sim}{x}$. For two groups these two approaches lead to the same function. For more than two groups, the linear compounds which maximize ϕ are solutions of the equation $(B - \gamma W)\underset{\sim}{\lambda} = \underset{\sim}{0}$.

This familiar equation has solutions which are the eigenvectors of $W^{-1}B$, and there are at most $\min(k-1,p)$ of them where k is the number of populations, and p the number of variables. Since the $\{\underset{\sim}{\lambda}'\underset{\sim}{x}\}$ are linear transformations of normal variates, they too are normal (conditional on $\underset{\sim}{\lambda}$), and it is easy to show that the optimal assignment rule when using r eigenvectors is to assign to π_i if

$$\sum_{\ell=1}^{r} (\underset{\sim}{\lambda}_\ell'(\underset{\sim}{x} - \underset{\sim}{\mu}_i))^2 = \min_j \sum_{1}^{r} (\underset{\sim}{\lambda}_\ell'(\underset{\sim}{x} - \underset{\sim}{\mu}_j))^2$$

or

$$(\underset{\sim}{y} - \frac{1}{2}\underset{\sim}{\nu}_i)'\underset{\sim}{\nu}_i = \max_j (\underset{\sim}{y} - \frac{1}{2}\underset{\sim}{\nu}_j)'\underset{\sim}{\nu}_j$$

where

$$\underset{\sim}{y}' = (\underset{\sim}{\lambda}_1'\underset{\sim}{x}, \underset{\sim}{\lambda}_2'\underset{\sim}{x}, \ldots, \underset{\sim}{\lambda}_r'\underset{\sim}{x})$$

and

$$\underset{\sim}{\nu}_i' = (\underset{\sim}{\lambda}_1'\underset{\sim}{\mu}_i, \ldots, \underset{\sim}{\lambda}_r\underset{\sim}{\mu}_i)$$

This method will be referred to as the eigenvector or EV method.

In either case the boundaries of the assignment regions are hyperplanes. For the MDF, the hyperplanes have dimension p-1 and for the EV they have dimension $r \leq \min(k-1,p) - 1$.

With the availability of computers, it is possible to use either method fairly easily. If p is large, it may be desirable to represent the data in fewer dimensions, which the EV approach does, although it is doubtful there is any great saving in computation since the compounds must be found. At any rate the EV method is used, particularly by psychometricians, so it does seem useful to compare its behavior to that of the MDF.

A number of questions may be asked about these functions.

1. Under what situation can the EV approach work well compared to the MDF method?
2. What is the effect of sample size on the behavior of the methods?
3. How does the number of groups affect their performance?
4. How does the number of variables affect their performance?
5. How would one asnwer questions 2-4 if one is concerned about the relative performance of the methods?

In the following pages we attempt to answer these questions using the estimated proportion of correct classification as our criterion. Further work will be needed to study the robustness of the methods when unequal covariances appear, or non-normal distributions are present.

2. *Configuration of Population Means*

In the two group case, it is possible to draw a line between the two means of the populations. In the k-group case, this will not be possible in general. The means will lie in a space of dimension less than or equal to the minimum of k-1 and p. The placement of the mean values will have a substantial effect on the performance of the EV classification rule when not all of the eigenvectors are used. In this paper, we consider two extreme placements. We assume $\Sigma = I$. In the first, the means are collinear (C case) and equally spaced. If the parameters are known, the first eigenvector contains all the information about the population and is equivalent to the MDF. In this case the eigenvalues other than the first are all 0. The second case has the means arranged so the pairwise distance is the same for any two groups. This implies that the means are the vertices of a regular simplex (S case). This is the least favorable case for the EV method in the sense that all eigenvalues are equal and each vector accounts for $1/(k-1)$ of the variability.

It is clear that we can assume the means are arranged as

$$\underset{\sim}{\mu}_1' = (0,..0),\ \underset{\sim}{\mu}_2' = (\delta_{21},0,...0),\ \underset{\sim}{\mu}_3' = (\delta_{31},\delta_{32},0,...,0),$$

etc. If $k > p + 1$, there may be k-p-1 means in which all components are used. For the case of collinear means, spaced δ units apart we may consider

$$\underset{\sim}{\mu}_1' = (0,...0),\ \underset{\sim}{\mu}_2' = (\delta,0,...0),\ \underset{\sim}{\mu}_3' = (2\delta,0,...,0)$$

The probability of correct classification if one uses the MDF procedure is

$$\Phi(\delta/2) - \Phi(-\delta/2) \quad \text{for} \quad \pi_2 \cdots \pi_{k-1}$$

and $\Phi(\delta/2)$ for π_1 and π_k. This can be seen as follows. An observation from π_i will be correctly classified if $f_i(\underset{\sim}{x})$ is a maximum. The assignment rule is assign to

$$\pi_1 \quad \text{if} \quad -\infty < x_1 \leq \delta/2$$

$$\pi_2 \quad \text{if} \quad \delta/2 < x_1 \leq (\delta/2)+\delta$$

$$\pi_3 \quad \text{if} \quad \delta/2+\delta < x_1 \leq \delta/2+2\delta$$

$$\vdots$$

$$\pi_{k-1} \quad \text{if} \quad \delta/2+(k-3)\delta < x_1 \leq \delta/2+(k-2)\delta$$

$$\pi_k \quad \text{if} \quad \delta/2+(k-2)\delta < x_1 < \infty$$

where x_1 is the first component of $\underset{\sim}{x}$. The following table gives some values of the probabilities.

TABLE A

Probabilities of Correct Classification for C Case

δ	$P_1=P_k$	$P_i\,(2\leq i\leq k-1)$	$\overline{P}_3$	$\overline{P}_5$	$\overline{P}_{10}$
1/2	.60	.20	.47	.36	.28
1	.69	.38	.59	.50	.44
2	.84	.68	.79	.74	.71
3	.93	.86	.91	.89	.87

The last three columns are the mean probabilities of correct classification for $k = 3,5,10$.

For this case, it is easy to obtain exact probabilities of correct classification. For more general placement of means, the problem becomes more difficult. This configuration of means is most favorable to the EV method since all of the between group variability occurs along one axis.

A less favorable configuration is one in which the means are placed at the vertices of a regular k-1 dimensional figure. This is an equilateral triangle in 2 dimensions, a pyramid in 3 dimensions, and in general is a regular simplex. One can show that the coordinates of the vertices of a regular simplex spaced δ units apart are $\underset{\sim}{\mu}_1' = (0,\ldots,0)$

$$\underset{\sim}{\mu}_\ell = \delta(a_{1\ell}, a_{2\ell},\ldots,a_{\ell\ell}, 0,\ldots,0)$$

where

$$a_{\ell\ell} = ((1+\ell)/2\ell)^{1/2}$$

$$a_{i\ell} = 1/2\binom{i+1}{2}^{1/2} \qquad i < \ell$$

$$a_{i\ell} = 0 \qquad i > \ell \ .$$

Table 1 gives coordinates for the vertices of a simplex spaced one unit apart for up to 10 dimensions. This is the least favorable case for the EV method since the eigenvalues are all equal and each vector accounts for 1/k-1 of the between group variability. The MDF procedure assigns an observation to π_i if

$$z_i = (\underset{\sim}{x}_i - \mu_i/2)' \, \Sigma^{-1} \underset{\sim}{\mu}_i$$

is a maximum (in our case we will have $\Sigma = I$). To find the probability of correct classification for observation from π_j we must find

$$P(z_i > z_j | x \in \pi_i)$$

for all j. We shall study this probability for observations from π_i, but our methods are applicable for all π_i. Suppose $x \in \pi_1$. We can show that the joint distribution of $\underset{\sim}{z}'=(z_1,\ldots,z_k)$ is normal with mean $\underset{\sim}{\mu}'_z = (0,-\delta^2/2, -\delta^2/2,\ldots,-\delta^2/2)$ covariance matrix

$$A = (a_{ij}) = \begin{bmatrix} 0 & 0 & \cdots & 0 \\ 0 & \delta^2 & \cdots & \delta^2/2 \\ \vdots & \vdots & \ddots & \vdots \\ 0 & \delta^2/2 & \cdots & \delta^2 \end{bmatrix}$$

$$a_{11} = 0$$

$$a_{ii} = \delta^2 \qquad i \geq 2$$

$$a_{1j} = a_{i1} = 0$$

$$a_{ij} = \delta^2/2 \qquad i \neq j,\ i > 1,\ j > 1$$

Thus, the probability of correct classification of $\underset{\sim}{x}$ is

$$P(z_1>z_2,\ z_1>z_3,\ldots,z_1>z_k) = P(z_2<0,\ z_3<0,\ldots,z_k<0)$$

Let $y_1 \ldots y_k$ be independent $N(0,\delta^2/2)$, and let $x_i = y_i + y_k$. Then, the event $(z_2<0,\ldots,z_k<0)$ is equivalent to the event $(x_1<\delta^2/2,\ldots,x_{k-1}<\delta^2/2)$. Since the y_i are independent, we have

$$P(x_1<\delta^2/2,\ldots,x_{k-1}<\delta^2/2) = P(y_1<\delta^2/2-y_k,\ldots,y_{k-1}<\delta^2/2-y_k)$$

$$= \int_{-\infty}^{\infty} P(y_1<\delta^2/2-y_k)\ldots P(y_{k-1}<\delta^2/2-y_k)d_{yk}$$

$$= \int_{-\infty}^{\infty} \left[\Phi\left(\frac{\delta^2/2-y_k}{\sqrt{\delta^2/2}}\right)\right]^{k-1} f(y_k)dy_k$$

This integral was mentioned by Bechhofer and Sobel (1954). Table 2 gives these probabilities for various values of δ and k. Note that the first column $k = 2$ is the value for two-group discriminant analysis.

Comparing the results of Table 2 with the sample results in Table 3, we see that the behavior is as expected. For the samples of size 25, the probability of correct classification is overestimated by more than for samples of size 100. This bias is not unexpected. For larger p, the bias is somewhat more severe.

The integration was done using the ten point Hermitian quadrature routine from the IBM Scientific Subroutine Package.

One may think of the mean pairwise distance as a measure of divergence among the k populations. This is not altogether satisfactory, but it provides a simple measure to compare the two cases. For the regular simplex, all means are δ units apart, so the mean pairwise distance is δ. For the collinear case, suppose the spacing is γ. Then the mean distance is

$$\frac{1}{\binom{k}{2}} \sum_{i>j} \gamma(i-j) = \frac{\binom{k}{2}\left[\frac{k+1}{3}\right]\gamma}{\binom{k}{2}} = \left[\frac{k+1}{3}\right]\gamma$$

So if $\gamma = \delta$, the mean pairwise distance is greater for the collinear case than for the regular simplex configuration. This, of course, is why better discrimination is possible with collinear means spaced evenly δ units apart, than with means at the vertices of a regular simplex.

It may be of some interest to compare the probability of correct classification for the same mean between group distance for the simplex and collinear cases. Table B gives some comparisons.

TABLE B

Probabilities of Correct Classification for the Same Mean Between Population Distance

k	C Spacing	S Spacing	Collinear P(Correct Classification)	Simplex P(Correct Classification)
3	1	1.33	.59	.617
	2	2.67	.79	.846
	3	4.00	.91	.959
5	1	2.00	.50	.627
	2	4.00	.74	.928
	3	6.00	.89	.995
10	1	3.67	.44	.829
	2	7.33	.71	.999
	3	11.00	.87	1.000

Thus, for large k, the regular simplex arrangement leads to higher probabilities of correct classification.

3. *Behavior of the Methods*

To evaluate the performance of these methods we used sampling experiments since the distribution theory for these procedures even for the two group case is exceedingly complicated. The estimated mean probability of correct classification was used as the criterion. The estimate was obtained by resubstituting the original observations in the calculated MDF or EV. This is known to be biased, and comparing the observed quantities with the theoretically calculable ones (namely, the collinear case) we can estimate the approximate extent of this bias. Our results are given for up to 3 eigenvectors as this is the maximum number that seem to be used. All programs were written by the author using the IBM Scientific Subroutine Package when appropriate.

The sampling experiments consisted of taking 10 samples for each value of k, p, the number of variables, n, the number of observations in each of the k groups, δ, the between mean distance, and the two configurations of means. The values of these were: $k = 5$ or 10, $p = 4$ or 10, $n = 25$ or 100, $\delta = 1,2,$ or 3, Type = collinear or simplex.

Since the simplex requires at least k-1 dimensions, it was not possible to arrange the means in a 9 dimensional simplex when $p = 4$. Thus, the simplex cases $p = 4$, $k = 10$, are not given.

For each combination, multivariate normal samples were taken, and estimates of the probability of correct classification by the MDF and EV methods were found. For the EV method, the use of 1, 2, or 3 eigenvectors was studied. The means and standard deviations of the probability of correct classification for each method are given in Table 3. The subscript on EV refers to the number of eigenvectors used. Because the EV method had almost the same standard deviation for all three cases, their standard deviations were pooled.

The following conclusions may be drawn from these data.

a) As expected, the EV approach did much better for the C cases than for the S cases. Averaged over all other factors, using one eigenvector correctly classified 0.70 of the C cases, but only .28 of the S cases. The differences between the MDF and EV methods are fairly small for the C groups, but considerably greater for the S cases. In the S cases no combination of eigenvectors was good for $p = 10$, for $p = 4$ occasionally 3 eigenvectors were satisfactory.

b) Increasing the sample size led to an observed decrease in the apparent probability of correct classification.

While an increase in sample size should improve the functions, and thus imply an increase in the probability of correct classification, the method of estimating error, namely resubstituting the observations in the functions is known to have a positive bias, and increasing the sample size reduces this bias. The effect is about the same for all groups. This problem could have been avoided by use of an index sample technique in which the functions are tested on an independent set of observations.

c) An increase in k, the number of groups, leads to a decrease in the mean probability of correct classification because there are more chances for erroneous assignments. Tables A and 2 show the situation. Comparing them to Table 3, we observe close agreement between the theoretical and sample cases, with the bias of the resubstitution method being noticeable.

d) An increase in the number of variables, p, gives an increase in the probability of correct classification. This is interesting since no further information is supplied in these extra variables. The explanation lies in the fact that a set of data can be fit more exactly when more variables are used. The monotone increase in R^2 in multiple-regression problems is a familiar example of the same phenomenon.

4. *Discussion*

If the means are collinear or nearly so, the EV method works about as well as the MDF method. If the number of variables is large, this may represent some economy in computation, particularly if the function is to be used in assigning future observations. However, if the above conditions do not hold, the MDF is much better than the EV and should be used.

Unfortunately, no firm statements on the effects of sample size can be made, because of the problems mentioned earlier. It appears that both the MDF and EV methods are affected to about the same degree.

The number of groups is inversely related to the performance of both methods and is about the same for both methods. The following table gives the performance figures. These are averaged over all other factors (i.e. δ, n, and p).

TABLE C

Probability of correct classification by number of groups, type of population, and method

		MDF	EV_1	EV_2	EV_3
C	k=5	.74	.72	.73	.73
	k=10	.71	.69	.70	.70
S	k=5	.64	.38	.50	.58
	k=10	.41	.18	.25	.30

The apparent improvement in performance as number of variables increases is greater for the MDF than any of the EV procedures. Table D gives the details.

TABLE D

Probability of correct classification by number of variables, type of population and method

		MDF	EV_1	EV_2	EV_3
C	p=4	.71	.70	.71	.71
	p=10	.74	.70	.72	.73
S	p=4	.47	.28	.37	.43
	p=10	.58	.28	.39	.46

The eigenvectors are often used to describe the populations graphically. Many of the computer programs offer plots of the first two eigenvectors (sometimes referred to as canonical vectors) as an option to the user. These can be valuable adjuncts to the analysis when the means are roughly in a two dimensional subspace. The extent to which this holds is usually measured by the proportion of the total variance "explained" by the first two eigenvectors.

This study has considered the behavior of the MDF

and EV methods under "ideal" conditions. That is, the variables were normally distributed and the covariance matrices were the same. Nothing has been done regarding the behavior of these methods when non-normality is present, the populations have different covariance matrices, or different sample sizes in the populations are used. These remain problems for future study.

ACKNOWLEDGEMENTS

This research was supported by Research Career Development Award HD-46344. Computing time was made possible by the Triangle Universities Computing Center and by support from the University of North Carolina. Discussions with Judith O'Fallon, M. L. Eaton, L. L. Kupper and P. K. Sen have been most helpful.

REFERENCES

Anderson, T. W. (1951), Classification by multivariate analysis, *Psychometrika*, 16, 31-50.

Anderson, T. W. (1973), Asymptotic evaluation of the probabilities of misclassification, this Volume.

Bechhofer, R. E. and Sobel, M. (1954), A single sample multiple decision procedure for ranking variances of normal populations, *Ann. Math. Stat.*, 25, 273-289.

Dunn, O. J. (1971), Some expected values for probabilities of correct classification in discriminant analysis, *Technometrics*, 13, 345-353.

Fisher, R. A. (1936), The use of multiple measurements in taxonomic problems, *Ann. Eugen.*, 7, 179-188.

Gilbert, E. (1968), On discrimination using qualitative variables, *J. Amer. Statist. Assoc.*, 63, 1399-1412.

Gilbert, E. (1969), The effect of unequal variance-covariance matrices on Fisher's linear discriminant function, *Biometrics*, 25, 505-515.

Glick, N. (1972), Sample based classification procedures derived from density estimators, *J. Amer. Statist. Assoc.*, 67, 116-122.

Hills, M. (1966), Allocation rules and their error rates, *J. Roy. Stat. Soc.*, B, 28, 1-31.

Lachenbruch, P. A. (1966), Discriminant analysis when the

initial samples are misclassified, *Technometrics*, 8, 657-662.

Lachenbruch, P. A. and Mickey, M. R. (1968), Estimation of error rates in discriminant analysis, *Technometrics*, 10, 1-11.

Lachenbruch, P. A., Sneeringer, C. and Revo, L. T. (1973), Robustness of the linear and quadratic discriminant functions to certain types of non-normality, *Communications in Statistics*, 1, 39-56.

Okamoto, M. (1963), An asymptotic expansion for the distribution of the linear discriminant functions, *Ann. Math. Stat.*, 34, 1286-1301.

TABLE 1

Coordinates of Simplicies

ℓ	Vector	i = 1	2	3	4	5	6	7	8	9	10
0	1	0	0	0	0	0	0	0	0	0	0
1	2	1	0	0	0	0	0	0	0	0	0
2	3	1/2	$\sqrt{3/4}$	0	0	0	0	0	0	0	0
3	4	1/2	$\frac{1}{2\sqrt{3}}$	$\sqrt{4/6}$	0	0	0	0	0	0	0
4	5	1/2	$\frac{1}{2\sqrt{3}}$	$\frac{1}{2\sqrt{6}}$	$\sqrt{5/8}$	0	0	0	0	0	0
5	6	1/2	$\frac{1}{2\sqrt{3}}$	$\frac{1}{2\sqrt{6}}$	$\frac{1}{2\sqrt{10}}$	$\sqrt{6/10}$	0	0	0	0	0
6	7	1/2	$\frac{1}{2\sqrt{3}}$	$\frac{1}{2\sqrt{6}}$	$\frac{1}{2\sqrt{10}}$	$\frac{1}{2\sqrt{15}}$	$\sqrt{7/12}$	0	0	0	0
7	8	1/2	$\frac{1}{2\sqrt{3}}$	$\frac{1}{2\sqrt{6}}$	$\frac{1}{2\sqrt{10}}$	$\frac{1}{2\sqrt{15}}$	$\frac{1}{2\sqrt{21}}$	$\sqrt{8/14}$	0	0	0
8	9	1/2	$\frac{1}{2\sqrt{3}}$	$\frac{1}{2\sqrt{6}}$	$\frac{1}{2\sqrt{10}}$	$\frac{1}{2\sqrt{15}}$	$\frac{1}{2\sqrt{21}}$	$\frac{1}{2\sqrt{28}}$	$\sqrt{9/16}$	0	0
9	10	1/2	$\frac{1}{2\sqrt{3}}$	$\frac{1}{2\sqrt{6}}$	$\frac{1}{2\sqrt{10}}$	$\frac{1}{2\sqrt{15}}$	$\frac{1}{2\sqrt{21}}$	$\frac{1}{2\sqrt{28}}$	$\frac{1}{2\sqrt{36}}$	$\sqrt{10/18}$	0
	Mean	1/2	$\frac{1}{2\sqrt{3}}$	$\frac{1}{2\sqrt{6}}$	$\frac{1}{2\sqrt{10}}$	$\frac{1}{2\sqrt{15}}$	$\frac{1}{2\sqrt{21}}$	$\frac{1}{2\sqrt{28}}$	$\frac{1}{2\sqrt{36}}$	$\frac{1}{2\sqrt{45}}$	0

TABLE 2

Correct Classification Probabilities

δ	K = 2	3	4	5	6	7	8	9	10	11
0.0	.500	.333	.250	.200	.166	.142	.125	.111	.100	.091
0.1	.520	.354	.269	.217	.182	.156	.137	.123	.111	.102
0.2	.540	.374	.288	.235	.198	.171	.151	.135	.123	.113
0.3	.560	.395	.308	.253	.215	.187	.166	.149	.136	.125
0.4	.579	.416	.328	.272	.233	.204	.182	.164	.149	.138
0.5	.599	.438	.349	.292	.252	.222	.199	.180	.164	.152
0.6	.618	.459	.370	.312	.271	.241	.216	.197	.180	.167
0.7	.637	.481	.392	.333	.291	.260	.235	.214	.197	.183
0.8	.655	.503	.414	.355	.312	.280	.254	.233	.215	.200
0.9	.674	.525	.436	.377	.333	.300	.274	.252	.234	.219
1.0	.691	.546	.458	.399	.355	.322	.295	.273	.254	.238
1.1	.709	.568	.481	.421	.377	.343	.316	.293	.274	.258
1.2	.726	.589	.504	.444	.400	.365	.338	.315	.295	.278
1.3	.742	.610	.526	.467	.423	.388	.360	.337	.317	.300
1.4	.758	.631	.549	.490	.446	.411	.383	.359	.339	.322
1.5	.773	.651	.571	.513	.469	.434	.405	.382	.361	.344
1.6	.788	.671	.593	.536	.493	.458	.429	.405	.384	.366
1.7	.802	.690	.615	.599	.516	.481	.452	.428	.407	.389
1.8	.816	.709	.636	.582	.539	.505	.476	.452	.431	.412
1.9	.829	.728	.657	.605	.563	.529	.500	.475	.454	.436
2.0	.841	.745	.678	.627	.586	.552	.524	.499	.478	.460
2.1	.853	.762	.698	.648	.609	.576	.548	.524	.502	.484
2.2	.864	.779	.717	.670	.631	.599	.572	.548	.527	.508

TABLE 2 (Continued)

δ	2	3	4	5	6	7	8	9	10	11
2.3	.875	.794	.736	.690	.653	.622	.595	.572	.551	.532
2.4	.885	.809	.754	.710	.674	.644	.618	.595	.575	.557
2.5	.894	.824	.771	.729	.695	.666	.641	.619	.599	.581
2.6	.903	.837	.788	.748	.715	.687	.663	.642	.623	.605
2.7	.911	.850	.804	.766	.735	.708	.685	.664	.646	.629
2.8	.919	.862	.819	.783	.753	.728	.706	.686	.668	.652
2.9	.926	.874	.833	.800	.771	.747	.726	.707	.690	.674
3.0	.933	.885	.847	.815	.788	.765	.745	.727	.711	.696
3.1	.939	.895	.859	.830	.805	.783	.764	.746	.731	.717
3.2	.945	.904	.871	.844	.820	.800	.781	.765	.750	.737
3.3	.951	.913	.883	.857	.835	.816	.798	.783	.769	.756
3.4	.955	.921	.893	.870	.849	.831	.814	.800	.786	.774
3.5	.960	.929	.903	.881	.862	.845	.830	.816	.803	.792
3.6	.964	.936	.912	.892	.874	.859	.844	.831	.819	.808
3.7	.968	.942	.921	.902	.886	.871	.858	.846	.834	.824
3.8	.971	.948	.929	.912	.897	.883	.871	.859	.849	.839
3.9	.974	.954	.936	.920	.907	.894	.883	.872	.862	.853
4.0	.977	.959	.943	.928	.916	.904	.894	.884	.875	.867
4.1	.880	.963	.949	.936	.924	.914	.904	.895	.887	.879
4.2	.982	.967	.954	.943	.932	.923	.914	.906	.898	.891
4.3	.984	.971	.959	.949	.939	.931	.923	.915	.908	.901
4.4	.986	.974	.964	.954	.946	.938	.931	.924	.918	.912
4.5	.988	.977	.968	.960	.952	.945	.938	.932	.926	.921
4.6	.989	.980	.972	.964	.957	.951	.945	.939	.934	.929
4.7	.991	.982	.975	.968	.962	.957	.951	.946	.941	.937
4.8	.992	.985	.978	.972	.967	.962	.957	.952	.948	.944

TABLE 2 (Continued)

δ	2	3	4	5	6	7	8	9	10	11
4.9	.993	.987	.981	.976	.971	.966	.962	.958	.954	.950
5.0	.994	.988	.983	.979	.974	.970	.966	.963	.959	.956

TABLE 3

Estimated Correct Classification Rates

$\delta = 1$

Type	p	k	n	Mean MDF	Mean EV_1	Mean EV_2	Mean EV_3	Standard Deviation MDF	Standard Deviation EV
C	4	5	25	.52	.50	.52	.53	.07	.06
		5	100	.50	.49	.50	.50	.03	.03
		10	25	.48	.45	.47	.47	.03	.03
		10	100	.45	.44	.44	.44	.02	.02
	10	5	25	.60	.53	.58	.61	.05	.03
		5	100	.52	.50	.51	.51	.02	.02
		10	25	.56	.47	.50	.52	.02	.03
		10	100	.49	.46	.47	.48	.01	.01
S	4	5	25	.42	.32	.39	.42	.02	.03
		5	100	.39	.30	.35	.38	.02	.02
		10	25	Not Applicable					
		10	100						
	10	5	25	.48	.32	.41	.46	.03	.04
		5	100	.41	.29	.35	.38	.01	.02
		10	25	.31	.16	.21	.25	.03	.02
		10	100	.26	.15	.18	.21	.01	.01

$\delta = 2$

Type	p	k	n	Mean MDF	Mean EV_1	Mean EV_2	Mean EV_3	Standard Deviation MDF	Standard Deviation EV
C	4	5	25	.75	.75	.76	.75	.05	.04
		5	100	.74	.74	.75	.75	.02	.02
		10	25	.73	.72	.73	.73	.03	.03
		10	100	.71	.70	.71	.71	.02	.01
	10	5	25	.80	.75	.78	.79	.03	.04
		5	100	.76	.75	.76	.75	.02	.02
		10	25	.77	.73	.75	.76	.02	.02
		10	100	.75	.73	.74	.74	.01	.01
S	4	5	25	.64	.39	.51	.59	.03	.03
		5	100	.63	.36	.48	.57	.01	.02
		10	25	Not Applicable					
		10	100						
	10	5	25	.68	.40	.55	.63	.05	.03
		5	100	.64	.36	.48	.57	.02	.02
		10	25	.53	.20	.29	.35	.03	.02
		10	100	.48	.18	.24	.29	.01	.01

$\delta = 3$

				Mean				Standard Deviation	
Type	p	k	n	MDF	EV_1	EV_2	EV_3	MDF	EV
C	4	5	25	.92	.91	.91	.92	.02	.02
		5	100	.90	.90	.91	.91	.02	.02
		10	25	.91	.90	.90	.91	.02	.02
		10	100	.90	.90	.90	.90	.01	.01
	10	5	25	.92	.92	.92	.93	.02	.02
		5	100	.88	.87	.87	.88	.01	.01
		10	25	.87	.86	.86	.86	.02	.02
		10	100	.90	.90	.90	.90	.01	.01
S	4	5	25	.84	.45	.62	.74	.04	.03
		5	100	.83	.42	.59	.72	.02	.03
		10	25	Not Applicable					
		10	100						
	10	5	25	.86	.46	.66	.77	.02	.04
		5	100	.84	.43	.61	.73	.02	.03
		10	25	.74	.23	.36	.46	.03	.03
		10	100	.72	.20	.31	.40	.01	.01

DISCRIMINATION AND THE AFFINITY OF DISTRIBUTIONS

Kameo Matusita
The Institute of Statistical Mathematics, Tokyo

1. Introduction

Discriminant analysis is usually carried out by projecting sample clusters in a multi-dimensional space onto a subspace of a lower dimension. To speak of the case of two distributions in the space R_k, for example, the linear discriminant function $c'X$ (c,X being k-dimensional vectors) is considered, where the vector c is determined usually by maximizing the ratio of the external to the internal variance of $c'X$, and here it is normally assumed that the covariance matrices of the distributions concerned are equal.

Now, the quantity called the affinity serves well as a measure of information for discriminating distributions (see Matusita, 1971). Employing the affinity, we shall present in this paper a method for determining a projection for discrimination, i.e., a method which minimizes the affinity of distributions concerned. This method will be seen to be reasonable from properties of the affinity. The linear discriminant function mentioned above is also obtained by this method, which shows that the usual way of finding the discriminant function is justified from the point of view of the affinity. Furthermore, it will be seen that we can get a discriminant function for the case of unequal covariance matrices by the same method.

In section 2 we shall state some properties of the affinity as a measure of information for discrimination, and in section 3 we shall give a method for determining a projection for efficient discrimination. Finally, in section 4 the Gaussian case will be treated.

2. *Affinity as a measure of information for discrimination*

First we shall consider the case of two distributions. Let F_1, F_2 be two distributions in the space R in which a measure m (lebesgue or counting or mixed) is defined, and let $p_1(x)$, $p_2(x)$ be density functions of F_1, F_2 with respect to m. Then we define the affinity between F_1 and F_2 by

$$\rho_2(F_1,F_2) = \int_R \sqrt{p_1(x)p_2(x)}\, dm.$$

This quantity is related to the distance between F_1 and F_2

$$d_2(F_1,F_2) = \left[\int_R (\sqrt{p_1(x)} - \sqrt{p_2(x)})^2 dm \right]^{1/2}$$

by the equation

$$d_2^2(F_1,F_2) = 2(1 - \rho_2(F_1,F_2)) .$$

In the case of several distributions we define the affinity as follows. Let $F_1,\ldots,F_r$ be distributions in R with density functions $p_1(x),\ldots,p_r(x)$ with respect to m. Then the quantity

$$\rho_r(F_1,\ldots,F_r) = \int_R (p_1(x),\ldots,p_r(x))^{1/r} dm$$

is called the affinity of $F_1,\ldots,F_r$. The value of this affinity can be shown to be independent of the choice of the underlying measure m. Furthermore, we have the following:

$$\rho_r(F_1,\ldots,F_r) \text{ is symmetric in } F_1,\ldots,F_r . \tag{1}$$

$$0 \leq \rho_r(F_1,\ldots,F_r) \leq \rho_{r-1}(F_1,\ldots,F_{r-1}) \leq \ldots \leq \rho_2(F_1,F_2) \leq 1 . \tag{2}$$

$$\rho_r(F_1,\ldots,F_r) = 1 \qquad F_1 = \ldots = F_r . \tag{3}$$

The affinity ρ_r is connected with the distance

$$d_r(F_1,F_2) = \left| \int_R \left(\sqrt[r]{p_1(x)} - \sqrt[r]{p_2(x)}\right)^r dm \right|^{1/r} ,$$

that is,

$$\rho_r \geqq 1 - (r-1)\delta \quad \text{when} \quad d_r(F_i,F_j) \leqq \delta \quad \text{for any pair} \quad (i,j),$$

and whenever

$$d_2(F_i,F_j) \leqq \delta_o \text{ for any pair } (i,j), \text{ and } \rho_o = \min_{(i,j)} \rho_2(F_i,F_j),$$

we have

$$1 - (r-1)2^{\frac{1}{r}} \delta_o^{\frac{1}{r}} \leqq \rho_r(F_1,\ldots,F_r) \leqq \rho_o^{\frac{1}{r}} . \tag{4}$$

Let $\{F_{i\nu}\}$ $(i=1,2,\ldots,r;\ \nu=1,2,\ldots)$ be sequences of distributions in R with density functions $p_{i\nu}(x)$ with respect to m and suppose $\lim_{\nu} p_{i\nu}(x) = p_{i0}(x)$ a.e.w.r.t. m in R, where $p_{i0}(x)$ is a density function, and let F_{i0} denote the distribution defined by $p_{i0}(x)$. Then we have

$$\rho_r(F_{1\nu},\ldots,F_{r\nu}) \underset{\nu\to\infty}{\to} \rho_r(F_{10},\ldots,F_{r0}) \tag{5}$$

$$\rho_r(F_1,\ldots,F_r) = \inf \sum_i (P(E_i|p_1)\ldots P(E_i|p_r)) \tag{6}$$

where inf is taken over all partitions $\{E_i\}$ of R such that $R = E_1 + \ldots + E_N$, $E_i \cap E_j = \emptyset$ $(i \neq j)$, and $P(E_i|P_k)$ denotes the probability of E_i according to the distribution F_k.

Let $T(\cdot)$ be a measurable transformation from the measure space $(\chi, \AA, m)$ to the space (T, β, μ), where for any $G \in \beta$, $T^{-1}(G) \in \AA$, $\mu(G) = m(T^{-1}(G))$. Let F_i $(i=1,2,\ldots,r)$ be distributions in χ with density functions $p_i(x)$ with respect to m, and let H_i be the distribution in T, induced from F_i by $T(\cdot)$. Then each H_i has a density function

$q_i(x)$ with respect to μ. Concerning these F_i and G_i we have

$$\rho_r(H_1,\ldots,H_r) \geqq \rho_r(F_1,\ldots,F_r) . \tag{7}$$

The equality holds when and only when

$$\frac{q_1(T(x))}{p_1(x)} = \ldots = \frac{q_r(T(x))}{p_r(x)} \quad \text{a.e.w.r.t. } m \text{ in } \chi - E_0$$

where $E_0 = \{x : p_i(x) = 0 \text{ for some } i, x \in \chi\}$.

Corollary. When $m(E_0) = 0$,

$$\rho_s(H_{\nu_1},\ldots,H_{\nu_s}) = \rho_s(F_{\nu_1},\ldots,F_{\nu_s}) ,$$

where $(\nu_1,\ldots,\nu_s)$ is any s-tuple from $1,2,\ldots,r$, is a necessary and sufficient condition that $T(\cdot)$ gives a sufficient statistic for discriminating $F_1,\ldots,F_r$.

As to proofs of these propositions, we refer to Matusita (1967a, 1971).

The affinity of distributions, which has properties as mentioned above, represents well the likeness of distributions. In other words, the affinity represents the discrepancy among distributions, so that it can serve as a measure of discriminating distributions. Actually, it has properties similar to those of information measures such as the Kullback-Leibler's information for discrimination. However, while the Kullback-Leibler's information is not symmetric in distributions, the affinity is symmetric in distributions and has a direct relationship with error probability in classification or discrimination.

For example, let us consider the following problem.

Given $F_1,\ldots,F_r$ over the space $(\chi,\mathring{A},m)$, assume that a sample has come from one of $\{F_i\}$. Furthermore, assume that the probability that a sample from F_i is known to be w_i $(i=1,2,\ldots,r;\ \sum_i w_i = 1)$. In this situation, decide from which F_i the observed sample has come.

For this problem, let $\{E_1,\ldots,E_r\}$ be a partition

of $\mathcal{X}$ such that

1) for $x \in E_i$, we have $w_i p_i(x) \geqq w_j p_j(x)$, $j \neq 1$, where $p_i(x)$ denotes the density function of F_i,

2) $E_i \cap E_j = \emptyset$ $(i \neq j)$; all E_i are m-measurable,

3) $E_1 + \ldots + E_r = \mathcal{X}$,

and let ϕ be the decision rule for the problem defined by $\{E_i\}$. Then ϕ is an optimum decision rule for discriminating $F_1,\ldots,F_r$ in the sense that it gives the smallest error rate

$$1 - \left(\int_{E_1} w_1 p_1(x)dm + \ldots + \int_{E_r} w_r p_r(x)dm \right) .$$

This error rate is bounded above by

$$\sqrt{w_1 w_2}\, \rho_2(F_1,F_2) + \ldots + \sqrt{w_{r-1} w_r}\, \rho_2(F_{r-1},F_r)$$

and bounded below by

$$\frac{(r - 1)(w_1 \ldots w_r)\, \rho_1^r(F_1,\ldots,F_r)}{r^{r-1}} .$$

As to transformations in the space, inequality (7) shows that they do not increase information for discrimination.

The affinity between two distributions is, as mentioned above, related to the distance $d_2(\cdot,\cdot)$, so that the distance $d_2(\cdot,\cdot)$ has the corresponding properties of affinity. Actually, Kirmani (1968, 1971) showed this fact.

3. *Projection for discrimination*

Suppose that distributions $F_1,\ldots,F_r$ are given in the k-dimensional space R_k, and the problem is to discriminate among $F_1,\ldots,F_r$ by a projection onto an s-dimensional subspace R_s. Let E denote a projection onto an s-dimensional space R_s. Further, let $\{e_1,\ldots,e_k\}$ be an orthonormal basis of R_k, i.e., $(e_i,e_j) = \delta_{ij}$, and let

$\{a_1,\ldots,a_s\}$ be an orthonormal basis of R_s, i.e. $(a_i,a_j) = \delta_{ij}$. Then each a_i can be written as a linear combination of e_j:

$$a_i = \sum_j \alpha_{ij} e_j .$$

The matrix $C = (\alpha_{ij})$ has rank s and it holds that

$$CC' = I_s \quad \text{(the s×s unit matrix)}.$$

The vector $x = \sum_j \gamma_j e_j$ in R_k is projected on the vector $\sum_i a_i(\sum_j \alpha_{ij}\gamma_j)$; symbolically,

$$x \to Cx .$$

The matrix C is a representation of the projection E and characterizes that projection. Let $E(F_i)$ denote the distribution induced by projection E from F_i $(i=1,2,\ldots,r)$. Then our problem is to find a projection E such that $E(F_1),\ldots,E(F_r)$ lie as far apart from one another as possible. For that, we propose here the following principle.

Find a projection E *onto an s-dimensional subspace which minimizes*

$$\rho_r(E(F_1),\ldots,E(F_r)) .$$

Since we have

$$\rho_r(F_1,\ldots,F_r) \leq \rho_r(E(F_1),\ldots,E(F_r)) ,$$

we get

$$\rho_r(F_1,\ldots,F_r) \leq \inf_E \rho_r(E)F_1),\ldots,E(F_r)) .$$

Let $\{E_n\}$ be a sequence of projections onto s-dimensional subspace such that

$$\rho_r(E_n(F_1),\ldots,E_n(F_r)) \to \inf_E \rho_r(E(F_1),\ldots,E(F_r)) .$$

Then we can select a subsequence of $\{E_n\}$, $\{E_{v_n}\}$, which converges to a projection. In fact, let $\{C_n\}$ be the sequence of representation matrices of $\{E_n\}$. Then, as each C_n satisfies $C_nC_n' = I_s$, each element C_n is contained in the interval $[-1,1]$, and we can take a subsequence $\{C_{v_n}\}$ out of $\{C_n\}$ which converges to a matrix of $(s\times k)$-type. Denote this matrix by C_0. Then C_0 satisfies the equation $C_0C_0' = I_s$, from which it follows that the rank of C_0 is s. Thus, the matrix C_0 can be regarded as a representation matrix of a projection onto an s-dimensional space. Denote this projection by E_0. Then the sequence $\{E_{v_n}\}$ converges to E_0, and $E_{v_n}(F_i)$ converges to $E_0(F_i)$ $(i=1, 2,\ldots,r)$. Consequently, we have

$$\rho_r(E_{v_n}(F_1),\ldots,E_{v_n}(F_r)) \to \rho_r(E_0(F_1),\ldots,E_0(F_r))$$

and

$$\rho_r(E_0(F_i),\ldots,E_0(F_r)) = \inf_E \rho_r(E(F_1),\ldots,E(F_r)) .$$

Thus the existence of a projection which minimizes $\rho_r(E(F_1),\ldots,E(F_r))$ is established.

If it holds generally that $\rho_r(E(F_1),\ldots,E(F_r))$ is invariant under non-singular linear transformations in the space onto which points in R_k are projected by E, we obtain

$$\rho_r(TE(F_1),\ldots,TE(F_r)) = \rho_r(E(F_1),\ldots,E(F_r)) ,$$

where T denotes a non-singular linear transformation in that space and $TE(F)$ is the induced distribution from $E(F)$ by T. Since a projection itself is a linear transformation in R_k, TE is also a linear transformation in R_k. Let C be a representation matrix on E. Then TC is a $s\times k$ matrix with rank s and we have

$$\rho_r(TC(F_1),\ldots,TC(F_r)) = \rho_r(C(F_1),\ldots,C(F_r)) ,$$

where $C(F)$ is the induced distribution from F by the transformation C, but this is the same as $E(F)$. On the other hand, any $s\times k$ matrix B with rank s can be written in the form $B = TC$ where C is a $s\times k$ matrix with rank s and T is a non-singular $s\times s$ matrix. Thus, we get

$$\inf_{B} \rho_r(B(F_1),\ldots,B(F_r)) = \inf_{E} \rho_r(E(F_1),\ldots,E(F_r)) ,$$

where B runs over the set of s×k matrices with rank s and E over the set of projections in R_k onto s-dimensional subspaces. Therefore:

When the above-mentioned condition holds, we need not always confine ourselves to the set of projections for our problem; we can seek a matrix B *from the set of* s×k *matrices with rank* s *which minimizes the affinity* $\rho_r(B(F_1),\ldots,B(F_r))$.

4. *Gaussian cases*

Let $F_1,\ldots,F_r$ be k-dimensional Gaussian distributions:

$$F_1 = N(a_1, A_1^{-1}),\ldots,F_r = N(a_r, A_r^{-1}) ,$$

where a_i are k-dimensional (column) vectors, and $A_1,\ldots,A_r$ are positive definite (symmetric) k×k matrices, Then we have

$$\rho_r(F_1,\ldots,F_r) = \frac{\prod_i |A_i|^{1/2r}}{|\frac{1}{r}\sum_i A_i|^{1/2}}$$

$$\times \exp\left[-\frac{1}{2r^2}\left\{\left(\sum_i A_i a_i,\ (\sum_i A_i)^{-1}\sum_i A_i a_i\right) - \sum_i (A_i a_i,\ a_i)\right)\right\}\right].$$

The right-hand side is invariant under a linear transformation $Tx + b$ with $|T| \neq 0$.

Especially, when $A_1 = \ldots = A_r = A$ we get

$$\rho_r(F_1,\ldots,F_r) = \exp\left[-\frac{1}{2r^2}\sum_{i,j}(A(a_i-a_j),\ (a_i-a_j))\right] ,$$

and, further, when $r = 2$,

$$\rho_2(F_1,F_2) = \exp\left[-\frac{1}{8}(A(a_1-a_2),\ (a_1-a_2))\right] .$$

$(A(a_1-a_2), (a_1-a_2))$ in the right-hand side is, as is known, the Mahalanobis distance between two distributions $N(a_1,A^{-1})$ and $N(a_2,A^{-1})$.

When $a_1 = \ldots = a_r$, the general form of ρ_r becomes

$$\rho_r(F_1,\ldots,F_r) = \frac{\prod_i |A_i|^{1/2r}}{|\frac{1}{r}\Sigma A_i|^{1/2}} .$$

Now, let B be an s×k matrix with rank s (s<k), and consider the transformation $y = Bx$. Then the induced distributions from $F_1,\ldots,F_r$ by this transformation are:

$$H_1 = N(Ba_1, BA_1^{-1}B'),\ldots, H_r = N(Ba_r, BA_r^{-1}B') ,$$

and we have

$$\rho_r(H_1,\ldots,H_r) = \frac{\prod_i |(BA_i^{-1}B')^{-1}|^{1/2r}}{|\frac{1}{r}\sum_i (BA_i^{-1}B')^{-1}|^{1/2}}$$

$$\times \exp\left[\frac{-1}{2r^2}\left\{(\sum_i (BA_i^{-1}B')^{-1}Ba_i,\right.\right.$$

$$\left.\left.(\sum_i (BA_i^{-1}B')^{-1})^{-1}\Sigma(BA_i^{-1}Ba_i) - \sum_i((BA_i^{-1}B')^{-1}Ba_i,Ba_i)\right\}\right] .$$

When $A_1 = \ldots = A_r = A$,

$$\rho_r(H_1,\ldots,H_r)$$

$$= \exp\left[\frac{-1}{2r^2}\sum_{i,j}((BA^{-1}B')^{-1}B(a_i-a_j), B(a_i-a_j))\right] ,$$

and when $a_1 = \ldots = a_r$,

$$\rho_r(H_1,\ldots,H_r) = \frac{\prod_i |(BA_i^{-1}B')^{-1}|^{1/2r}}{|\frac{1}{r}\sum_i (BA_i^{-1}B')^{-1}|^{1/2}} .$$

Our problem is to find an s×k matrix B which minimizes $\rho_r(H_1,\ldots,H_r)$, because the affinity in the Gaussian case is invariant under non-singular linear transformations.

For example, when $r = 2$, $s = 1$, and the covariance matrices are equal, we have

$$\rho_2(H_1,H_2) = \exp\left[-\frac{1}{8}((c'A^{-1}c)^{-1}(c'a_1-c'a_2),\ (c'a_1-c'a_2))\right]$$

where c is a column vector. To minimize $\rho_2(H_1,H_2)$ means to maximize

$$\frac{(c'(a_1 - a_2))^2}{8c'A^{-1}c}$$

(see Matusita, 1967b). This is the usual way for finding a line onto which to project distributions for discrimination.

For illustration, let us consider the case where

$$A_1 = I_k,\quad A_2 = \begin{pmatrix} \lambda_1 & & & 0 \\ & \lambda_2 & & \\ & & \ddots & \\ 0 & & & \lambda_k \end{pmatrix},\quad a_1 = \begin{pmatrix} 0 \\ 0 \\ \vdots \\ 0 \end{pmatrix},\quad a_2 = \begin{pmatrix} d_1 \\ d_2 \\ \vdots \\ d_k \end{pmatrix}.$$

Let $c = \begin{pmatrix} \gamma_1 \\ \gamma_2 \\ \vdots \\ \gamma_k \end{pmatrix}$. Then the problem is to find c which

minimize

$$\frac{2^{k/2}((\gamma_1^2 +\ldots+ \gamma_k^2)(\lambda_1\gamma_1^2 +\ldots+ \lambda_k\gamma_k^2))^{1/4}}{[(\gamma_1^2 +\ldots+ \gamma_k^2)+(\lambda_1\gamma_1^2 +\ldots+ \lambda_k\gamma_k^2)]^{1/2}}$$

$$\times \exp\left[-\frac{1}{4} \frac{(\gamma_1\alpha_1 +\ldots+ \gamma_k\alpha_k)^2}{(\gamma_1^2 +\ldots+ \gamma_k^2)+(\lambda_1\gamma_1^2 +\ldots+ \lambda_k\gamma_k^2} \right].$$

In computation we can put the condition

$$(\lambda_1 + 1)\gamma_1^2 +\ldots+ (\lambda_k + 1)\gamma_k^2 = 1 .$$

When A_i, a_i are unknown, we, of course, employ sample means and sample covariance matrices.

REFERENCES

Kirmani, S. N. U. A. (1968), Some results on Matusita's measure of distance, *Jour. Indian Statist. Assoc.*, 6, 89-98.

Kirmani, S. N. U. A. (1971), Some limiting properties of Matusita's measure of distance, *Ann. Inst. Statist. Math.*, 23, 157-162.

Matusita, K. (1967a), On the notion of affinity of several distributions and some of its applications, *Ann. Inst. Statist. Math.*, 19, 181-192.

Matusita, K. (1967b), Classification based on distance in multivariate Gaussian cases, *Proc. Fifth Berkeley Symposium on Math. Statist. and Prob.*, 1, 299-304.

Matusita, K. (1971), Some properties of affinity and applications, *Ann. Inst. Statist. Math.*, 23, 137-155.

SIMULATION EXPERIMENTS WITH MULTIPLE GROUP LINEAR AND QUADRATIC DISCRIMINANT ANALYSIS

J. Michaelis
University of Mainz

Summary

A simulation program is described which can be performed to obtain estimates of the different types of misclassification probabilities for multiple group linear and quadratic discriminant analysis. The program can be used to study how these errors depend on sample sizes and the different parameters of the multivariate normal distribution. Examples for several simulation experiments are given and possible conclusions are discussed.

1. *Introduction, notation*

The simulation experiments considered here illustrate some of the questions which arise in the context of practical application of discriminant analysis.

The special practical application I am dealing with is the automated analysis of electrocardiograms. A computer program extracts from each patient's electrocardiogram a vector of measurements. On the basis of this vector of observed measurements, a person has to be classified as belonging to a normal or one of a certain number of disease groups. Large samples of "normals" and patients from whom such measurements were taken were kindly provided to us by Prof. Pipberger, Washington. The data have been described elsewhere in more detail, Pipberger (1970), where also further references to the data are given. The samples consisted of individuals, for whom the diseases were clearly determined by several valid, external criteria, not including any of the ECG measurements. On the basis of this practical application, I want to deal with some of the questions which were already mentioned in Lachenbruch (1973).

a) What is the dependence of the estimated frequency of correct classifications on the number of variables, sample sizes and the number of populations (diseases) taken into account? In a practical application this may be the question: How well the discrimination in new, independent samples compares to the classification achieved by the resubstitution method for the first samples.

b) How large is the expected difference of correct classifications if one uses the linear and quadratic discriminant analysis when differences in the covariance matrices are present? For the multiple group discriminant analysis this question is of interest when the parameters are known as well as when they are estimated. Different covariance matrices are very common in medical data, e.g. the variability tends to be greater for "abnormals" than for "normals". Frequently therefore the question arises, whether one should use the linear or the quadratic discriminant analysis, especially with respect to deviations from the multivariate normal distribution.

After raising this set of questions which partly depend on each other it seems useful to distinguish between the different types of misclassification probabilities and their estimations which one can consider. Before that I want to introduce the following notation:

Let there be k populations π_i, $i=1,2,\ldots,k$ whose elements are associated with a p-dimensional multivariate normally distributed random vector X. θ_i is the known parameter set of X in π_i. We assume that all prior probabilities and the costs of all types of misclassifications are equal. Then we choose an allocation rule δ so that an observed vector X_o shall be allocated to π_i if $f(X_o, \theta_i) > f(X_o, \theta_j)$, $j \neq i$, $j = 1,\ldots,k$, where $f(X_o, \theta_j)$ denotes the value of the multivariate normal density function $f(X,\theta_j)$ corresponding to X_o.

If the covariance matrices are equal for all populations, δ is called "linear discriminant" rule, otherwise, "quadratic discriminant" rules.

Following partly the notation of Hills (1966), one can distinguish between:

1. $\alpha_i(\delta)$ = probability that a randomly chosen element of π_i will be misclassified when the allocation rule is based on the known parameters. This could be called the "optimal probability".

2. $\alpha_i(\hat{\delta})$ = (conditional) probability, that a randomly chosen element of π_i will be misclassified when the allocation rule is based on parameters which have been estimated from certain reference samples. This could be called the "actual probability".

3. $\hat{\alpha}_i(\hat{\delta})$ = estimate of the actual probability.

4. $E[\alpha_i(\hat{\delta})]$=(unconditional) probability, that a randomly chosen element of π_i will be misclassified when the allocation rule is based on parameters which have been estimated from a randomly chosen set of reference samples.

5. $E[\alpha_i(\hat{\delta})]$=estimate of the probability $E[\alpha_i(\hat{\delta})]$.

For the two-group linear discriminant analysis the calculation of $\alpha_i(\delta)$ is straightforward (see e.g. Anderson 1958); for the corresponding two-group quadratic discriminant analysis and for the multiple group analysis the calculation of the involved integrals becomes very complicated so that asymptotic expansions have been given, Anderson (1972), Hildebrandt et al (1973), Okamoto (1963), Press (1966), and Shah (1963).

The actual probability $\alpha_i(\delta)$ is frequently estimated by the "resubstitution" method: The estimated parameters $\hat{\theta}_i$ are either substituted in the calculation of the optimal probability as recommended by Fisher (1936), or the frequency of correct classifications is evaluated after the reference samples have been reclassified according to $\hat{\delta}$ - as recommended by Smith (1947). We will denote the latter estimation, which also can easily be applied to the quadratic and multiple group discriminant analysis, by $\hat{\alpha}_i^*(\hat{\delta})$.

For the estimation of $E[\alpha_i(\hat{\delta})]$ Lachenbruch and Mickey (1968) have proposed the so called jackknife technique which is also suited for quadratic and multiple-group discriminant analysis.

After giving a brief outline of the computer program which was designed to carry out different types of simulation experiments, I will describe some of the results of these experiments and discuss some possible conclusions.

2. *Description of the computer program*

k samples of a p-dimensional multivariate normally distributed random vector X can be generated with sample sizes n_j, j=1,2,...,k after specifying the corresponding model parameters μ_j, Σ_j in the following manner:

In the first step p-dimensional random vectors Z_{ij}, i=1,2,...,n_j are generated which have the multivariate normal distribution N(0,I). Then these vectors are transformed in the following way:

$$X_{ij} = T_j \cdot Z_{ij} + \mu_j$$

where T_j is chosen so that $T_j T_j' = \Sigma_j$. The program uses a Cholesky-factorisation of Σ_j to obtain T_j.

We checked the performance of the random number generator and of the overall procedure with respect to the following:

a) Univariate normal distributions of the components of Z.

b) Multivariate normal distribution of Z according to I (Bartlett's test of "total independence" as described in Morrison, 1967).

After generation of the samples, linear and quadratic discriminant analysis were used to classify the samples in three different ways:

a) The specified model-population parameters are used to classify each member of the samples. This gives an estimation of the "optimal probability" which could be denoted as $\hat{\alpha}_i(\delta)$ in accordance with the previously given notation.

b) The members of each sample are classified by the resubstitution method which we denoted as $\hat{\alpha}_i^*(\hat{\delta})$ previously and which shall also later be referred to as "internal classification", because the samples are classified based on the parameters which are estimated from the samples themselves.

c) The members of each sample are classified on the basis of parameters which have been estimated from another sample. This corresponds to the definiton of $\hat{\alpha}_i(\hat{\delta})$ if only two sets of samples are considered but leads to the estimation of $E[\alpha_i(\hat{\delta})]$ if more replications are performed as will be described below. We will refer to this method also as "external classification". Linear discriminant analysis was performed by computing a pooled covariance matrix disregarding the previously specified differences between these matrices.

For a given model the difference between the frequencies of correct classifications by internal and external classification depends on the sample sizes. This reflects, of course, in some way how well the parameters are estimated according to sample sizes. But in practical applications of discriminant analysis one might be more interested in looking at the differences between the results of internal and external classification which depend not only on the accuracy of estimation of the parameters but also on the parameter structure of the chosen model.

For each set of specified model populations the described generation and classification of samples has in most of our simulation runs been repeated ten times. Thereafter, mean values and standard deviations of the observed frequencies of correct classifications have been calculated. The complement of the mean value for the internal classification can be regarded as an estimate of $E[\hat{\alpha}_i^*(\hat{\delta})]$.

3. *Results*

Table 1 shows an example of several simulation runs. The number of populations was 5, the number of variates 8. Instead of arbitrarily choosing or constructing the different model-population parameters, we took those parameters which had been estimated from the Pipberger data which I mentioned before. It will be pointed out, that by choosing these parameters it was not the aim of the simulation experiments to give direct answers to the questions raised before with respect to these special data. It was rather the idea to choose the parameters somewhat close to a really observed situation.

The sample sizes were taken equal for all groups, in the upper half of Table 1 30 and in the lower half 100 for each group. Frequencies of correct classifications are given for the different types of classification described before. Linear and quadratic discriminant analysis were always applied to the same samples. Whereas the internal classification gives constantly an underestimation with respect to $\hat{\alpha}_i(\hat{\delta})$, the external classification gives a constant overestimation. With sample sizes increasing to infinity all three estimations should converge to $\alpha_i(\delta)$ as it was theoretically shown by Glick (1972).

One can notice that the quadratic analysis gives a better discrimination than the linear analysis as one would expect, because different covariance matrices have been chosen for the model. Gilbert (1969) has shown for the two-group case that the "optimal probabilities" of misclassification are only markedly lowered by application of the quadratic analysis compared to the use of the linear analysis if differences of the covariance matrices are present and the Mahalanobis distance has a relatively small value. This effect was already illustrated for the univariate case by Smith (1947) and holds equivalently for the multiple-group case. In Table 1 the difference between the results of internal and external analysis is substantially larger for the quadratic than for the linear discriminant analysis, especially for the smaller sample sizes. This is presumably due to the fact that the number of estimated parameters is much smaller in the linear discriminant analysis.

It is interesting to note that for all simulated larger samples the external quadratic discriminant analysis gives better results than the corresponding linear analysis, although the estimation of the parameters is not yet very good, as can be seen from the differences between internal and external analysis - especially for the quadratic discriminant analysis. Even with the smaller sample sizes, where the differences between internal and external analysis are very large, in most samples external estimation for the quadratic analysis gives better results than in the corresponding linear analysis.

We - by the way - never saw this in a corresponding analysis of the original Pipberger data which might be due to the fact that the distributions of the simulated data were in a good accordance with the multivariate normal distribution, whereas the original data showed some deviations from multivariate normality. As Anderson and Bahadur (1962) already have pointed out, deviations from multivariate normality may affect the results of quadratic discriminant analysis much more than those of the linear analysis.

Figure 1 gives a graphical summary of several simulation experiments. It shows frequences of correct classifications which were obtained for the three types of classifications and different sample sizes. Again as model-parameters were chosen those which were estimated from the Pipberger data. The first model (left diagrams) consisted of 5 populations with $p = 10$. The second model (diagrams in the middle of Fig.1) was a modification of the first one by using the same parameters but deleting 2 variables of the random vector which in the original data contributed least to the discrimination. The third model consisted of a subset of the second one by deleting one population. Although the second and third model can be regarded as a subset of the first one, different samples were generated for each of the models. Linear and quadratic discriminant analysis, however, were always applied to the same samples. For every chosen model one can observe how the frequencies of correct classification by internal and external classification converge against the value which is estimated on the basis of the known parameters with increasing sample sizes. These observations, however, always hold only for the chosen model, which is also demonstrated by Figure 1: Although the structures of the 3 models have very much in common, the types of convergence differ markedly. The way in which the differences between the results of internal and external classification become smaller with increasing sample sizes depends ont only on the number of populations and the dimension of X but also on the defined population parameters.

For the linear 2-group discriminant analysis Lachenbruch (1968) and Cornfield (1967) have shown independently how the overestimation of the Mahalanobis' distance and therefore the underestimation of $\alpha_i(\hat{\delta})$ depends on the value of the known distance. One could also notice from the data which Lachenbruch presented in his paper (see present Volume) for the multiple group linear discriminant function with equal covariance matrices, that the overestimation of the correct classification rates depended on the pairwise group distances which he had chosen.

However, several other simulation studies, similar to the ones shown in Figure 1, have not so far pointed to a certain kind of dependence on the parameter structure. This is possibly due to the fact that the models, which I chose on the basis of the observed data, were very complicated and that the inherent problems - e.g. computing of a pooled covariance matrix when the matrices are different - require a much more extensive systematical variation of the parameters than I could perform so far to achieve some results which can be generalized.

Figure 2 shows some results of simulation experiments with different model-parameters. Again the model-parameters are chosen equal to some estimated parameters from really observed data. Here I only want to demonstrate (left diagram for the quadratic discriminant analysis) that $\hat{\alpha}^*(\hat{\delta})$ can become numerically very close to $\hat{\alpha}(\delta)$ simply due to the fact that the model is chosen in such a way, that nearly complete separation of the populations is possible. That the parameters are poorly estimated from the small sample sizes is made apparent only by the external classification. The large oscillations which one can notice on the right diagrams are due to the fact that here very small sample sizes have been generated, in which case the number of simulation runs should have been increased to get better estimates of the probabilities.

The difference between the results of internal and external classifications has a different meaning - depending on which part of the percentage scale they are observed. This becomes apparent also in the diagrams of Figure 2. For systematic studies it might therefore be useful to choose a transformation of these differences which allows comparisons which do not depend on the part of the scale where the observations are made. The logit-transformation seems to be well suited to allow such comparisons.

4. *Discussion*

What kind of conclusions can be drawn from simulation experiments like the ones described before?

At first it seems important to state again that the simulation technique can give estimates of $\alpha_i(\delta)$. This can be useful in cases where, e.g., numerical integration or asymptotic expansions would require more computing time than the simulation experiments.

From the simulation examples which were shown I would like to recommend both an internal and external analysis in each practical application of multiple group discriminant analysis. This can be done by the resubstitution method recommended by Smith and the technique which was proposed by Lachenbruch and Mickey. For both methods, however, one must keep in mind that the obtained estimation is only unbiased if the original sample of reference groups is unbiased. This sample is no longer unbiased, e. g., if a stepwise variable elimination has been performed based on parameters which were estimated from the sample itself. The differences between these two classification methods indicate the interval in which the "optimal probability" can be expected to lie. For practical applications this may be of interest in the following case: If there is a great difference between $\hat{\alpha}_i^*(\hat{\delta})$ and $E[\alpha_i(\hat{\delta})]$, one can expect to achieve a better discrimination of independent samples by further increasing the sample sizes of the reference groups.

The observation of these differences for linear and quadratic discriminant analysis may also lead to a choice of one of the two methods. E.g., one might, for practical applications, prefer the linear to the quadratic discriminant analysis if there is an indication that the "optimal probabilities" do not differ very much for both methods, because the required computations are much simpler for the linear discriminant analysis. This is not an important aspect if a computer can be used but may become important, e.g., if a physician will apply a diagnostic algorithm.

A general drawback to all simulation experiments is that the obtained results only hold for the specific model definition. One can, however, hope that a systematic variation of model specifications may give hints to more general and perhaps explicit solutions to some of the questions which were raised at the beginning of this paper.

By choosing the model parameters equal to parameters which have been estimated from empirical samples one cannot transfer directly the results obtained from the simulation experiments to the original populations. It is, however, interesting to observe whether there exist any larger differences of internal and external classifications between the simulated model and the really observed data for corresponding sample sizes. Such deviations could be due, e.g., to the fact that the original data are not normally distributed. If such deviations do not exist, the results of simulation experiments can be taken as an additional information about the sample sizes which are required to achieve a certain level of correct classification when the derived formulae are applied to independent samples.

REFERENCES

Anderson, T. W. (1958), *An Introduction to Multivariate Statistical Analysis*, Wiley: New York.

Anderson, T. W. (1972), Asymptotic evaluation of the probabilities of misclassification by linear discriminant functions, this Volume.

Anderson, T. W. and Bahadur, R. R. (1962), Classification into two multivariate normal distributions with different covariance matrices, *Ann. Math. Stat.*, 33, 420-431.

Cornfield, J. (1967), Discriminant functions, *Rev. Int. Stat. Inst.*, 35, 142-153.

Fisher, R. A. (1936), Use of multiple measurements in taxonomic problems, *Ann. Eugen.*, 7, 179-188.

Gilbert, E. S. (1969), The effect of unequal variance - covariance matrices on Fisher's linear discriminant function, *Biometrics*, 25, 505-515.

Glick, N. (1972), Sample-based classification procedures derived from density estimators, *J. Amer. Stat. Assoc.*, 67, 116-122.

Hildebrandt, B., Michaelis, J., and Koller, S. (1973), Die Häufigkeit der Fehlklassifikation bei der quadratischen Diskriminanzanalyse, *Biom. Z.* (in

press).
Hills, M. (1966), Allocation rules and their error rates, *J. Roy. Stat. Soc.* B, 28, 1-31.

Lachenbruch, P. A. (1968), On expected probabilities of misclassification in discriminant analysis, necessary sample size and a relation with the multiple correlation coefficient, *Biometrics*, 24, 323-334.

Lachenbruch, P. A. and Mickey, M. R. (1968), Estimation of error rates in discriminant analysis, *Technometrics*, 10, 1-11.

Lachenbruch, P. A. (1973), Some results on the multiple group discrimination problem, this Volume.

Michaelis, J. (1972), Zur Anwendung der Diskriminanzanalyse für die medizinische Diagnostik, Habilitationsschrift, Mainz.

Morrison, D. F. (1967), *Multivariate Statistical Methods*, McGraw-Hill: New York.

Naylor, T., Baltintey, J. L., Burdick, D. S., and Chu, R. (1967), *Computer Simulation Techniques*, New York.

Okamoto, M. (1963), An asymptotic expansion for the distribution of linear discriminant function, *Ann. Math. Stat.*, 34, 1286-1301.

Pipberger, H. V. (1970), Computer analysis of electrocardiograms, *Clinical Electrocardiography and Computers* (Caceres, C. A. and Dreifus, L. S., Eds.), Academic Press: New York.

Press, S. (1966), Linear combinations of non-central chi-squares variates, *Ann. Math. Stat.*, 37, 480-487.

Shah, B. K. (1963), Distribution of definite and of indefinite quadratic forms from a non-central normal distribution, *Ann. Math. Stat.*, 34, 186-190.

Smith, C. A. B. (1947), Some examples of discrimination, *Ann. Eugen.*, 13, 272-282.

TABLE 1

Example for two simulation series for linear and quadratic dsicriminant analysis. Parameters for the model-populations have been chosen according to those which have been estimated from really observed data (automatic analysis of vectorcardiograms, data provided by Prof. Pipberger, Washington).

$$k = 5 \ , \quad p = 8$$

I = Classification of the generated samples according to the model parameters.
II = Classification by "internal parameter estimation".
III = Classification by "external parameter estimation".

a) sample sizes: 30 cases per group

	Frequency (%) of correct classifications					
	linear discriminant analysis			quadratic discriminant analysis		
	I	II	III	I	II	III
	62,7	68,0		76,7	85,3	
	67,3	71,3	69,3	70,7	83,3	64,0
	70,0	74,7	68,0	78,7	88,0	69,3
	70,0	73,3	69,3	67,7	90,0	70,0
	70,0	70,7	67,3	81,3	87,3	64,7
	60,0	60,0	58,7	74,0	87,3	62,7
	60,7	64,7	58,0	71,3	80,7	66,7
	65,3	69,3	65,3	78,7	88,0	70,0
	72,0	73,3	69,3	79,3	90,0	70,0
	66,0	73,3	66,0	74,7	86,7	68,0
$\overline{Q}$	66,4	69,9	65,7	75,3	86,7	67,3
s_{emp}	4,2	4.6	4,4	4,4	2,9	2,9

TABLE 1 (Continued)

b) sample sizes: 100 cases per group

	I	II	III	I	II	III
	67,4	68,6		77,4	79,2	
	67,8	69,2	68,8	75,8	78,4	71,2
	68,4	68,6	65,8	76,6	78,8	74,2
	70,0	69,6	67,8	75,8	79,8	73,2
	67,8	69,2	66,4	72,6	77,6	72,2
	68,4	69,0	67,8	74,4	77,6	69,6
	66,4	68,0	66,2	76,8	81,2	72,4
	66,8	67,4	63,4	72,6	76,6	72,4
	67,8	66,8	65,8	76,2	79,9	71,6
	64,6	66,0	63,2	75,8	79,6	71,2
$\overline{Q}$	67,5	68,2	66,1	75,4	78,8	72,1
s_{emp}	1,4	1,2	1,9	1,7	1,3	1,4

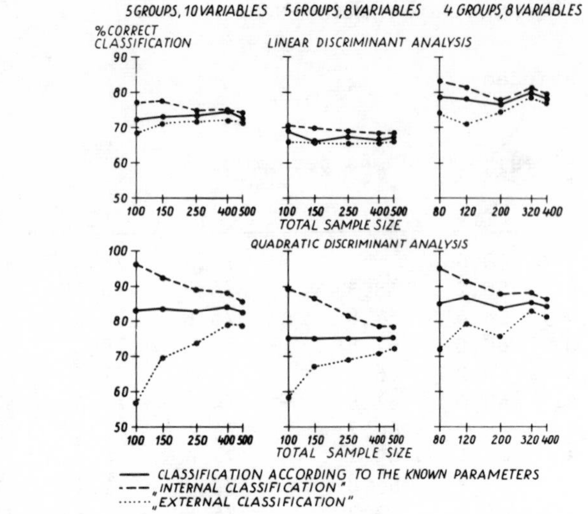

Fig. 1 Examples of some simulation experiments. The parameters of the model have been specified according to estimations from really observed data. Further description of the models is given in the text.

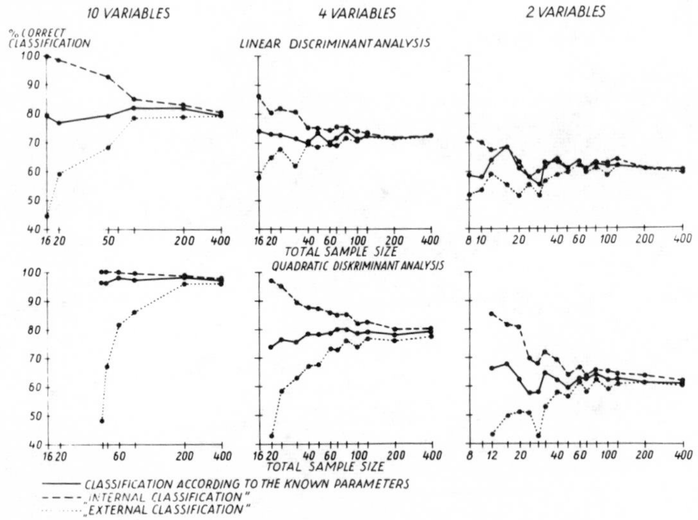

Fig. 2 Examples of further simulation experiments: Number of populations = 4.
Parameters were specified according to a stepwise variable elimination with really observed data.

FINITE AND INFINITE MODELS FOR GENERALIZED GROUP-TESTING WITH UNEQUAL PROBABILITIES OF SUCCESS FOR EACH ITEM*

Elliott Nebenzahl and *Milton Sobel*

California State University at Hayward
and
University of Minnesota

1. *Introduction*

In group-testing we are allowed to test any number of units from the same or different sources but each test of a batch of (say, x) units gives us one of two possible results: i) either all x units are good or ii) at least one of the x units is defective and we don't know which one(s). In this paper our goal is again to classify units efficiently in the sense of minimizing the number of tests required. This paper is a generalization of previous work in group-testing, Sobel and Groll (1959) and Sobel (1960), in the sense that we allow units to have different probabilities q_i $(i=1,2,\ldots,k)$ of being good whereas previous work assumed a common value of q for all units. Each value of i $(i=1,2,\ldots,k)$ corresponds to a different source or stream of units and each stream represents an assembly line type of operation with an unending number of units coming forth to be tested. Each unit from the ith stream can only be good (with probability q_i) or defective (with probability $p_i = 1-q_i$). All units from the same or different streams represent independent binomial chance variables.

The one restriction that we put on the plan or strategy for testing units is that no stream should be held

*Sponsored partially by U.S. Army Grant DA-ARO-D-31-124-72-6187 at the University of Minnesota, Minneapolis, Minn., and partially by contract NIH-E-71-2180 at the University of California Medical School in San Francisco, California.

up indefinitely, i.e., any unit in any of the k streams will be classified in some finite number of tests.

For convenience we sometimes group the units into sets of size k, where the jth set consists of the jth unit from each of the k sources. Let c denote the total number of such sets; we consider both the case $c = \infty$ and the case in which c is large but finite. For convenience and in order to make meaningful comparisons we rephrase our goal as the minimization of the expected number of tests per set of units classified, subject to the restriction mentioned above.

For any set of q_i-values we can always write these as powers of one q (say, the smallest q). We do not treat the most general case of unequal q_i-values since we assume that the q_i-values are integer powers of the smallest q. In this case we can interpret the unit with probability (say) q^3 as being a set of 3 units, each with probability q, and we only want to know whether this set contains all good units or at least 1 defective unit. The advantage of this interpretation is that we can bring to bear on the problem information from other papers such as Sobel and Groll (1959). Below we refer to this idea as group-testing with groups or the G-point of view as opposed to the individual or I-point of view.

2. *Vertical vs. Horizontal Procedures*

A vertical procedure is one for which the probability that a unit is not classified until after m units from subsequent sets have been classified tends to zero as $m \to \infty$. This insures us that no one source will be held up indefinitely. The problem of finding the optimal vertical procedure is not easy and it turns out to be useful to find the optimal procedure in another class of procedures that we call 'horizontal'. One of the main results of this paper is to point out that for every horizontal procedure we can find a vertical procedure which is equivalent in the sense of having the same expected number of tests per set classified.

To explain the horizontal procedure we assume a large finite c and later let $c \to \infty$. Rather than give a formal definition of a horizontal procedure, we illustrate it by an example. Suppose $k = 4$ and let x_i denote a unit from the ith source which has probability q^i of being good $(i=1,2,3,4)$. To be specific, we take $q = .9$. A horizontal procedure S_h gives us an ordered list of preferred batch structures for testing a plan (or tree) for testing each of these batches. Suppose a particular horizontal procedure $S_h^{(4)}$ gives us the list

$$S_h^{(4)} = \begin{array}{ll} \textit{Stage 1:} & (x_3,\ x_4) \\ \textit{Stage 2:} & (x_1,\ x_2,\ x_4) \\ \textit{Stage 3:} & (x_1,\ x_2,\ x_2,\ x_2) \\ \textit{Stage 4:} & (x_1,\ x_1,\ x_1,\ x_1,\ x_1,\ x_1,\ x_1). \end{array} \tag{2.1}$$

Then in Stage 1 we continue to test pairs (x_3,x_4) according to a specific given plan specified by the procedure until the units x_3 are all classified. A particular plan (later shown to be optimal for the example above) is the following, where arrows to the left (right) indicate success (failure):

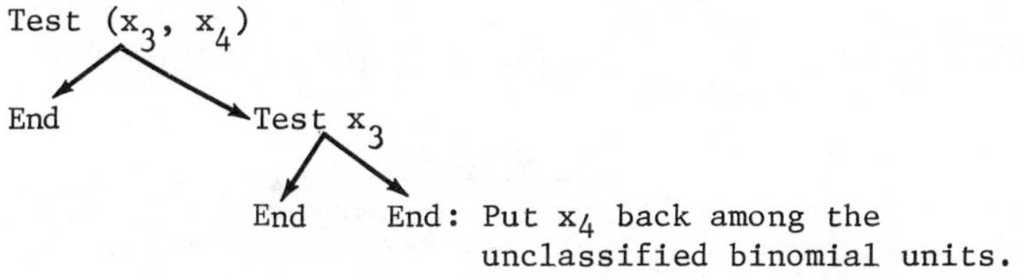

Figure 1. Plan (or Tree) for Testing $(x_3,\ x_4)$.

Since we start with the same number of x's of each type it is easy to see that the x_3's will be depleted (i.e., classified) before the x_4's under the above scheme.

In the group-testing terminology we only use plans that take us from one H-situation (where the units are all binomially distributed) to the very next H-situation.

In Stage 2 of (2.1) we continue testing triples $(x_1,\ x_2,\ x_4)$ according to a specific plan given by the procedure $S_h^{(4)}$, until another type of unit is depleted; in this case the x_4 is depleted earlier. In Stage 3 we continue until the x_2's are depleted and finally in Stage 4 we test seven units of type x_1 according to a specific plan until all the remaining units are classified.

It should be carefully noted that when we let $c \to \infty$ the limit of the horizontal procedure is not a procedure at all. However we are interested in the limit L_∞ as $c \to \infty$ of the expected number of tests per set classified since, as we shall see later, there is a vertical procedure with the same value $L = L_\infty$ where L is the expected number of tests per set classified.

Furthermore we can find the horizontal procedure with the smallest L_∞-value. We have some reasons to believe that these equivalent vertical procedures are optimal in the entire class of vertical procedures.

A vertical procedure is defined to be equivalent to a horizontal procedure if it has an expected number of tests per set classified that is equal to the limiting L-value of the horizontal procedure.

3. *An Example of a Vertical Procedure Equivalent to a Horizontal Procedure*

Using group-testing terminology we define a G-situation to be one in which we have information that some set of units contains at least one defective. The restriction to nested procedures in which we try to locate 1 defective unit in the smallest possible number of tests then gives us an R_1-type procedure (cf. Sobel and Groll (1959)] in which the G and H-situations are the only two that are possible.

For the example illustrating equivalence suppose $k = 3$ and for the three types of units x_1, x_2, x_3 we take $q_1 = .9$ and $q_2 = (.9)^2$ and $q_3 = (.9)^4$ respectively. From the G-point of view we have an equal number of groups of sizes 1, 2 and 4; all units have the common probability .9 of being good and we are only interested in knowing whether or not each entire group is good. From the I-point of view, in the sequel we shall refer to the unit that has probability q^i as type i or simple as the i-unit (i=1,2, ...) when the i-values are different; hence for our example above a test on (1,2,4) is equivalent to a test on (x_1, x_2, x_3). We now define a horizontal procedure for the above problem and find its L_∞-value; we then derive the equivalent vertical procedure and prove the equivalence by showing that $L = L_\infty$.

A tree is a rule or plan, as in Figure 1 above, which specifies what units to test i) initially in the H-situation and ii) in each succeeding G-situation until the very next H-situation is reached. For convenience we refer to a particular tree by the units tested initially, e.g., the particular tree for testing (x_1, x_2, x_3) can be referred to as the (1,2,4)-tree.

Let the horizontal procedure S_h be defined by

$$S_h = \begin{cases} (1,\ 2,\ 4) \\ (2,\ 4) \\ (4,\ 4) \end{cases}, \tag{3.1}$$

where the (1,2,4), (2,4) and (4,4) trees are given by

$$x = (1,\ 2,\ 4) \begin{array}{l} \nearrow \text{End} \\ \rightarrow G(1,2,4) \\ \quad x=(1,2) \end{array} \begin{array}{l} \nearrow \text{End} \\ \rightarrow G(1,2) \\ \quad x=(1) \end{array} \begin{array}{l} \nearrow H(4) \\ \rightarrow H(2,4) \\ \ \end{array}, \tag{3.2}$$

$$x = (2,\ 4) \begin{array}{l} \nearrow \text{End} \\ \rightarrow G(2,4) \\ \quad x=(2) \end{array} \begin{array}{l} \nearrow \text{End} \\ \rightarrow H(4) \\ \ \end{array}, \tag{3.3}$$

$$x = (4,\ 4) \begin{array}{l} \nearrow \text{End} \\ \rightarrow G(4,4) \\ \quad x=(4) \end{array} \begin{array}{l} \nearrow \text{End} \\ \rightarrow H(4) \\ \ \end{array}. \tag{3.4}$$

In each of the above trees the horizontal (resp., slanted) arrows corresponds to a failure (resp., success) on that particular trial. The expression "H(4)" (say) in (3.2) (say) means that from the (1,2,4) grouping that we tested on the first step of the tree, a single 4-unit has been left unclassified at the end of the tree and returns to the binomial state; the word "End" in (3.2) means that the entire (1,2,4) grouping has been classified. We start testing with c sets of (1,2,4)'s and work (1,2,4)-trees until we exhaust the 1-units. The number of such trees is, of course, c and the number of tests involved is $cE\{T|T_{124}\}$, where $E\{T|T_{124}\}$ is the expected number of tests per (1,2,4)-tree. For large c $(c\to\infty)$, the number of 4's (resp., 2's) remaining after the 1's have been exhausted is $c(1-q^3)$ (resp., $c(1-q)$). A single 2-unit is used up on each (2,4)-tree and thus the number of such trees until the 2's are exhausted is $c(1-q)$; since q^2 4's are classified per (2,4)-tree, the number of 4-units remaining is $c(1-q)^3 - c(1-q)q^2 = c(1-q^2)$. Also, $(1+q^4)$ 4's are classified per (4,4) tree and thus it takes $\frac{c(1-q^2)}{1+q^4}$ such trees to exhaust the 4's, with the number of tests involved equal to $\frac{c(1-q^2)}{1+q^4}E\{T|T_{44}\}$. It is easy to verify that

$$\begin{aligned} E\{T|T_{124}\} &= 3 - q^3 - q^7 \\ E\{T|T_{24}\} &= 2 - q^6 \\ E\{T|T_{44}\} &= 2 - q^8 \quad . \end{aligned} \tag{3.5}$$

Therefore,

$$L_\infty = \frac{c(3-q^3-q^7)+c(1-q)(2-q^6)+\dfrac{c(1-q^2)}{1+q^4}(2-q^8)}{c} \approx 2.1196 \tag{3.6}$$

where L_∞ is written both as a ratio of polynomials for q close to .9 and also numerically at $q = .9$.

We now define the equivalent vertical procedure. Let the S_i $(i=1,2,\ldots,5)$ which represent the five possible H-situations be defined by

$$\begin{gathered} S_1 = H(\ldots), \quad S_2 = H(4,\ldots), \quad S_3 = H(4,4,\ldots), \\ S_4 = H(2,4,\ldots), \quad S_5 = H(2,4,4,\ldots) \ , \end{gathered} \tag{3.7}$$

where H(....) represents a binomial state with an equal number of 1's, 2's, and 4's; H(4,...) represents one where there is an extra 4, etc. Our vertical procedure can then be described by the trees

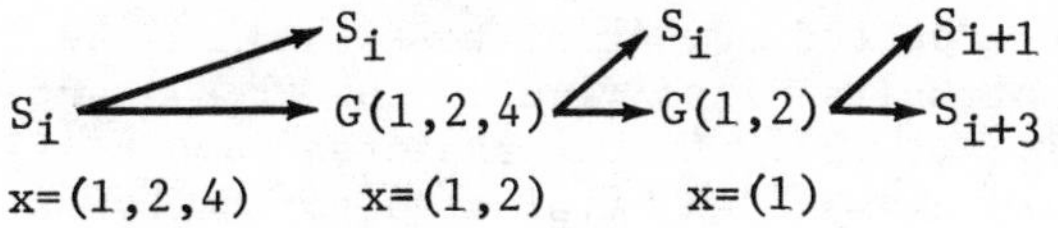

$$
S_i \longrightarrow \begin{array}{c} S_i \\ G(1,2,4) \end{array} \longleftarrow \begin{array}{c} S_i \\ G(1,2) \end{array} \longleftarrow \begin{array}{c} S_{i+1} \\ S_{i+3} \end{array} \quad (i=1,2) , \qquad (3.8)
$$
$$
x=(1,2,4) \qquad x=(1,2) \qquad x=(1)
$$

$$
S_3 \longrightarrow \begin{array}{c} S_1 \\ G(4,4) \end{array} \longleftarrow \begin{array}{c} S_1 \\ S_2 \end{array} \quad , \qquad (3.9)
$$
$$
x=(4,4) \qquad x=(4)
$$

$$
S_i \longrightarrow \begin{array}{c} S_{i-3} \\ G(2,4) \end{array} \longleftarrow \begin{array}{c} S_{i-3} \\ S_{i-2} \end{array} \quad (i=4,5) . \qquad (3.10)
$$
$$
x=(2,4) \qquad x=(2)
$$

It is useful to characterize our system as a Markov chain with state space $\{S_1, S_2, S_3, S_4, S_5\}$. If $P = (p_{ij})$ is defined as the matrix of transition probabilities, i.e., the matrix of probabilities of switching in a single tree from state S_i to state S_j $(i,j=1,2,\ldots,5)$, then its value for q near .9 is easily seen to be equal to

$$
P = \begin{bmatrix}
q^3, & q(1-q^2), & 0 , & 1-q, & 0 \\
0, & q^3 , & q(1-q^2), & 0 , & 1-q \\
q^4, & 1-q^4 , & 0 , & 0 , & 0 \\
q^2, & 1-q^2 , & 0 , & 0 , & 0 \\
0, & q^2 , & 1-q^2 , & 0 , & 0
\end{bmatrix} \qquad (3.11)
$$

Using (3.11) and the equations

$$
\pi_j = \sum_{i=1}^{5} \pi_i p_{ij} \qquad (j=1,\ldots,5) \qquad (3.12)
$$

to solve for the π_j, we find that

$$
\pi_1 = \frac{q^4}{D}, \quad \pi_2 = \frac{1}{D}, \quad \pi_3 = \frac{1-q^2}{D}, \quad \pi_4 = \frac{q^4(1-q)}{D}, \quad \pi_5 = \frac{1-q}{D},
$$
$$
\qquad (3.13)
$$

where

$$
D = 3 - q - q^2 + 2q^4 - q^5 . \qquad (3.14)
$$

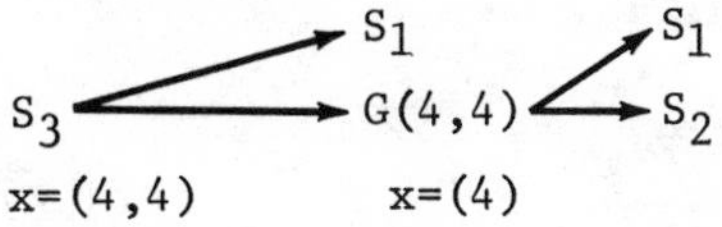

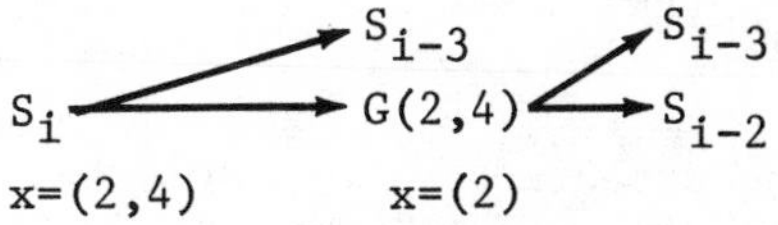

Next, let ET_i be the expected number of tests and EN_i be the expected number of units classified in the tree starting in state S_i $(i=1,2,\ldots,5)$. L, the expected number of tests per set of units classified, evaluated for the vertical procedure is then given by

$$L = 3 \frac{\sum_{i=1}^{5} \pi_i ET_i}{\sum_{i=1}^{5} \pi_i EN_i} , \tag{3.15}$$

i.e., L is equal to 3 times the expected numbers of tests per unit classified. Evaluating ET_i and EN_i $(i=1,2,\ldots,5)$ for q close to .9 yields

$$\begin{aligned} ET_i &= 3 - q^3 - q^7, \quad EN_i = 1 + q + q^3 \quad (i=1,2), \\ ET_3 &= 2 - q^8, \quad EN_3 = 1 + q^4, \\ ET_i &= 2 - q^6, \quad EN_i = 1 + q^2 \quad (i=4,5) . \end{aligned} \tag{3.16}$$

Using the above together with (3.13), (3.14) and (3.15) results in (3.6), i.e.,

$$L = \frac{7-2q-2q^2-q^3+5q^4-2q^5-q^6-q^7-q^8}{1 + q^4} \approx 2.1196 \tag{3.17}$$

at $q = .9$.

Notice that the expression in (3.17) is the same as in (3.6). This result is not surprising since it is easy to see the equivalence of the horizontal and vertical approaches. In Section 8, it is proved that every horizontal procedure is equivalent to a vertical one.

4. *Finding the Optimal Procedure: Trial and Error Method*

In searching for the optimal horizontal procedure we use the G-point of view which helps us to bring to bear information from previous work on group-testing. It was seen in Section 6 of Sobel (1960) that in the case of an infinite number of units with the same q-value we get an 'optimal' tree i) by maximizing the (Shannon) information in each H-situation and ii) by using the R_1-procedure [cf. Sobel and Groll (1959)] for each G-situation. We use these guidelines as a first step to eliminate lots of possibilities in finding the optimal horizontal procedure. For example, if $k = 4$ and let $q_i = q^i$ (i=1,2,3,4) where q=.9. By Sobel and Groll (1959) we know (using the G-point of view) that we need 7 'units' to maximize the information; thus we can use (3,4), (1,2,4), (1,2,2,2), (1,1,1,1,1,1,1), etc. Furthermore if the collection of 7 'units' is defective then we would like to select 3 'units' to test in this G-situation, e.g., from a (1,2,4) defective set we can test the combination (1,2) on the very next test. Generalizing the above example, let $k = 2,3,4$ and 5 with $q_i = q^i$ $(i=1,2,\ldots,k)$ and $q = .9$. Using the above mentioned guidelines the optimal horizontal procedures $S_h^{(k)}$, found by trial and error, are given by

$$S_h^{(2)} = \begin{cases}(1,2,2,2)\\(1,1,1,1,1,1,1)\end{cases}, \quad S_h^{(3)} = \begin{cases}(3,2,2)\\(3,1,1,1,1)\\(1,1,1,1,1,1,1)\end{cases}, \tag{4.1}$$

$$S_h^{(4)} = \begin{cases}(3,4)\\(1,2,4)\\(1,2,2,2)\\(1,1,1,1,1,1,1)\end{cases}, \quad S_h^{(5)} = \begin{cases}(3,4)\\(1,2,4)\\(1,2,2,2)\\(1,1,1,1,1,1,1)\\(5)\end{cases}.$$

The last row of $S_h^{(5)}$ has an extremely simple tree, the (5)-tree, since it tests the units one-at-a-time. We find a theoretical basis for these results in the next section.

5. *Theoretical Results with Special Reference to* $k = 2$

A group of units $(x_1,x_2,\ldots,x_i)$ will be called *coarser* than another group $(y_1,y_2,\ldots,y_j)$ if $i < j$ and there exist ordered, unequal integers $\ell_1,\ell_2,\ldots,\ell_{i-1}$ such that

$$x_1 = \sum_{\alpha=1}^{\ell_1} y_\alpha \;;\; x_2 = \sum_{\alpha=\ell_1+1}^{\ell_2} y_\alpha \;;\; \ldots,\; x_i = \sum_{\alpha=\ell_{i-1}+1}^{j} y_\alpha \tag{5.1}$$

where at least one sum contains 2 or more elements.

Unless stated otherwise, we assume that all of the

trees used in this paper utilize in addition to properties i) and ii) of Section 4 the assumption iii): In a G-situation if there is more than one combination that satisfies ii) we take the coarsest of these combinations. For example, suppose $q_1 = .9$, $q_2 = (.9)^2$ and we use (1,1,1,2,2) in the first test. Then, if it turned out to be defective, we could test (1,1,1) or (1,2) in the resulting G-situation. According to assumption iii) we prefer (1,2) since it is coarser than (1,1,1). This assumption has not been proved but has proved to be the right attack in a number of numerical examples. We have proved that if this plan is used in the G-situation then the same plan of using the coarsest combination of those that satisfy property ii) should also be used in the H-situation.

A tree T_1 is said to be coarser than another tree T_2 if from the G-point of view the total number of 'units' at each step of T_1 is the same as the corresponding step in T_2 and from the I-point of view the group tested under T_1 is coarser than the corresponding group tested in T_2; clearly, the finer tree T_2 can continue beyond T_1. If a family of trees exhibits the property that any pair of trees (in the family) can be compared as to coarseness, then we refer to it as a linear (coarse-fine) family of trees.

We emphasize the case $k = 2$, in which there are only 2 types of units. The two types are ℓ_i-units with probability q^{ℓ_i} of being good, where $\ell_2 = d\ell_1 > \ell_1$ and d is a positive integer. Let F be a linear family of trees containing trees which range in coarseness from some coarsest tree in F to the finest tree T_{ℓ_1}, where $T_{\ell_1} = (\ell_1, \ell_1, \ldots, \ell_1)$ is also included in F.

For a given tree T_0, let $E\{T|T_0\}$ denote the expected number of tests under T_0 and let $E\{N(\ell_i)|T_0\}$ denote the expected number of ℓ_i-units classified under T_0 (i=1,2). As above, we use T_{ℓ_2} to denote the tree $(\ell_2, \ell_2, \ldots, \ell_2)$ consisting only of ℓ-units. Let $T(d,\ell_1)$ denote the tree in F which starts by testing d of the ℓ_2-units; recall that $\ell_2 = d\ell_1$.

Theorem 1

A necessary and sufficient condition that a horizontal procedure restricted to a linear family F does best by choosing the coarsest possible group of units is that

$$\frac{E\{T|T_{\ell_2}\}}{E\{N(\ell_2)|T_{\ell_2}\}} \geq E\{N(\ell_1)|T(d,\ell_1)\}\,\frac{E\{T|T_{\ell_1}\}}{E\{N(\ell_1)|T_{\ell_1}\}} - (E\{T|T(d,\ell_1)\}-1) \ . \qquad (5.2)$$

Illustration of the Result

Let $\ell_1 = 1$ and $\ell_2 = 2\ell_1 = 2$ and suppose x_i has probability q^i of being good (i=1,2) where q = .9. The linear family F of trees consists of the four trees (1,2, 2,2), (1,1,1,2,2), (1,1,1,1,1,2) and (1,1,1,1,1,1,1), which are listed in the order of coarsest to finest. Below we give each of the four trees in detail:

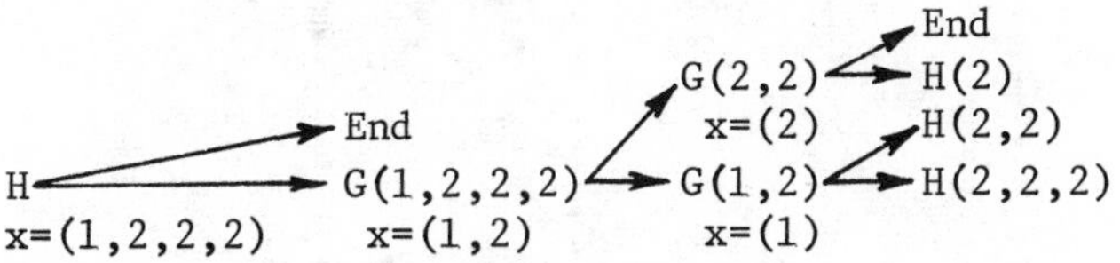

(5.3)

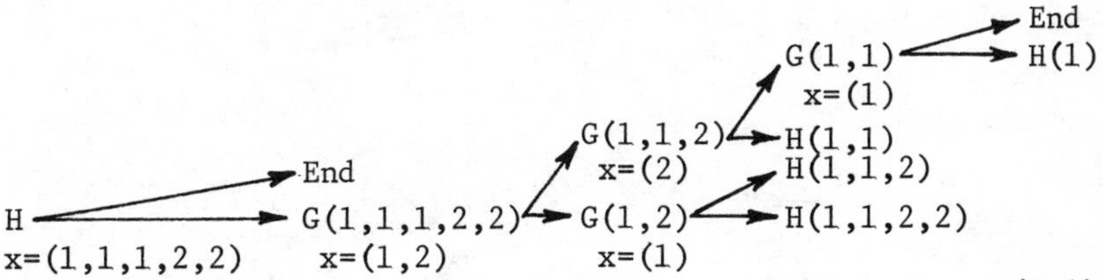

(5.4)

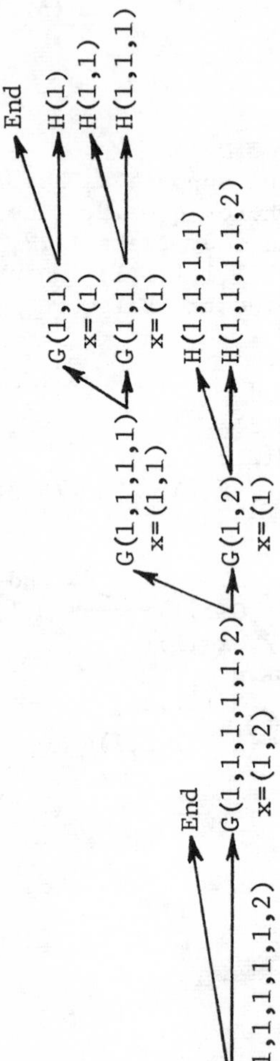
H
x=(1,1,1,1,1,1,2)
End
G(1,1,1,1,1,2)
x=(1,2)
G(1,1,1,1,1)
x=(1,1)
G(1,2)
x=(1)
G(1,1)
x=(1)
G(1,1)
x=(1)
H(1,1,1,1)
H(1,1,1,1,2)
End
H(1)
H(1,1)
H(1,1,1)

(5.5)

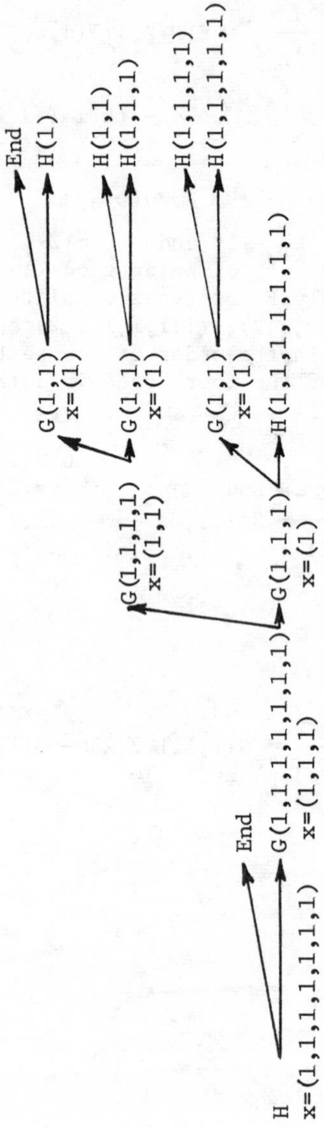
H
x=(1,1,1,1,1,1,1)
End
G(1,1,1,1,1,1,1)
x=(1,1,1)
G(1,1,1,1)
x=(1,1)
G(1,1,1)
x=(1)
G(1,1)
x=(1)
G(1,1)
x=(1)
G(1,1)
x=(1)
H(1,1,1,1,1,1,1)
End
H(1)
H(1,1)
H(1,1,1)
H(1,1,1,1)
H(1,1,1,1,1)

(5.6)

The tree T_{ℓ_1} (or T_1) is given in (5.6); the tree T_{ℓ_2} (or T_2) and the tree $T(d,\ell_1)$ (or $T(2,1)$) are respectively given by

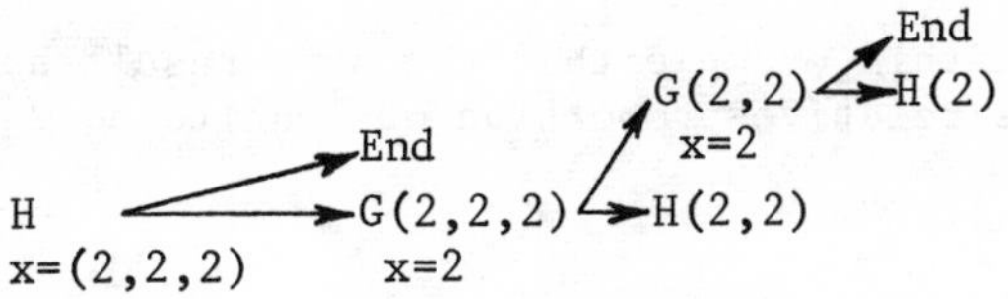

, (5.7)

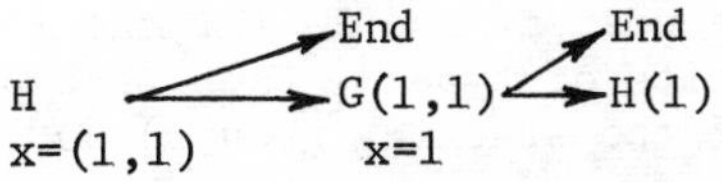

. (5.8)

(We always begin a tree in the H-situation, denoted by H.)

By elementary calculations we obtain from (5.6), (5.7) and (5.8) for $q = .9$

$$\frac{E\{T|T_1\}}{E\{N(1)|T_1\}} = \frac{3+q-3q^7}{1+q+q^2+q^4+q^5+q^6} = \frac{2.4651}{5.2170} = .4725 , \tag{5.9}$$

$$\frac{E\{T|T_2\}}{E\{N(2)|T_2} = \frac{2+q^2-2q^6}{+q^2+q^4} = \frac{1.7472}{2.4661} = .7085 , \tag{5.10}$$

$$\begin{aligned} E\{T|T(2,1)\} &= 2 - q^2 = 1.19 ; \\ E\{N(1)|T(2,1)\} &= 1 + q = 1.9 . \end{aligned} \tag{5.11}$$

The condition (5.2) of Theorem 1 is satisfied when $q = .9$, $\ell_1 = 1$ and $\ell_2 = 2$ since

$$.7085 \geq (1.9)(.4725) - (1.19-1) = .7078 . \tag{5.12}$$

Hence according to Theorem 1 we conclude that the best horizontal procedure S_h^* is

$$S_h^* = S_h^{(2)} = \begin{Bmatrix} (1,2,2,2) \\ (1,1,1,1,1,1,1) \end{Bmatrix} . \tag{5.13}$$

Thus the result obtained in Section 4 by trial and error (cf. (4.1)) can be obtained from Theorem 1.

Remark

We note in this example that the same result holds regardless of the relative proportion of 1-units to 2-units at the outset.

Proof of Theorem 1

Consider 2 successive trees in the linear family F, which we write as T_c for the coarser and T_F for the finer. If the coarser tree has a units of type ℓ_1 then we can write the trees (in terms of the starting groups) as

$$T_c = (\underbrace{\ell_1, \ell_1, \ldots, \ell_1}_{a},\ \underbrace{\ell_2, \ell_2, \ldots, \ell_2}_{b})$$
$$T_F = (\underbrace{\ell_1, \ell_1, \ldots, \ell_1}_{a},\ \underbrace{\ell_1, \ell_1, \ldots, \ell_1}_{d},\ \underbrace{\ell_2, \ell_2, \ldots, \ell_2}_{b-1}) \tag{5.14}$$

where $a \geq 0$, $b \geq 1$ and $d \geq 2$. We now show that T_c gives better results than T_F if and only if (5.2) holds. Our proof is in three parts; in part A we assume that the ratio $p > 0$ of ℓ_1-type to ℓ_2-type units is sufficiently small so that the coarsest tree in F will exhaust the ℓ_1-type unit first, if it is repeatedly applied to the units. It follows that all the trees of F, and in particular T_c and T_F, will have the same property. In comparing T_c and T_F we will be interested in the events (with reference to the T_c trees):

$$B_1 = \{\text{all } \ell_2\text{-units are classified and the last one classified is good}\}, \tag{5.15}$$

$$B_2 = \{\text{all } \ell_2\text{-units are classified and the last one classified is defective}\}. \tag{5.16}$$

We write Q $(0 \leq Q \leq 1)$ for the probability that all ℓ_2-units are classified and this enables us to write for the events B_1 and B_2

$$P\{B_1\} = Qq^{\ell_2} \ ; \qquad P\{B_2\} = Q(1-q^{\ell_2}) \ . \tag{5.17}$$

The T_c tree differs from the T_F tree only when all the ℓ_2-units are classified and only if the last ℓ_2-unit is defective (cf. (5.3)-(5.6)). This observation leads to the relation

$$E\{T|T_F\} = E\{T|T_C\} + Q(1-q^{\ell_2})\Sigma i\theta_i \tag{5.18}$$

where θ_i is the conditional probability, given a $G(\ell_1,\ell_1,\ldots,\ell_1)$ situation as a starting point, that it will take i tests to return to the very next H-situation. From the $T(d,\ell_1)$ tree (cf. (5.8)) we note that

$$E\{T|T(d,\ell_1)\} = 1 + (1-q^{\ell_2})\Sigma i\theta_i \tag{5.19}$$

and hence we can write (5.18) in the form

$$E\{T|T_F\} = E\{T|T_C\} + Q[E\{T|T(d,\ell_1)\} - 1] \ . \tag{5.20}$$

Similarly, comparing the units classified, we use (5.16) and obtain

$$E\{N(\ell_1)|T_F\} = E\{N(\ell_1)|T_C\} + dQq^{\ell_2} + Q(1-q^{\ell_2})\Sigma j\theta'_j \tag{5.21}$$

where θ'_j is the conditional probability, given a $G(\ell_1,\ell_1,\ldots,\ell_1)$ as a starting point, that j units of type ℓ_1 are classified by the time we reach the very next H-situation. From the $T(d,\ell_1)$ tree (cf. (5.8)) we note that

$$E\{N(\ell_1)|T(d,\ell_1)\} = dq^{\ell_2} + (1-q^{\ell_2})\Sigma j\theta'_j \tag{5.22}$$

and hence we can write (5.21) in the form

$$E\{N(\ell_1)|T_F\} = E\{N(\ell_1)|T_C\} + QE\{N(\ell_1)|T(d,\ell_1)\} \ . \tag{5.23}$$

The two results (5.20) and (5.23) will be used later.

To compare the trees T_C and T_F we compare the two horizontal procedures or strategies S_1 and S_2 defined by

$$S_1 = \begin{Bmatrix} T_C \\ T_{\ell_2} \end{Bmatrix} , \quad S_2 = \begin{Bmatrix} T_F \\ T_{\ell_2} \end{Bmatrix} \tag{5.24}$$

both based on the same numbers of units, M_i of type ℓ_i (i= 1,2). Under the assumption that $p = M_1/M_2$ is small, the units of type ℓ_1 will be depleted first and the expected number of tests under S_i (i=1,2) needed to classify all these units is given by

$$E\{T|S_1\} = \frac{M_1}{E\{N(\ell_1)|T_C\}} E\{T|T_C\} + \left[\frac{M_2 - \dfrac{M_1 E\{N(\ell_2)|T_C\}}{E\{N(\ell_1)|T_C\}}}{E\{N(\ell_2)|T_{\ell_2}\}} \right] E\{T|T_{\ell_2}\} , \tag{5.25}$$

and the same result holds for S_2 if we replace T_C by T_F. To show that $E\{T|S_1\} \leq E\{T|S_2\}$ for p sufficiently small is now equivalent to showing that $\Delta_C \leq \Delta_F$ where

$$\Delta_C = \frac{1}{E\{N(\ell_1)|T_C\}} \left[E\{T|T_C\} - \frac{E\{N(\ell_2)|T_C\}}{E\{N(\ell_2)|T_{\ell_2}\}} E\{T|T_{\ell_2}\} \right] , \tag{5.26}$$

and the same holds for Δ_F if T_C is replaced by T_F.

Using (5.20) and (5.23) and the additional fact that

$$E\{N(\ell_2)|T_F\} = E\{N(\ell_2)|T_C\} - Q , \tag{5.27}$$

we obtain for Δ_F the result (in terms of T_C only)

$$\Delta_F = \frac{1}{E\{N(\ell_1)|T_C\} + QE\{N(\ell_1)|T(d,\ell_1)\}} \times \left[E\{T|T_C\} + Q[E\{T|T(d,\ell_1)\} - 1] - \frac{[E\{N(\ell_2)|T_C\} - Q]}{E\{N(\ell_2)|T_{\ell_2}\}} E\{T|T_{\ell_2}\} \right] . \tag{5.28}$$

For convenience we use F, G, H, I, J to denote as follows

$$F = \frac{E\{T|T_{\ell_2}\}}{E\{N(\ell_2)|T_{\ell_2}\}} \;;\quad G = E\{T|T_C\} - E\{N(\ell_2)|T_C\}F \;;$$

$$H = E\{N(\ell_1)|T_C\} \;;\quad I = E\{T|T(d,\ell_1)\} - 1 + F \;; \tag{5.29}$$

$$J = E\{N(\ell_1)|T(d,\ell_1)\} \;.$$

Then by straightforward algebra the inequality $\Delta_C \leq \Delta_F$ becomes

$$\frac{G}{H} \leq \frac{G + QI}{H + QJ} \quad \text{or} \quad GJ \leq HI. \tag{5.30}$$

If we solve the latter inequality for F, then we obtain

$$F \geq \frac{E\{N(\ell_1)|T(d,\ell_1)\}E\{T|T_C\} - E\{N(\ell_1)|T_C\}[E\{T|T(d,\ell_1)\}-1]}{E\{N(\ell_1)|T_C\} + E\{N(\ell_1)|T(d,\ell_1)\}E\{N(\ell_2)|T_C\}} \;. \tag{5.31}$$

To simplify the above result we consider the numerator in (5.31) (call it $N(T_r)$) and the denominator in (5.31) (call it $D(T_r)$) separately with the tree T_C replaced by an arbitrary tree T_r in F. It is easily seen by using (5.20), (5.23) and (5.27) that

$$N(T_C) = N(T_F) = N(T_r) \quad \text{and} \quad D(T_C) = D(T_F) = D(T_r) \tag{5.32}$$

for any pair T_C and T_F (contiguous or not) in F. Hence for any T_C in F

$$N(T_C) = N(T_{\ell_1}) \quad \text{and} \quad D(T_C) = D(T_{\ell_1}) \;. \tag{5.33}$$

Using these results in (5.31) and noting that $E\{N(\ell_2)|T_{\ell_1}\} = 0$ we obtain the final result

$$F \geq \frac{N(T_{\ell_1})}{D(T_{\ell_1})} = \frac{E\{N(\ell_1)|T(d,\ell_1)\}E\{T|T_{\ell_1}\}}{E\{N(\ell_1)|T_{\ell_1}\}} - (E\{T|T(d,\ell_1)\}-1). \tag{5.34}$$

This proves Theorem 1 for small p.

Part B of Theorem 1

For given trees T_C and T_F as in (5.14) we define p^* as the supremum of values of p for which both trees exhaust the ℓ_1-type unit first. We define p^{**} as the infimum value of p for which both tests exhaust the ℓ_2-type first; then $p^* \leq p^{**}$ and we can write

$$p^* = \frac{E\{N(\ell_1)|T_C\}}{E\{N(\ell_2)|T_C\}}, \quad p^{**} = \frac{E\{N(\ell_1)|T_F\}}{E\{N(\ell_2)|T_F\}}. \tag{5.35}$$

In Part B of our proof we assume that $p^* < p^{**}$ and consider this interval $p^* < p \leq p^{**}$, where the tree T_C exhaust the ℓ_2-type unit first and T_F exhausts the ℓ_1-type unit first. We consider the two horizontal procedures or strategies S_1' and S_2' defined by

$$S_1' = \begin{Bmatrix} T_C \\ T_{\ell_1} \end{Bmatrix}, \quad S_2' = \begin{Bmatrix} T_F \\ T_{\ell_2} \end{Bmatrix}. \tag{5.36}$$

For $p^* < p \leq p^{**}$ and M_i (i=1,2) as defined above we have as in (5.25)

$$E\{T|S_1'\} = \frac{M_2 E\{T|T_C\}}{E\{N(\ell_2)|T_C\}} + \left[M_1 - \frac{M_2 E\{N(\ell_1)|T_C\}}{E\{N(\ell_2)|T_C\}}\right] \frac{E\{T|T_{\ell_1}\}}{E\{N(\ell_1)|T_{\ell_1}\}}, \tag{5.37}$$

$$E\{T|S_2'\} = E\{T|S_2\}, \tag{5.38}$$

where the latter is the analogue of (5.25). Since our result clearly depends only on the ratio p, we can without loss of generality set $M_1 = p$ and $M_2 = 1$. Then for $p^* < p \le p^{**}$ we can write (5.37) and (5.38), respectively, as

$$E(T|S_1') = \frac{p^*}{E\{N(\ell_1)|T_C\}} E\{T|T_C\} + (p-p^*)\frac{E\{T|T_{\ell_1}\}}{E\{N(\ell_1)|T_{\ell_1}\}}. \tag{5.39}$$

$$E\{T|S_2'\} = \frac{p}{E\{N(\ell_1)|T_F\}} E\{T|T_F\} + \left[1 - p\frac{E\{N(\ell_2)|T_F\}}{E\{N(\ell_1)|T_F\}}\right]\frac{E\{T|T_{\ell_2}\}}{E\{N(\ell_2)|T_{\ell_2}\}}. \tag{5.40}$$

We now regard $E\{T|S_i'\}$ and $E\{T|S_i\}$ as functions of p and write $E_p\{T|S_i'\}$ and $E_p\{T|S_i\}$ $(i=1,2)$. For $p = p^*$ the second part of (5.37) (in brackets) vanishes and we can write for $p^* < p \le p^{**}$

$$E_p\{T|S_1'\} = E_{p^*}\{T|S_1\} + (p-p^*)\frac{E\{T|T_{\ell_1}\}}{E\{N(\ell_1)|T_{\ell_1}\}}, \tag{5.41}$$

$$E_p\{T|S_2'\} = E_{p^*}\{T|S_2\} + (p-p^*) \times\left[\frac{E\{T|T_F\}}{EN(\ell_1)|T_F} - \frac{E\{N(\ell_2)|T_F\}}{E\{N(\ell_1)|T_F\}}\frac{E\{T|T_{\ell_2}\}}{E\{N(\ell_2)|T_{\ell_2}\}}\right]. \tag{5.42}$$

Since we have already shown that $E_{p^*}\{T|S_1\} \le E_{p^*}\{T|S_2\}$ in Part A of the proof and because of the linearity in p it is sufficient for us to show that

$$E_{p^{**}}\{T|S_1'\} = p^*\frac{E\{T|T_C\}}{E\{N(\ell_1)|T_C\}} + (p^{**}-p^*)\frac{E\{T|T_C\}}{E\{N(\ell_1)|T_{\ell_1}\}} \le$$

$$\leq p^{**} \frac{E\{T|T_F\}}{E\{N(\ell_1)|T_F\}} \tag{5.43}$$

$$= E_{p^{**}}\{T|S\} ,$$

in order to show the inequality for $p^* < p \leq p^{**}$. Using (5.35) to replace p^* and p^{**} and (5.20), (5.23) and (5.27), to get it all in terms of the T_C tree, we can write (5.43) as

$$E_{p^{**}}\{T|S_1'\} = \frac{G'}{H_2} + \left[\frac{H_1+QJ}{H_2-Q} - \frac{H_1}{H_2}\right]F' \leq \frac{G'+QI'}{H_2-Q} = E_{p^{**}}\{T|S_2'\} \tag{5.44}$$

where

$$F' = \frac{E\{T|T_{\ell_1}\}}{E\{N(\ell_1)|T_{\ell_1}\}}, \quad G' = E\{T|T_C\}, \quad H_i = E\{N(\ell_i)|T_C\} \ (i=1,2)$$

$$I' = E\{T|T(d,\ell_1)\} - 1 , \quad J = E\{N(\ell_1)|T(d,\ell_1)\}. \tag{5.45}$$

Straightforward algebra now shows that (5.44) holds for $p^* < p \leq p^{**}$ if and only if

$$\frac{E\{T|T_{\ell_1}\}}{E\{N(\ell_1)|T_{\ell_1}\}} = F' \leq \frac{I'H_2 + G'}{JH_2 + H_1} . \tag{5.46}$$

As in the proof of (5.32) we can easily show that the numerator $N'(T)$ in (5.46), regarded as a function of the tree T, is the same for $T = T_C$, as for $T = T_F$ and the same holds for the denominator $D'(T)$ in (5.46). Thus we can take for T any tree in the family F and we can replace T_C in (5.46) by T_{ℓ_1}, obtaining (since H_2 is replaced by 0)

$$I'H_2 + G' = E\{T|T_{\ell_1}\} ; \quad JH_2 + H_1 = E\{N(\ell_1)|T_{\ell_1}\} . \tag{5.47}$$

Thus we have equality in (5.46) and Part B is proved, i.e., Theorem 1 holds for $p^* < p \leq p^{**}$. Moreover we attain equality in (5.2) when $p = p^{**}$.

In Part C of the proof we consider the range $p > p^{**}$ where T_C and T_F both exhaust the ℓ_2-type unit first. Defining the horizontal procedures or strategies S_1'' and S_2'' by

$$S_1'' = \begin{Bmatrix} T_C \\ T_{\ell_1} \end{Bmatrix}, \qquad S_2'' = \begin{Bmatrix} T_F \\ T_{\ell_1} \end{Bmatrix}, \tag{5.48}$$

we find, as in (5.41), and (5.42) that

$$E_p\{T|S_i''\} = E_{p^{**}}\{T|S_i'\} + (p-p^{**})\frac{E\{T|T_{\ell_1}\}}{E\{N(\ell_1)|T_{\ell_1}\}} \quad (i=1,2). \tag{5.49}$$

Since we have already that $E_{p^{**}}\{T|S_1'\} = E_{p^{**}}\{T|S_2'\}$ it follows from (5.49) that for all $p > p^{**}$

$$E_p\{T|S_1''\} = E_p\{T|S_2''\} \tag{5.50}$$

and this completes the proof of Theorem 1.

Remarks on Desirable Extensions of Theorem 1

It would be desirable to find a condition as in Theorem 1 and show that it holds for the case of three types of units, say ℓ_1, ℓ_2, and ℓ_3. Consider the case $\ell_1 = 1$, $\ell_2 = 2$, and $\ell_3 = 3$ as an example; we assume that the three types each have proportion 1/3. The trees of interest no longer form a linear coarse-fine set; in addition to the coarse-fine set for the 2-type case with $\ell_1 = 1$ and $\ell_2 = 2$, we now wish to consider trees for (3,2,2), for (3, 2,1,1) and for (3,1,1,1,1). Tree structures for (3,2,2) and (3,1,1,1,1) for $q = .9$ would be

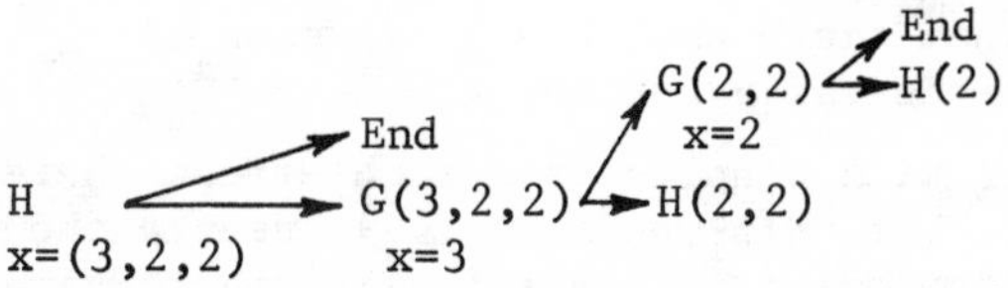

(5.51)

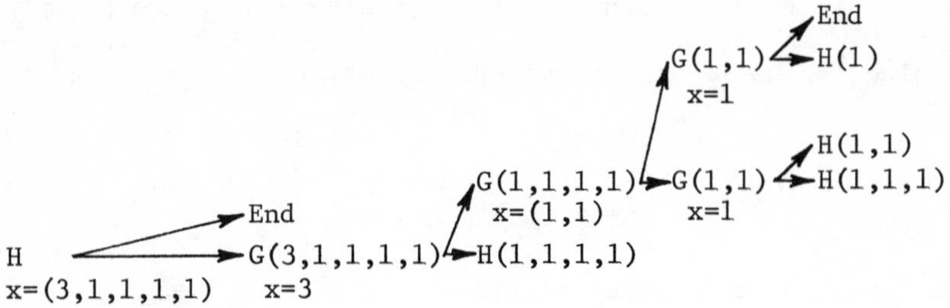

(5.52)

The (3,3,1)-tree (with a total of 7) also maximizes the information on the first step when $q = .9$ but it is eliminated from consideration because any strategy beginning with the (3,3,1)-tree always does worse than the horizontal procedure or strategy

$$\left\{\begin{matrix} (3,3) \\ (1,1,1,1,1,1,1) \end{matrix}\right\} = \left\{\begin{matrix} (1,1,1,1,1,1,1) \\ (3,3) \end{matrix}\right\} . \quad (5.53)$$

In other words (3,3,1) is not a good combination for $q = .9$ because it is better to test the 1's and 3's separately, in either order. The optimal strategy for $q = .9$ is $S_h^{(3)}$ given in (4.1).

Similar extensions to the case 4 types and to any number of types of units would, of course also be desirable and the trial and error results of Section 4 will then will become special applications of the general theorem.

6. *Using Recursive Equations to Find the Optimal Horizontal Procedure*

In this section, we illustrate a recursive equation technique for finding the optimal horizontal procedure. It is done for the case when there are $k = 5$ streams of units but the technique is valid for any k.

We begin testing with c units from each stream where c is large but finite. Let $H(p_1,p_2,p_3,p_4,p_5)$ denote the expected number of tests needed to classify all the remaining units if we start from an H-situation in which cp_i units from the ith stream ($i=1,2,\ldots,5$) are still unclassified. We assume that at least one of the p_i $(i=1,2,\ldots,5)$ is positive. Then $H(p_1,p_2,p_3,p_4,p_5)$ for the optimal horizontal procedure satisfies the following equations:

$$H(p_1,p_2,p_3,p_4,p_5) = \min_{T_0 \varepsilon F(p_1,p_2,\ldots,p_5)} \{b^*(T_0)E\{T|T_0\} + H(c_1(T_0),c_2(T_0),\ldots,c_5(T_0))\}, \tag{6.1}$$

$$b_i(T_0) = \frac{cp_i}{a_i(T_0)}, \quad b^*(T_0) = \min_{i=1,2,\ldots,5} \{b_i(T_0)\},$$
$$c_i(T_0) = p_i - \frac{a_i(T_0)b^*(T_0)}{c}, \tag{6.2}$$

where $a_i(T_0)$ is the expected number of units from the ith stream classified per T_0-tree. For given $(p_1,p_2,\ldots,p_5)$, we restrict $F(p_1,p_2,\ldots,p_5)$ to the class of all possible trees which maximize the information obtained from their very first step. This restriction can be justified by reasoning similar to that given at the beginning of Section 4. We note that since for some i

$$c_i(T_0) = 0 , \tag{6.3}$$

the H expression on the right-hand-side of the equations in (6.1) contains at least one more zero than the one on the left-hand-side; we thus define

$$H(0,0,0,0,0) = 0 , \tag{6.4}$$

as a boundary condition for the set of equations in (6.1). Let us also note that $c_i(T_0)$ and $b_i(T_0)$ are functions of p_i as well as of T_0; this functional dependence is not explicitly shown in these expressions because of their resulting cumbersomeness, but should be understood.

We illustrate the above technique in finding the optimal horizontal procedure for the case when q_i, the probability of the ith unit being good, is equal to q^i $(i=1,2,\ldots,5)$, where q is close to .9. From the recursive equations, we find that

$$H(1,1,1,1,1) = b^*(T_1)E\{T|T_1\}+H(c_1(T_1),c_2(T_1),\ldots,c_5(T_1)) \;, \tag{6.5}$$

where T_1 is the (3,4)-tree, where

$$\begin{gathered} b^*(T_1) = b_3(T_1) = c \;, \quad E\{T|T_1\} = 2 - q^7 \\ c_1(T_1) = c_2(T_1) = c_5(T_1) = 1, \quad c_3(T_1) = 0, \quad c_4(T_1) = 1-q^3, \end{gathered} \tag{6.6}$$

and where $F(1,1,1,1,1)$ consists of the trees

$$\left\{ \begin{array}{l} (1,2,4),(1,1,1,4),(3,4),(3,2,2),(3,3,1), \\ (3,2,1,1),(3,1,1,1,1),(1,1,5),(1,1,1,2,2),(1,1,1,1,1,2), \\ (1,2,2,2),(1,1,1,1,1,1,1),(2,5) \end{array} \right\}. \tag{6.7}$$

In finding (6.5), we also find that

$$\begin{aligned} &H(c_1(T_1),c_2(T_1),\ldots,c_5(T_1)) \\ &\quad = b^*(T_2)E\{T|T_2\} + H(c_1(T_2),\ldots,c_5(T_2)) \;, \end{aligned} \tag{6.8}$$

where T_2 is the (1,2,4)-tree, where

$$\begin{gathered} b^*(T_2) = b_4(T_2) = \frac{c(1-q^3)}{q^3} \;, \quad E\{T|T_2\} = 3-q^3-q^7 \;, \\ c_3(T_2) = c_4(T_2) = 0 \;, \quad c_1(T_2) = 1 - \frac{1-q^3}{q^3} \;, \\ c_2(T_2) = 1 - \frac{(1-q^3)q}{q^3} \;, \quad c_5(T_2) = 1 \;, \end{gathered} \tag{6.9}$$

and where $F(c_1(T_1),\ldots,c_5(T_1))$ contains the trees

$$\left\{ \begin{array}{l} (1,2,4),(1,1,1,4),(1,1,5),(1,1,1,2,2),(1,1,1,1,1,2), \\ (1,2,2,2),(1,1,1,1,1,1,1),(2,5) \end{array} \right\} ; \tag{6.10}$$

$$H(c_1(T_2),\ldots,c_5(T_2)) = b^*(T_3)E\{T|T_3\}+H(c_1(T_3),\ldots,c_5(T_3)), \tag{6.11}$$

where T_3 is the (1,2,2,2)-tree, where

$$b^*(T_3) = b_2(T_3) = \frac{c\left(1 - \frac{(1-q^3)q}{q^3}\right)}{q + q^3 + q^5}, \quad E\{T|T_3\} = 3-2q^7,$$

$$c_2(T_3) = c_3(T_3) = c_4(T_3) = 0, \quad c_1(T_3) = c_1(T_2) - \frac{b^*(T_3)}{c},$$

$$c_5(T_3) = 1, \tag{6.12}$$

and where $F(c_1(T_2),c_2(T_2),\ldots,c_5(T_2))$ contains the trees

$$\left\{\begin{matrix}(1,1,5),(1,1,1,2,2),(1,1,1,1,1,2),(1,2,2,2),\\(1,1,1,1,1,1,1),(2,5)\end{matrix}\right\}; \tag{6.13}$$

$$H(c_1(T_3),\ldots,c_5(T_3)) = b^*(T_4)E\{T|T_4\}+H(c_1(T_4),\ldots,c_5(T_4)), \tag{6.14}$$

where T_4 is the (1,1,1,1,1,1,1)-tree, where

$$b^*(T_4) = b_1(T_4) \doteq \frac{c[c_1(T_3)]}{1+q+q^2+q^3+\ldots+q^6}, \quad E\{T|T_4\} = 3+q-3q^7 \tag{6.15}$$

$$c_1(T_4) = c_2(T_4) = c_3(T_4) = c_4(T_4) = 0, \quad c_5(T_4) = 1,$$

and where $F(c_1(T_3),c_2(T_3),\ldots,c_5(T_3))$ contains the trees

$$\{(1,1,5),\ (1,1,1,1,1,1,1)\}; \tag{6.16}$$

and finally

$$H(0,0,0,0,1) = cE\{T|T_5\}, \tag{6.17}$$

where T_5 is the (5)-tree and

$$E\{T|T_5\} = 1 . \tag{6.18}$$

After finding the optimal strategy at each step we conclude that the optimal horizontal procedure for q in the neighborhood of .9 is given by

$$S_h^{(5)} = \begin{Bmatrix} (3,4) \\ (1,2,4) \\ (1,2,2,2) \\ (1,1,1,1,1,1,1) \\ (5) \end{Bmatrix} ; \tag{6.19}$$

as stated at the end of Section 4; the expected number of tests per set of units classified [from (6.5), (6.8), (6.11), (6.14), 6.17)] for $q = .9$ is given to at least 4 decimals by

$$\begin{aligned} \frac{H(1,1,1,1,1)}{c} &= 1.521703 + (.371742)(1.792703) \\ &\quad +(.299813)(2.043406)+(.062956)(2.465109)+1 \\ &= 1.521703 + .666423 + .612640 + .155193 + 1 \\ &= 3.955959 = 3.9560 \text{ (to four decimal places).} \end{aligned} \tag{6.20}$$

The (3,4), (5), (1,2,2,2) and (1,1,1,1,1,1,1)-trees are given at the beginning of Section 2, at the end of Section 4 and at the beginning of Section 5, respectively. The (1,2,4)-tree is given by

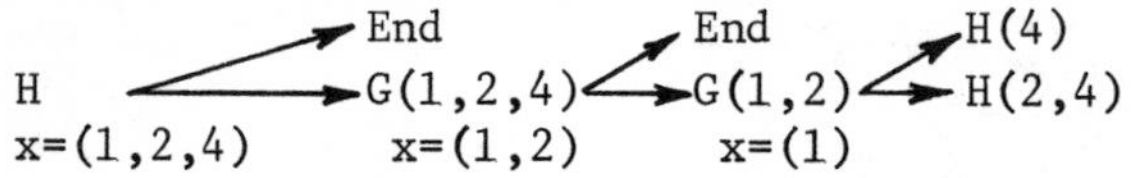

. (6.21)

We can similarly show that for $k = 2,3,4$ and $q_i = q^i$ for $i = 1,2,\ldots,k$, that the optimal horizontal procedures is given by $S_h^{(k)}$ (i=2,3,4) defined at the end of Section 4, and the expected number of tests per set of units classified for these procedures is

$$\frac{H(1,1,0,0,0)}{c} \approx 1.1803 ,$$

$$\frac{H(1,1,1,0,0)}{c} \approx 2.0188 ,$$
$$\frac{H(1,1,1,1,0)}{c} \approx 2.9560 , \qquad (6.22)$$

respectively. Note that the last result is exactly one less than the result in (6.20) where the 5-units are tested one-at-a-time.

The above calculations were done by hand. The recursive equation technique is particularly important because it enables us to use the computer in more complicated situations.

7. *Comparison with Finite Models*

This work on finite models was motivated by an unpublished table of R. R. Coveyou (1971). His table can be used for the batch group-testing problem if we knew a priori that there is exactly one defective unit contained in one of several batches. Then the problem can be handled by known techniques in information theory, coding theory, and/or search theory.

For example, we might have 10 objects grouped into batches of size 1,2,3,4 and we want to know which of these four batches is the one that contains a particular unit which we call defective. All the 10 objects (or units) have the same chance of being the defective unit; thus we can call this a homogeneous model. The problem is to find in the smallest expected number of steps (i.e., tests) which of these four batches (or groups) contains the bad unit. Here we never break up the batches and at each step we select one or more batches and see if all the units contained therein are good or if the bad unit is among them. In this problem the optimal procedure is unique and starts by testing the single batch of size 4 (not two batches adding to five or four). The expected number of tests is 1.9 and the lower bound by information theory is 1.846 according to the table of Coveyou. One of Coveyou's main interest was in systematically arranging a large class of such problems.

Of course, the above model has little in common with

our problem since we do not know at the outset whether *any* units or how many are defective. In our case all the units have a common probability q of being good and $p = 1-q$ of being defective and we assume mutual independence at the outset. Nevertheless some numerical comparisons between our problem and that of Coveyou (in addition to the fact that he provided a strong motivation) may also be of interest. We apply our general method to the same problem as above with four batches of sizes 1,2,3 and 4; special attention is given to the case $q = .9$. We distinguish an ordered and an unordered problem; in the former we apply the restriction that some ordering is given for the four batches (say 1,2,3,4) and we cannot classify any batch strictly before one that precedes it; this is usually called the "first come-first served" property in queueing theory. Clearly the unordered problem must give the best results but a basic property of group-testing holds for the fixed-ordering problem and not for the general unordered case.

The derivation of these procedures (due to space problems) will be published separately. To illustrate the results let $H(1,2,3,4)$ denote the total expected number of tests required in the unordered problem for different values of q. The value of $H(1,2,3,4)$ is given below by seven polynomials strung together to form a continuous non-increasing function, running constant at 4 for q close to zero and decreasing to 1 as $q \to 1$; the value at $q = .9$ is 3.0202.

R_1-*Procedure Results for the Unordered* $H(1,2,3,4)$-*Problem:*

Polynomials	*Range*	*Initial Strategy*
4	$0 < q < .6823$	1 or 2 or 3 or 4
$5-q-q^3$	$.6823 \leq q < .8087$	(1,2) or 3 or 4
$6-2q-q^2-q^5+q^6$	$.8087 \leq q < .8518$	(1,2) or 4
$7-2q-q^2-q^3-q^5$	$.8518 \leq q < .8679$	(1,2,3) or 4
$8-2q-q^2-3q^3-q^6+q^{10}$	$.8679 \leq q < .9057$	(1,2,3)
$8-2q-q^2-2q^3-q^6-q^{10}$	$.9057 \leq q < .9566$	(1,2,3,4)
$9-2q-q^2-3q^3-2q^6$	$.9566 \leq q < 1$	(1,2,3,4)

(7.1)

In this unordered case the optimal strategy in the G-situation does not depend only on the defective set but also on the binomial set. Thus a basic result that holds for batches of size 1 (cf. Theorem 1 of Sobel and Groll (1959)) does not hold here. In the corresponding fixed ordering problem the optimal strategy in the G-situation does depend only on the defective set and this simplifies the theory. Actually we give the numerical answers for several problems related to those used in other sections of this paper. For the unordered problem with $q = .9$ some optimal results in the class of NAR procedures (nested procedures that recombine binomial sets) are

$$H(1,2,3,4,5) = 3.9588 \ , \quad H(1,2,3,4) = 3.0202 \ , \quad H(1,2,4) = 2.1296 \ . \tag{7.2}$$

For the fixed order problem we use the notation H_f and put the arguments in the given fixed order. Then we obtain

$$H_f(1,2,3,4,5) = 3.9936 \ , \quad H_f(1,2,4,3,5) = 3.9644 \ , \tag{7.3}$$

and the latter is the best result among the $5! = 120$ possible fixed-orderings of (1,2,3,4,5). For the fixed-ordering problem with batch sizes (1,2,3,4) we obtain

$$H_f(1,2,3,4) = 3.0311 \ , \quad H_f(3,1,2,4) = 3.0202 \ ; \tag{7.4}$$

the latter result is the same as for the unordered problem and hence is the best result among the $4! = 24$ possible orderings of (1,2,3,4). For the fixed-order problem with batches of size (1,2,4) we obtain

$$H_f(1,2,4) = 2.1296 \tag{7.5}$$

and this is already the minimum of the $3! = 6$ possible orderings of (1,2,4).

8. *The Asymptotic Equivalence of Horizontal and Vertical Procedures*

Let c denote the number of sets of units; the sets all have a similar composition, e.g., (1,2,3,4) is a set of size 4 with respective probabilities q, q^2, q^3, and q^4 of being good. We note that for c finite the horizontal type scheme are bona-fide procedures but when $c = \infty$ and the strategy contains more than one type of tree, they are not bona-fide procedures. Hence it is desirable to show for $c = \infty$ that for every horizontal procedure there is an equivalent vertical procedure (i.e., one that will eventually classify any given unit in the infinite collection of sets) which has the same "limiting efficiency" as the horizontal procedure it is associated with.

Before we prove the above (as Theorem 3) let us give a formal definition of the "limiting efficiency" of both a horizontal and a vertical procedure. For the horizontal procedure, we let $T_H^{[c]}$ be the number of tests used in classifying c sets of units and we write its *limiting efficiency* (denoted by EFF_h) as

$$\text{Eff}_h = \lim_{c\to\infty} \frac{ET_h^{[c]}}{c} . \tag{8.1}$$

For the vertical procedure, we let $T_v^{[N]}$ (resp., $S_v^{[N]}$) be the number of tests used (resp., sets classified) in working through N trees and we write its *limiting efficiency* (denoted by EFF_v) as

$$\text{Eff}_v = \lim_{N\to\infty} \frac{ET_v^{[N]}}{ES_v^{[N]}} . \tag{8.2}$$

Theorem 3

For every horizontal 'procedure' S_h with $c = \infty$ there is an equivalent vertical procedure S_v whose group-testing efficiency can be obtained by analyzing the corresponding efficiency of S_h.

Proof:

The basic idea of the proof is that we can crystal-
lize certain structural properties of the horizontal 'procedure' S_h that determine the asymptotic proportion of the
time (i.e., of the total number of trees used) that we use
the trees of each type. In the definition of the horizontal procedure the various trees are listed in preferential
order to ensure that with probability one there is no infinite delay in the classification of any unit, i.e., no unit
gets classified after an infinite number of units in the
same level with positive probability. We then argue that
the associated vertical procedure must have the same asymptotic structure and hence it is a bona-fide procedure with
the same efficiency as S_h.

In the course of the proof we carry along the so-
called (1,2,3,4) example which starts with equal proportions of units of type i which have probability q^i (i=1,2,
3,4) of being good and, in particular, for q = .9. The
general nature of our proof, however, can be carried over
to any other such case. In the above case we start with
the horizontal procedure

$$S_h^{(4)} = \left\{\begin{array}{l} (3,4) \\ (1,2,4) \\ (1,2,2,2) \\ (1,1,1,1,1,1,1) \end{array}\right\} = \left\{\begin{array}{l} T_1 \\ T_2 \\ T_3 \\ T_4 \end{array}\right\} \qquad (8.3)$$

which has 4 trees, three of which are given in Figure 1,
(5.3) and (5.6) and the fourth is, as in (6.21),

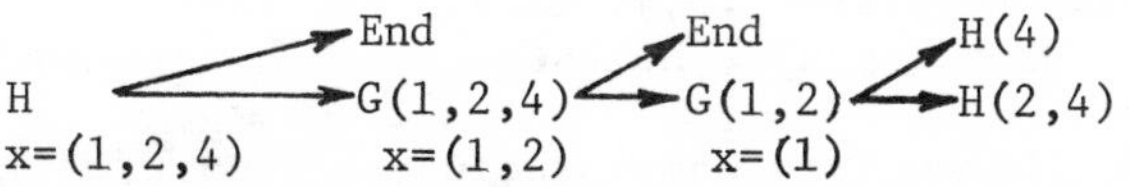

$$\begin{array}{ccccc} & \nearrow \text{End} & \nearrow \text{End} & \nearrow \text{H}(4) \\ \text{H} & \rightarrow \text{G}(1,2,4) \leftarrow & \rightarrow \text{G}(1,2) \leftarrow & \rightarrow \text{H}(2,4) \\ \text{x}=(1,2,4) & \text{x}=(1,2) & \text{x}=(1) \end{array} \qquad (8.4)$$

The order of these 4 trees in (8.3) has been purposely arranged so that for any (large) finite number c of sets with
probability approaching one as c → ∞, i) the first tree
if repeated over and over again will eliminate the 3-units
(before the 4-units), ii) the second tree, if repeated,
will eliminate the 4-units (before the 1-units and before
the 2-units), iii) the third tree, if repeated, will elim-

269

inate the 2-units (before the 1-units, and hence iv) the fourth tree will be used infinitely often in the limiting case ($c \to \infty$).

For any horizontal procedure S_h we can calculate the proportion of trees of each type that are used (in the limit as $c \to \infty$) and thus calculate its limiting ($c \to \infty$) efficiency. The particular calculations for $S_h^{(4)}$ are deferred until later when we use them to illustrate this fact.

Let the limiting proportion for the T_i tree be denoted by p_i ($i=1,2,3,4$). We first define a vertical procedure $S_v^{(4)}$, then show it has these same limiting proportions and from this fact show that it has the same limiting efficiency as $S_h^{(4)}$.

Recall that a set (1,2,3,4) consists one of each of the types $\ell_i = i$ ($i=1,2,3,4$). A *broken* set is one for which at least one unit has been classified. Since the 3-unit always get used up when we break a set, this is equivalent to saying that at least one unit has been involved in a group-test. Any unit in a broken set is a *free* unit; if the set is not broken then it is a *bound* unit.

We define a vertical procedure S_v associated with S_h by establishing a preference scheme among the different H- (i.e., binomial) situations that can arise with respect to free and bound units. We denote a unit of type i as an i-unit ($i=1,2,3,4$) and a tree that starts by testing (3,4) as a (3,4)-tree. For $S_v^{(4)}$ this preference scheme is

Preference 1: If none of the following 3 preference situations arise then test (3,4) (i.e., use the (3,4)-tree) by breaking up a new set. (8.5a)

Preference 2: If the free units include a 1-unit, a 2-unit and a 4-unit, then use the (1,2,4)-tree. (8.5b)

Preference 3: If the free units include a 1-unit and three 2-units, but no 4-units, then use the (1,2,2,2)-tree. (8.5c)

Preference 4: If the free units include seven 1-

units, but no 4-units and at most two 2-units, *or* one 4-unit and no 2-units, then use the (1,1,1,1, 1,1,1)-tree. (8.5d)

The event in Preference 4 dealing with one 4-unit and no 2-units can be neglected asymptotically since we are eliminating 4-units before 2-units with probability approaching one as $c \to \infty$; in fact we claim that this event cannot occur at all.

These four preference rules implicitly define a vertical procedure S_v; we can describe a small part of the procedure $S_v^{(4)}$ by

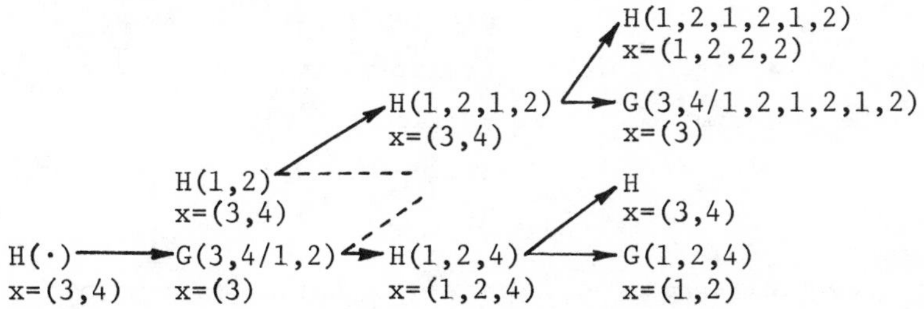

(8.6)

where $H(\cdot)$ indicates a binomial situation with no free units; $H(1,2)$ [say] indicates that a 1- and 2-unit are free, etc., and where the dashed lines lead to states not filled in.

We now note that both the vertical procedure $S_v^{(4)}$ defined by (8.3) and the horizontal 'procedure' $S_h^{(4)}$ defined by (8.1) have certain properties in common that characterize the asymptotic structure. These are

1) Every 3-unit is tested with a 4-unit.
2) Every 4-unit that is not classified by 1) is tested simultaneously with both a 1-unit and a 2-unit.
3) Every 2-unit that is not classified by 2) is tested simultaneously with a 1-unit and two other 2-units.
4) Every 1-unit that is not classified by 2) or 3) is tested simultaneously with six other 1-units.

Implicit in these properties is the fact that with probability $\to 1$ as $c \to \infty$ the 3-units get depleted first, then the 4-units, then the 2-units and finally the 1-units. These four properties determine the relative proportion of the trees that are (3,4)-trees, (1,2,4)-trees, (1,2,2,2)-trees and (1,1,1,1,1,1,1)-trees which must of course add to one. Since $S_v^{(4)}$ and $S_h^{(4)}$ have the same asymptotic structure and since $S_h^{(4)}$ has been constructed so that with probability $\to 1$ as $c \to \infty$ there is no infinite delay in classifying any one unit, the same property must also hold for $S_v^{(4)}$. This proves that $S_v^{(4)}$ is indeed a vertical procedure and justifies our notation for it. Moreover the above four asymptotic ($c \to \infty$) relative proportions for $S_h^{(4)}$ must also hold for $S_v^{(4)}$. In particular, with probability $\to 1$ as $c \to \infty$ the (1,1,1,1,1,1,1)-tree will be used infinitely often in the vertical procedure $S_v^{(4)}$.

We claim that the above result holds for any associated pair of horizontal procedure S_h and vertical procedure S_v. The properties that we want for the vertical procedure S_v, such as no infinite delay, are built beforehand into the horizontal procedure S_h. The horizontal procedure S_h then becomes simply a mathematical convenience for calculating the asymptotic ($c \to \infty$) properties of the vertical procedure S_v.

Illustration

After studying the efficiency of various horizontal schemes (which are easier to work with than vertical schemes) by trial and error methods as in Section 4 we decided for $q = .9$ to first use the (3,4)-tree and finally arrived at the horizontal procedure $S_h^{(4)}$ in (8.3). Not only can we say that the 3-units are depleted but the proportion of 4-units remaining (assuming a large number c of sets of units) is

$$\text{Prop. (4's after Step I)} = 1 - q^3 . \qquad (8.18)$$

A simple analysis of the (1,2,4)-tree in (8.4), which makes up our Step II, shows that for $q = .9$ the 4-units are depleted before the 1-units or the 2-units. Since the probability of classifying the 4-unit is q^3 it follows that we use the (1,2,4)-tree $N(1,2,4) = c(1-q^3)/q^3$ times compared to the (3,4)-tree being used $N(3,4) = c$ times. Since the 2-unit is classified in Step II with probability q it follows that the proportion of 2 units after Step II is

$$\text{Prop. (2's after Step II)} = \left(1 - \frac{1-q^3}{q^2}\right) = \frac{q^2+q^3-1}{q^2}. \tag{8.19}$$

In Step III we analyze the (1,2,2,2)-tree given in (5.3) and find that for $q = .9$ the 2-units get depleted before the 1-units. Also the expected number of 2-units classified by the (1,2,2,2)-tree is $q + q^3 + q^5$ and hence the number of these trees used is

$$N(1,2,2,2) = \frac{c(q^2+q^3-1)}{q^3(1+q^2+q^4)} = \frac{c(q^2+q^3-1)(1-q^2)}{q^3(1-q^6)}. \tag{8.20}$$

Note that we definitely use up one 1-unit per tree in Steps II and III. Hence the proportion of 1-units after Step II is $1-(1-q^3)/q^3 = (2q^3-1)/q^3$ and after Step III it is, using (8.20),

$$\begin{aligned}\text{Prop. (1's after Step III)} &= \frac{2q^3-1}{q^3} - \frac{(q^2+q^3-1)(1-q^2)}{q^3(1-q^6)} \\ &= \frac{q+q^2+q^3+q^4-2q^7-2}{q(1-q^6)}.\end{aligned} \tag{8.21}$$

Since each tree in Step IV classifies $(1-q^7)/(1-q)$ units of Type 1, it follows that the number of trees used in Step IV is

$$\begin{aligned}&N(1,1,1,1,1,1,1) \\ &= \frac{cp(-2+q+q^2+q^3+q^4-2q^7)}{q(1-q^6)(1-q^7)} = \frac{c(-2+3q-q^5-2q^7+2q^8)}{q(1-q^6)(1-q^7)}.\end{aligned} \tag{8.22}$$

The sum of these four N-values (divided by c) is found to be

$$\frac{1}{c}\,\Sigma N = \frac{4-q-q^2-q^3-q^4}{1-q^7} = D \text{ (say)} \tag{8.23}$$

and hence the four desired relative proportions are

$$\frac{1}{D},\ \frac{1-q^3}{q^3 D},\ \frac{(q^2+q^3-1)(1-q^2)}{q^3(1-q^6)D},\ \frac{-2+3q-q^5-2q^7+2q^8}{q(1-q^6)(1-q^7)D} \tag{8.24}$$

for the trees used in Step I, II, III, and IV, respectively. For $q = .9$ the numerical values for these four proportions are .576531, .214321, .172852, and .036295, respectively.

An interesting result about the computation in (8.23) and (8.24) is that $D = \frac{1}{c}\Sigma N$ represents the asymptotic ($c \to \infty$) expected number of trees required to classify one set. Since there are no infinite delays, D/4 is the asymptotic ($c \to \infty$) expected number of trees per unit classified. Hence 4/D is the asymptotic ($c \to \infty$) expected number of units classified per tree used; this holds for both procedures $S_h^{(4)}$ and $S_v^{(4)}$.

Hence the use of D as a normalizing constant puts our calculations on a per set basis and if we multiply through by 4 in (8.24) the proportions will be on a per unit basis. In other words, for the number of units U that are classified per tree used when c sets of units are classified, we can write the identity, using (8.24)

$$\begin{aligned}\lim_{c\to\infty} E\{U\} = \frac{1}{D}\{&(1+q^3)1 + (1+q+q^3)\left(\frac{1-q^3}{q^3}\right)\\ &+ (1+q+q^3+q^5)\frac{(q^2+q^3-1)}{q^3(1+q^2+q^4)}\\ &+ \left(\frac{1-q^7}{1-q}\right)\frac{(-2+3q-q^5-2q^7+2q^8)}{q(1-q^6)(1-q^7)}\} = \frac{4}{D}\ ;\end{aligned} \tag{8.25}$$

the identity being obtained by cancelling D in the last two parts of (8.25). The four quantities $(1+q^3)$, $(1+q+q^3)$, $(1+q^2+q^3+q^4+q^5+q^6)$ used in (8.25) represent the expected number of units classified in each of the four trees that make up our strategy $S_h^{(4)}$ in (8.3) and each is easily obtained by a simple analysis of the appropriate tree.

The asymptotic $(c \to \infty)$ efficiency is the ratio of the asymptotic expected number of tests to the asymptotic expected number of units classified. For the (1,2,3,4)-problem using the result (8.25), we can write the asymptotic efficiency (on a per unit classified basis) of any strategy S with no infinite delays as

$$\mathrm{Eff}(S) = \frac{\lim_{c\to\infty} E\{T|S\}}{\lim_{c\to\infty} E\{U|S\}} = \frac{D}{4} \lim_{c\to\infty} E\{T|S\} \tag{8.26}$$

and we multiply this by 4 to put it on a per set basis. For the strategy $S_h^{(4)}$ the 'per-set' efficiency Eff reduces to

$$\begin{aligned}\mathrm{Eff}(S_h^{(4)}) &= (2-q^7)+(3-q^3-q^7)\frac{(1-q^3)}{q^3} + \frac{(3-2q^7)(q^2+q^3-1)}{q^3(1+q^2+q^4)} \\ &\quad + \frac{(3+q-3q^7)(-2+3q-q^5-2q^7+2q^8)}{q(1-q^6)(1-q^7)} .\end{aligned} \tag{8.27}$$

This has some nice properties. Suppose we consider the corresponding (1,2,3,4,5)-problem using the same procedure $S_h^{(4)}$ and testing each 5-unit separately; call this $S_h^{(5)}$. For the vertical procedure $S_v^{(5)}$ the 5-units are tested when they are free and none of the preferences (8.5b)-(8.5d) are possible; again there is no infinite delay. Then the efficiency on a set basis for $S_h^{(5)}$ or $S_v^{(5)}$ is obtained by simply adding 1 to the result in (8.27).

For $q = .9$ the numerical value for $S_h^{(4)}$ and $S_v^{(4)}$ in (8.14) is 2.9560 and, as mentioned above in Section 4, this was found to be optimal. For the (1,2,3,4,5)-problem with $q = .9$, the optimal procedure is the one described above, i.e., the procedure $S_v^{(5)}$ associated with $S_h^{(5)}$, and hence the optimal efficiency is 3.9560 for $q = .9$.

9. *Lower Bounds*

For each of these problems one can obtain lower bounds exactly as was done in previous papers (cf. Sobel and Groll (1959), Sobel (1960) and Sobel (1967)); we omit their derivations. One lower bound is based on the information lower bound (we denote it by ILB) and it is given by

$$\text{ILB} = -\Sigma Q_i \log_2 Q_i \tag{9.1}$$

where the sum is over all possible states of nature for each set, Q_i is the probability of the ith possible state and $\Sigma Q_i = 1$. For example, in the $H(1,2,3,4)$ problem (finite or infinite model) we consider the 16 possible states of nature: all four batches are good with probability $Q_1 = q^{10}$, the 4-unit is good and the others are bad with probability $Q_2 = q^4(1-q)(1-q^2)(1-q^3)$, etc., so that i runs up to 16 in (9.1).

For $q = .9$ we obtain for the three problems $H(1, 2,3,4,5)$, $H(1,2,3,4)$ and $H(1,2,4)$, respectively,

$$\text{ILB}(1,2,3,4,5) = 3.9181 \ ; \quad \text{ILB}(1,2,3,4) = 2.9419 \ ; \quad \text{ILB}(1,2,4) = 2.0990 \ . \tag{9.2}$$

Recall that we obtained for the infinite model

$$H_\infty(1,2,3,4,5) = 3.9560 \ ; \quad H_\infty(1,2,3,4) = 2.9560 \ ; \quad H_\infty(1,2,4) = 2.1196 \tag{9.3}$$

and for the finite model we obtained

$$H(1,2,3,4,5) = 3.9588 \ ; \quad H(1,2,3,4) = 3.0202 \ ; \quad H(1,2,4) = 2.1296 \ . \tag{9.4}$$

The other lower bound is always superior to, i.e., larger than, the ILB. It is the expected length of the most economical code due to Huffman for the given probabilities Q_i; we denote it by HLB. The procedure of deriving it is described in Sobel (1960); there is no explicit formula for this lower bound. For $q = .9$ we obtain in the above three problems

$$\text{HLB}(1,2,3,4,5) = 3.9469 \;; \quad \text{HLB}(1,2,3,4) = 3.0035 \;; \quad \text{HLB}(1,2,4) = 2.1174 \;. \tag{9.5}$$

It is interesting to note that while the ILB is a lower bound for both models the HLB is a lower bound only for the finite model; this further justifies the consideration of both bounds. The HLB (ILB) is generally not attainable by any nested group-testing rrocedure in the finite (in either) model.

In the finite model version of the (1,2,4)-problem it is interesting to note that a non-nested procedure does exist that attains the HLB at the right end of (9.5). The following non-nested procedure R^*

```
          End        End      End          End      End            End
H(1,2,4) ← G(1,2,4) ← G(1,2) → J(1,2;1,4) ← H(2,4) ← G(2,4) → H(4) ← End
test       (1,2)      (1,4)    (1)          (2,4)    (2)        (4)
```

(9.6)

consisting of a single chain of length 7 and terminanting arrows elsewhere, does attain the HLB for $q = .9$. The fact that we tested (1,4) after the event '(1,2) was bad' (and had to write something other than G or H, namely, J(1, 2;1,4) in (9.6)) shows that the procedure R^* is not a nested one. In fact, the expected number of tests is easily shown to be (compare with the HLB in (9.5))

$$\begin{aligned} E\{T|R^*\} &= 7-3q-q^2-q^3-q^5-q^6+q^7 \\ &= 1+12p-8p^2-4p^3+15p^4-14p^5+6p^6-p^7 \end{aligned} \tag{9.7}$$

and the latter is 2.1174 for $p = .1$. The same procedure can be applied in an obvious manner to the infinite model, simply by repeating the operation on successive sets, and the HLB is the resulting measure of efficiency.

Some discussion of when the HLB is attained for the case of a common (known) q is given in Sobel (1967) but the general case of unequal q's has not been thoroughly investigated. In general, if we start with four or more batches the HLB is not attainable; we conjecture that it is not attainable in the other two problems, (1,2,3,4) and (1,2,3,4,5), discussed above for non-trivial q. In particular, for $q = .9$ we conjecture that these two proposed nested procedures are optimal among all group-testing procedures.

For the (1,2,3,4)-problem it should be clear from (9.3) and (9.5) that even if there existed a procedure R that attained the HLB = 3.0035 in the finite model, the analogue or repetition of R for the infinite model would still not compete with the proposed nested procedure $S_v^{(4)}$ (cf. (8.3) through (8.6)) which has efficiency $H_\infty(1,2,3,4) = 2.9560 <$ HLB = 3.0035. This whole question of when there exists a non-nested procedure that is better than the proposed nested procedure will have to be treated in a separate paper.

ACKNOWLEDGEMENTS

The authors wish to acknowledge a number of conversations with Professor S. Kumar of the Canadian Government, Department of Indian and Northern Affairs. Thanks are also due to Professor R. A. Elashoff since the basic ideas were developed while working on contract NIH-E-71-2180 at the University of California Medical School in San Francisco, California. The authors also wish to acknowledge with thanks the help of Dr. W. E. Lever of the Oak Ridge National Laboratories, Oak Ridge, Tenn., in deriving the exact form of the procedure R_1 for the (1,2,3,4)-problem as given in (7.1).

REFERENCES

Coveyou, R. R. (1971), Tables for an incomplete search problem, Oak Ridge National Laboratory (personal communication).

Hwang, F. K. (1973), On finding a single defective in binomial group-testing. (To appear)

Sobel, M. and Groll, P. A. (1959), Group-testing to eliminate efficiently all defectives in a binomial sample, *Bell System Tech. Jour.*, 38, 1179-1252.

Sobel, M. (1960), Group-testing to classify all defectives in a binomial sample, *Information and Decision Processes* (R. E. Machol, Ed.), McGraw-Hill: New York, 127-161.

Sobel, M. (1967), Optimal Group-Testing, *Proceedings of the Colloquium on Information Theory Organized by the Bolyai Mathematical Society*, Debrecen, Hungary, 411-488.

Appendix to Section 5

In this appendix, we give a partial proof (indicating where it is not complete) that if we restrict our attention to trees which maximize the information on their very first step (in the H-situation), then Condition (5.2) for the application of Theorem 1 is always satisfied. (We also assume, as usual, that all trees under consideration satisfy (ii) of Section 4 above.) Since we are often interested in trees which maximize this information, Theorem 1 provides us with a valuable tool in our search for the optimal procedure.

Let $q_{a,a+1}$ denote the root of $1-x^a-x^{a+1} = 0$; it is easily seen that these roots increase with a and converge to 1 as $a \to \infty$. For any given q, this sequence determines an integer $m = m(q)$ such that $q_{m-1,m} < q < q_{m,m+1}$ (with equality on either side, say $q \leq q_{m,m+1}$). It is shown on p.1215 of Sobel and Groll (1959) that for trees which only use one type of unit, each with probability q of being good, the choice of (the above) $m = m(q)$ such units to test on the initial H-situation maximizes the information. The integer m in turn determines two other integers when we write

$$m = 2^{\alpha} + \beta \quad (0 \leq \beta < 2^{\alpha}); \tag{5.54}$$

thus for $m = 4$, we have $\alpha = 2$ and $\beta = 0$. Let $H^*(m)$ denote the expected number of tests to get from one H-situation to the next, if we use m units in our first test. Now (5.54) is in accord with equations (20a) and (23) on p.1190 of Sobel and Groll (1959), and from (23) of the same reference we easily obtain

$$H^*(m) = (\alpha+1)(1-q^m) + q^{m-2\beta} , \tag{5.55}$$

where m is chosen to maximize the information from the first test in the initial H-situation. The result given in (5.55) was shown to hold in an interval ending at $q = 1$ but the fact that this interval always starts to the left of $q_{m-1,m}$ has been verified up to $m = 16$ (and beyond) but not shown to hold for all m; all the calculations to date support this conjecture. This conjecture has been proved by F. K. Hwang (1973).

Suppose we are working with two types of units, say $e_1 = 1$ and $e_2 = d$ (d is a positive integer), i.e., q and q^d are the probabilities of being good for the two types of units. Let $m = 2^{\alpha} + \beta$ be the maximizing-information value for all q in the interval $q_{m-1,m} < q < q_{m,m+1}$ and define T_1 to be the tree that initially tests m of the 1-units. It follows that $E\{T|T_1\} = H^*(m)$ as given in (5.55). It is also easily seen that

$$E\{N(1)|T_1\} = mq^m + mq^{m-1}p + (m-1)q^{m-2}p + \ldots + 1q^0 p = \frac{1-q^m}{1-q} . \tag{5.56}$$

Let us first consider the case when $d = 2$. Then $T(2,1)$ starts with $x = (1,1)$ and

$$E\{T|T(2,1)\}-1 = (1-q^2) \; ; \quad E\{N(1)|T(2,1)\} = 1+q . \tag{5.57}$$

Hence for the right side of (5.2), we have

$$(1+q)\frac{[(\alpha+1)(1-q^m)+q^{m-2\beta}]}{(1-q^m)/(1-q)} - (1-q^2) = \frac{1-q^2}{1-q^m}[\alpha(1-q^m)+q^{m-2\beta}] \ . \tag{5.58}$$

Let T_2 be the tree which uses only the 2-units and maximizes the information on its first step. Suppose m is odd and hence (for $m>1$) cannot be a power of two. At this point we are not sure whether $\frac{m-1}{2}$ or $\frac{m+1}{2}$ two's maximize the above information. Consider T_2 first with $\frac{m-1}{2}$ two's and then (later) with $\frac{m+1}{2}$ two's. We use the same formula as in (5.55) above with q replaced by q^2 and (m,α,β) replaced by $(\frac{m-1}{2}, \alpha-1, \frac{\beta-1}{2})$, respectively. This gives for procedure T_2 starting with $\frac{m-1}{2}$ two's

$$E\{T|T_2\} = \alpha(1-q^{m-1}) + q^{m-1-2(\beta-1)} \ . \tag{5.59}$$

Similarly, using (5.56) with (q,m) replaced by $(q^2, \frac{m-1}{2})$, respectively,

$$E\{N(2)|T(2)\} = \frac{1-q^{m-1}}{1-q^2} \tag{5.60}$$

and hence the left side of (5.2) is given by

$$\frac{E\{T|T_2\}}{E\{N(2)|T_2\}} = \frac{1-q^2}{1-q^{m-1}}[\alpha(1-q^{m-1}) + q^{m+1-2\beta}] \ . \tag{5.61}$$

To show the inequality in (5.2) is equivalent to showing that

$$\frac{q^{m+1-2\beta}}{1-q^{m-1}} \geq \frac{q^{m-2\beta}}{1-q^m} \tag{5.62}$$

and this in turn is equivalent to showing that

$$q(1-q^m) \geq (1-q^{m-1}) \tag{5.63}$$

or that

$$1 - q^{m-1} - q^{m} \leq 0 . \tag{5.64}$$

The latter inequality holds for $q > q_{m-1,m}$.

We now consider T_2 with $(\frac{m+1}{2})$ two's and break things up into two cases according as (i) $0 < \beta < 2^{\alpha}-1$ or (ii) $\beta = 2^{\alpha}-1$. If $0 < \beta < 2^{\alpha}-1$, we have from (5.55) and (5.56) with (q,m,α,β) replaced by $(q^2, \frac{m+1}{2}, \alpha-1, \frac{\beta+1}{2})$, respectively,

$$E\{T|T_2\} = \alpha(1-q^{m+1}) + q^{m+1-2(\beta+1)} , \tag{5.65}$$

$$E\{N(2)|T_2\} = \frac{1 - q^{m+1}}{1 - q^2} . \tag{5.66}$$

Then the left side of (5.2) becomes

$$\frac{E\{T|T_2\}}{E\{N(2)|T_2\}} = \frac{1-q^2}{1-q^{m+1}}[\alpha(1-q^{m+1}) + q^{m-1-2\beta}] . \tag{5.67}$$

Hence, comparing with (5.8), we have to show that

$$\frac{q^{m-1-2\beta}}{1-q^{m+1}} \geq \frac{q^{m-2\beta}}{1-q^{m}} \tag{5.68}$$

and this is easily seen to be equivalent to

$$1 - q^{m} - q^{m+1} \geq 0 . \tag{5.69}$$

The latter holds for $q \leq q_{m,m+1}$. If $\beta = 2^{\alpha}-1$, then from (5.55) and (5.56), with (q,m,α,β) replaced by $(q^2, \frac{m+1}{2}, \alpha, 0)$, it follows that

$$E\{T|T_2\} = \alpha(1-q^{m+1}) + 1 , \tag{5.70}$$

$$E\{N(2)|T_2\} = \frac{1 - q^{m+1}}{1 - q^2} . \tag{5.71}$$

Again in this case (namely when $\beta = 2^{\alpha}-1$), (5.2) is true if and only if

$$1 - q^{m} - q^{m+1} \geq 0 . \qquad (5.72)$$

Hence the inequalities (5.64), (5.69) and (5.72) all hold for $q_{m-1,m} < q \leq q_{m,m+1}$, which is exactly the interval we assumed at the outset. The case m even must also be considered but it is easy to see that this leads to equality in (5.2). We have thus proved for the case $d = 2$ the following theorem:

Theorem 2

Let the two types of units be the $e_1 = 1$-unit and $e_2 = d$-unit. If we restrict ourselves to trees which maximize the information on their initial step (and also satisfy (ii) of Section 4) then condition (5.2) always holds.

It is our conjecture, though, that Theorem 2 is true for all $d \geq 2$. In line with this conjecture, we extend the proof of Theorem 2 by virtue of the following lemma:

Lemma 1

For any two divisors of m, say $a' > a$,

$$\frac{E\{T|T_a\}}{E\{N(2)|T_a\}} \leq \frac{E\{T|T_{a'}\}}{E\{N(a')|T_{a'}\}} , \qquad (5.73)$$

where T_a (resp., $T_{a'}$) is the tree which maximizes the information on its initial step and works only with a-units (resp., a'-units), i.e., with units having probability q^a (resp., $q^{a'}$). We prove Lemma 1 for $a = 1$ and 2 and conjecture that it is true for all a. The case $a = 2$ is the one needed to show that Theorem 2 holds for all values of $d = a' > a = 2$, which are multiples of 2 and divisors of m. This is so because by proving Theorem 2 for $d = 2$, we have shown that

$$\frac{E\{T|T_2\}}{E\{N(2)|T_2\}} \geq \text{right side of (5.2)}, \qquad (5.74)$$

and, together with (5.73), this implies that

$$\frac{E\{T|T_{a'}\}}{E\{N(a')|T_{a'}\}} \geq \text{right side of (5.2)}, \tag{5.75}$$

which in turn proves Theorem 2 for $d = a'$.

We now prove Lemma 1 for $a = 1$ and 2. Let $m = 2^{\alpha} + \beta$ $(0 \leq \beta < 2^{\alpha})$, where we are writing α for $\alpha(m)$ and β for $\beta(m)$. Let m be chosen so that $q_{m-1,m} < q \leq$ $\leq q_{m,m+1}$. The number of units we test on the first step of the $T_{a'}$ tree is thus m/a' (recall that a' is a divisor of m). Replacing (q,m,α,β) by $(q^{a'}, m/a', \alpha(m/a'), \beta(m/a'))$ in (5.55) and (5.56) gives us

$$\frac{E\{T|T_{a'}\}}{E\{N(a')|T_{a'}\}} = (1-q^{a'})\{\alpha(\tfrac{m}{a'}) + 1 + \frac{q^{m-2a'\beta(m/a')}}{1 - q^{m}}\} . \tag{5.76}$$

Let $\alpha(m/a) = \alpha^*$; for $a = 1$ or 2 or any power of 2, $\beta(m/a) = \beta/a$. It thus follows that

$$\frac{E\{T|T_{a}\}}{E\{N(a)|T_{a}\}} = (1-q^{a})\{\alpha^* + 1 + \frac{q^{m-2\beta}}{1-q^{m}}\} . \tag{5.77}$$

We say x and y belong to the same power cycles if they have the same α-value, i.e., $\alpha(x) = \alpha(y)$.

Part 1: Assume m/a and m/a' are in the same power cycle.

Suppose first that $a = 1$ and the above assumption holds; then to prove Lemma 1, we have to show (by (5.76) and (5.77)) that

$$(1-q)\{\alpha^* + 1 + \frac{q^{m-2\beta}}{1-q^{m}}\} \leq (1-q^{a'})\{\alpha^* + 1 + \frac{q^{m-2a'\beta(m/a')}}{1 - q^{m}}\}. \tag{5.78}$$

Since $a' > 1$, $1-q < 1-q^{a'}$ and since $\beta(m/a') \geq 0$

it suffices to show that

$$1 \leq q^{2\beta}(1+q+\ldots+q^{a'-1}) \; . \tag{5.79}$$

Since $a' \geq 2$ we have at least two terms on the right side of (5.79) and we can disregard the remaining terms. Since $\beta < m/2$ it follows that $2\beta \leq m-1$ and hence the result follows since

$$1 - q^{2\beta} - q^{2\beta+1} \leq 1 - q^{m-1} - q^{m} \leq 0 \; . \tag{5.80}$$

Suppose $a = 2$ and the assumption of Part 1 holds. We have to show that

$$(1-q^2)\{\alpha^* + \frac{q^{m-2\beta}}{1-q^m}\} \leq (1-q^{d'})\{\alpha^* + \frac{q^{m-2a'\beta(m/a')}}{1 - q^m}\} \tag{5.81}$$

or that

$$\alpha^* q^2(1-q^{a'-2})(1-q^m) + (1-q^{a'})a^{m-2a'\beta(m/a')} \geq (1-q^2)q^{m-2\beta} \; . \tag{5.82}$$

Since $\alpha(m/a) = \alpha - 1$ for $a = 2$, we have by the assumption of Part 1

$$\frac{m}{a'} = \frac{2}{a'}\, 2^{\alpha-1} + \frac{\beta}{a'} \geq 2^{\alpha-1} = \alpha(m/a') \tag{5.83}$$

and hence, since $a' > 2$,

$$\beta(m/a') = \frac{m}{a'} - 2^{\alpha-1} = \frac{\beta}{a'} - (1 - \frac{2}{a'})2^{\alpha-1} > \frac{\beta}{a'} \; . \tag{5.84}$$

Thus we can replace $m - 2a'\beta(m/a')$ in (5.82) by $m - 2\beta$ and it suffices to prove that

$$\alpha^* q^2(1-q^{a'-2})(1-q^m) + (1-q^{a'-2})q^{m-2\beta+2} \geq 0 \; , \tag{5.85}$$

which obviously holds; this completes the proof under the Part 1 assumption.

Part 2: Assume m/a and m/a' are not in the same power cycle.

We will show that it suffices to assume adjoining power cycles, i.e., $\alpha(m/a) = \alpha^* = 1 + \alpha(m/a')$.

Suppose $a = 1$ and that m/a and m/a' are in adjoining power cycles. Using (5.76), (5.77) and the fact that $\beta(m/a') \geq 0$, it is sufficient to show that

$$\alpha^* q^a(1-q^{a'-a})(1-q^m)+(1-q^{a'})q^m \geq q^{m-2\beta}(1-q^a)+(1-q^m)(1-q^a). \tag{5.86}$$

For $m = 2$ we take $a = 1$ and $a' = 2$ and the inequality reduces to $1-q-q^2 \leq 0$, which holds for $m = 2$; hence we can assume $m \geq 3$. For $a = 1$, $\alpha^* \geq 1$ we first treat

Case 1: $a' \geq 2a = 2$ and $m > 2\beta + 1$.

Dividing (5.86) by $1-q^a$, and using the facts that $1 - q^{a'-a} \geq 1 - q^a$ and

$$1 - q^{a'} \geq 1 - q^{2a} = (1-q^a)(1+q^a) \tag{5.87}$$

it suffices to show that

$$q^a(1-q^m) + q^m(1+q^a) \geq q^{m-2\beta} + (1-q^m) . \tag{5.88}$$

Since $1-q^m \leq q^{m-1}$, it suffices to show that

$$q^a - q^{m-2\beta} \geq q^{m-1}(1-q) \tag{5.89}$$

or equivalently that

$$1 + q + \ldots + q^{m-2\beta-2} \geq q^{m-2} . \tag{5.90}$$

Here we use the fact that $m > 2\beta + 1$ or $m - 2\beta - 2 \geq 0$.

For the case $m = 2\beta + 1$ we disregard the possibility $a' = 2$, since m is now odd.

Case 2: $m = 2\beta + 1$, $a' \geq 3$.

From (5.86) by substitution and straighforward algebra we now have to show that

$$q^2(1-q^{a'-2}) \geq (1-2q^{2\beta+1})(1-q) \tag{5.91}$$

or equivalently that

$$q^2 + q^3 + \ldots + q^{a'-1} \geq 1 - 2q^{2\beta+1} . \tag{5.92}$$

Since there is at least 1 term on the left side of (5.92) we have to show that

$$1 - q^2 - q^{2\beta+1} - q^{2\beta+1} \leq 0 \tag{5.93}$$

which holds for $m = 2\beta+1 \geq 3$ since the first three terms alone are nonpositive.

If m/a and m/a' are (k+1) cycles apart (k=1,2,...) then $\alpha^* \geq (k+1)$, $a' \geq 3$ and $m \geq 3$. Using (5.76), (5.77) and the fact that $\beta(m/a') \geq 0$, it is sufficient to prove that

$$\begin{aligned}&(k+1)q^a(1-q^{a'-a})(1-q^m) + (1-q^{a'})q^m \\ &\geq q^{m-2\beta}(1-q^a) + (k+1)(1-q^m)(1-q^a) .\end{aligned} \tag{5.94}$$

Using the fact that (5.86) is true for $\alpha^* = 1$, proving (5.94) amounts to proving that

$$1 - 2q^a + q^{a'} \leq 1 - 2q + q^3 \leq 0 \tag{5.95}$$

or that

$$(1-q)(1-q-q^2) \leq 0 ; \tag{5.96}$$

this holds even for $m \geq 2$.

Since a' cannot be less than 2 for $a = 1$, we have completed the proof for $a = 1$.

Now we consider the case $a = 2$ with m/a and m/a' in adjoining power cycles, i.e., under Part 2. Again we wish to show (5.86). Since $a \leq m/2$, it follows that $m/a \geq 2$ and hence $\alpha^* \geq 1$; it suffices to prove (5.86) with $\alpha^* = 1$ as before.

Case 1: $a' \geq 2a = 4$ and $m \geq 4$.

From (5.86) we have to show that

$$q^a(1-q^{a'-a}) \geq (q^{m-2\beta}+1-2q^m)(1-q^a) . \tag{5.97}$$

Since $1 - q^{a'-a} \geq 1 - q^a$ for $a' \geq 2a$, it suffices to show that

$$q^2 \geq 1 + q^{m-2\beta} - 2q^m \tag{5.98}$$

or, replacing $1 - q^m$ by q^{m-1}, that

$$q^2(1-q^{m-2\beta-2}) \geq q^{m-1}(1-q) \tag{5.99}$$

or equivalently that

$$q^2 + q^3 + \dots + q^{m-2\beta-1} \geq q^{m-1} . \tag{5.100}$$

For $m \geq 2\beta + 3$ the result follows from (5.100); hence we need only consider the two remaining possibilities $m = 2\beta + 2$ and $m = 2\beta + 1$. Since $a = 2$ we rule out $m = 2\beta+1$ since m must be even to be a multiple of a. For $m = 2\beta+2$ we need only consider $\beta \geq 1$ since $m \geq 4$.
[Remark: We note that the smallest m of this type is $m = 6$ with $a' = 6$ and after that we have $m = 14$ with $a' = 7$ or 14; we note that $\beta(m/a') = 0$ in these cases.]

Since $m = 2^\alpha + \beta = 2\beta + 2$ it follows that β is an integer and hence $m = 2\beta + 2$ contains a factor of 2 but not 4. Hence $a' > a$ must contain a factor other than 2 and since $a' \geq 4$ we can assume that $a' \geq 5$ or that $a' - a \geq 3$. It now follows from (5.97) that it is sufficient to show that

$$q^2(1+q+q^2) \geq (q^2+1-2q^m)(1+q) \tag{5.101}$$

or, using the fact that $1-q^m \leq q^{m-1}$, that

$$q^4 \geq q^{m-1}(1-q^2) \tag{5.102}$$

and this clearly holds for $m \geq 5$. If $a' < 2a = 4$ then we need only consider $a' = 3$ and $m \geq 6$.

Case 2: $a' = 3$, $a = 2$ and $m \geq 6$.

For $\beta = 0$ the inequality (5.86) with $\alpha^* = 1$ reduces to $1-q-q^m \leq 0$ which clearly holds for $m \geq 2$; hence we can assume $\beta \geq 1$. In this case ($\beta \geq 1$) we need the exponent $2a'\beta(m/a')$ in (5.82). Since $a' < 2a$ (strictly)

$$2a'\beta(m/a') = 2(m-a'2^{\alpha^*-1}) = 2m - \frac{3}{2}(m-\beta) = \frac{m+3\beta}{2} \qquad (5.103)$$

and hence $m-2a'\beta(m/a') = (m-3\beta)/2$. Using (5.76), (5.77) and the above result (5.103) and then setting $\alpha^* = 1$, it suffices to show that

$$q^{m-3\beta/2}(1+q+q^2) \geq (1+q)q^{m-2\beta} + (1-q^m)(1+q-q^2). \qquad (5.104)$$

Replacing $1-q^m$ by q^{m-1} and dividing by $q^{m-2\beta}$, it suffices to show that

$$q^{\beta-m/2}(1+q+q^2) \geq 1 + q + q^{2\beta-1} + q^{2\beta} - q^{2\beta+1}. \qquad (5.105)$$

Since $\beta \leq (m-1)/2$, and $m \geq 6$, the powers on the left side of (5.105) are at most $(-m-1)/4 + 2 \leq 1/4$ and since $q^{1/4} > q$, $q^{-7/4} \geq 1$ and $q^{-3/4} \geq 1$, it suffices to show that

$$1 - q^{2\beta-1} \geq q^{2\beta}(1-q) \qquad (5.106)$$

or equivalently that

$$1 + q + \ldots + q^{2\beta-2} \geq q^{2\beta} \qquad (5.107)$$

and this clearly holds for $\beta \geq 1$.

[Remark: Although we assumed $m \geq 6$ in the above, the assumption that m/a and m/a' are in adjoining cycles actually requires that m be at least 18, in which case $m/a = 9$ and $m/a' = 6$. If m/a and m/a' are more than one cycle apart, then the same argument as in (5.94), (5.95) and (5.96) shows that the result holds a fortiori.]

This completes the proof of Lemma 1 for $a = 1$ and 2.

QUASI-LINEAR DISCRIMINATION

J. Tiago de Oliveira
University of Lisbon

1. *Introduction*

We will begin by stating the basic ideas of discrimination which will be dealt with in this paper. Suppose that P and Q are distinct populations, from which we did observe independent and identically distributed samples $(x_1,\ldots,x_m)$ and $(y_1,\ldots,y_n)$, respectively, x and y being possibly vectors. Finally let z be the observation relative to some individual to be assigned to P or Q, on the basis of the previous information (samples).

If we knew, exactly, the distribution of P and Q, the problem would be one of hypothesis testing, eventually with equalization of the misclassification errors (the errors of 1st and 2nd kinds). Since, in fact, the distributions are not exactly known, previous samples are used to supply information on the parameters so that, as a rule, estimators of the parameters are substituted for the real (unknown) values, changing then, in some way, the misclassification errors.

Essentially, since in the distribution of the random variables there appear location and dispersion parameters, the technique used avoids their estimation, reducing the problem to the consideration of the essential parameters (shape, correlation, etc.). For simplicity of the exposition we will suppose the observations to be one-dimensional and we will use x for $(x_1,\ldots,x_m)$ and y for $(y_1,\ldots,y_n)$.

It should be noted that, in some cases, owing to the nature of the observations, location or dispersion may have effects that are overlooked in this approach.

2. The basic ideas for known populations

Let $p(\cdot)$ and $q(\cdot)$ be the density functions of populations P and Q. Then the likelihood of the sample is

$$L(x,y) = \prod_1^n p(x_i) \cdot \prod_1^n q(y_j) .$$

We want to decompose the sample space $R^{m+n+1} = \{x_1,\ldots,x_m, y_1,\ldots,y_n, z\}$ where z denotes the not yet assigned observation, in two complementary regions A and B or equivalently to decompose the real line R in two complementary random regions $A(x,y)$ and $B(x,y)$ such that if $(x,y,z) \varepsilon A$ [or $z \varepsilon A(x,y)$] we discriminate z as belonging to P, if $(x,y,z) \varepsilon B$ [or $z \varepsilon B(x,y)$] we discriminate z as belonging to Q.

The misclassification errors are:

$$\begin{aligned} M_Q &= \text{Prob}\{\text{discriminate } z \text{ for } P | z \text{ belongs to } Q\} \\ &= \int_A L(x,y) q(z)\, dx dy dz \end{aligned}$$

$$\begin{aligned} M_P &= \text{Prob}\{\text{discriminate } z \text{ for } Q | z \text{ belongs to } P\} \\ &= \int_B L(x,y) p(z)\, dx dy dz = 1 - \int_A L(x,y,z) p(z)\, dx dy dz. \end{aligned}$$

If we denote by π and $\chi (\pi+\chi = 1)$ the overall proportions of the populations P and Q and by c_P and c_Q the losses resulting from assigning an individual of P to Q and of Q to P, respectively, then the average loss is

$$\pi c_P M_P + \chi c_Q M_Q .$$

The loss is minimum if we take:

$$A : \chi c_Q\, q(z) \leq \pi c_P\, p(z)$$

$$B : \chi c_Q\, q(z) \geq \pi c_P\, p(z)$$

if equality has zero probability. Notice that, in this case, both A and B are independent of x and y, previous samples not giving thus additional information because

$p(\cdot)$ and $q(\cdot)$ do not depend on any parameters.

As c_P and c_Q can, in general, be known, should we know π and χ the approach given would be the best one. Note that, in this case, the misclassification errors are not minimized even if we take $c_P = c_Q$ (=1); in the last case only the overall misclassification error is minimized.

Consequently we will impose *the fairness assumption* $M_P = M_Q$ and minimize (equal) misclassification errors.

The fairness assumption can be written

$$\int_A L(x,y)[p(z) + q(z)]dxdydz = 1$$

and we have to minimize

$$M_Q = \int_A L(x,y)q(z)dxdydz .$$

The Neyman-Pearson theorem leads to

$$A : \frac{q(z)}{p(z) + q(z)} \leq c' \quad (< 1)$$

$$A : \frac{p(z)}{q(z)} \geq \frac{1-c'}{c'} = c .$$

The region A is, then, of the same form as before, c being computed from the fairness assumption. The result has evidently the same form for vector random variables. It is clear how we can shift from the second formulation to the first one: instead of computing c by the fairness assumption we shall take $c = \chi c_Q/\pi c_P$.

Consequently we will, in what follows, consider only the second approach. It should be noted, already, that although not using bayesian methods, the results obtained are, in many cases, bayesian-like. The same can be said if, instead of the fairness condition, we require that the misclassification errors be in a given ratio.

Note, moreover, that if the border $p(z) = cq(z)$ is not of zero measure, randomization is necessary.

We can also obtain an interesting relation between the common misclassification error and the Kolmogoroff distance between the two densities.

As it is well known, the Kolmogoroff distance is

$$\Delta(p,q) = \sup_S \left| \int_S p(z)dz - \int_S q(z)dz \right| = \frac{1}{2} \int_R |p(z)-q(z)| dz$$

$$= \int_D (p(z)-q(z))dz = \int_{D^c} (q(z)-p(z))dz$$

where $D = \{z: p(z) \geq q(z)\}$.

For fair regions A' we have for the misclassification error M:

$$1 - 2M = 1 - 2 \int_{A'} q(z)dz = \int_{A'} (p(z)-q(z))dz$$
$$\leq \int_D (p(z)-q(z)) = \Delta(p,q) \quad ,$$

thus giving the general lower bound for any fair region

$$M \geq \frac{1 - \Delta(p,q)}{2} .$$

It is noted that the misclassification errors for D and for D^c (taken as A) verify respectively the equalities $M_P + M_Q = 1 - \Delta$ and $M_P + M_Q = 1 + \Delta$.

Other inequalities can be obtained. Integrating the relation defining A $(p(z) \geq cq(z))$ over A we get

$$1 - M \geq cM$$

and integrating the reverse inequality, defining $B = A^c$, over A^c, we get

$$M \leq c(1-M) ,$$

so that

$$\frac{M}{1-M} \leq c \leq \frac{1-M}{M}$$

implying thus $M \leq 1/2$ as would be expected; otherwise it should be sufficient to interchange A and A^c to improve the procedure. Also

$$M \leq \frac{\min(1,c)}{1+c} \quad (\leq 1/2) .$$

It must be noted that the bounds

$$\frac{1-\Delta}{2} \leq M \leq \frac{\min(1,c)}{1+c} \leq 1/2$$

cannot be improved.

In fact, if $p(z) = 0$ if $z < 0$, $= e^{-z}$ if $z \geq 0$ and $q(z) = 0$ if $z < 0$, $= (2\theta-1)e^{-z} + (1-\theta)e^{-z/2}$ if $z \geq 0$, $0 \leq \theta \leq 1$, we get

$$\Delta(\theta) = \frac{1-\theta}{2}, \qquad M(\theta) = \frac{1 + \theta^2 - \theta - (1-\theta)\sqrt{(1+\theta^2)}}{2\theta^2},$$

$$c(\theta) = \frac{\sqrt{(1+\theta^2)}-(1-\theta)}{1-\theta+(2\theta-1)\sqrt{(1+\theta^2)}} \geq 1$$

$$\frac{\min(1,\ c(\theta))}{1 + c(\theta)} = \frac{1 - \theta + (2\theta-1)\sqrt{(1+\theta^2)}}{2\theta\sqrt{(1 + \theta^2)}} ;$$

thus $M(\theta)$ is equal to the lower bound if $\theta = 0$ and equal to both bounds if $\theta = 1$; for other values of θ, $M(\theta)$ is strictly between the bounds.

It should, also, be noted that if, instead of the minimization of the equal misclassification errors, we did minimize the overall misclassification error

$$M_P + M_Q = 1 - \int_A L(x,y)(p(z)-q(z))dxdydz ,$$

we would get the region

$$A = \{z : p(z) \geq q(z)\} = D \quad \text{and have}$$

$$M_P + M_Q = 1 - \Delta .$$

3. *Quasi-linear discrimination with only accidental parameters*

Let the densities of P and Q be of the form

$$\frac{1}{\delta}\, p(\frac{x-\lambda}{\delta}) \quad \text{and} \quad \frac{1}{\delta}\, q(\frac{y-\lambda}{\delta})$$

where $p(\cdot)$ and $q(\cdot)$ are completely known.

As decision regarding discrimination is to be independent of changes of λ or of δ (with the corresponding transformation for the observations), region A (and B) must be independent of those changes, that is,

$$(x,y,z) \in A \quad \text{if} \quad (ax+b,\ ay+b,\ az+b) \in A$$

for any $a > 0$ and b; A will be called a quasi-linear discriminant region (for P).

In what follows we will use the technique developed in Tiago de Oliveira (1966) related to the ideas of Pitman (1939).

In this way the discrimination problem is analogous to the test of a simple versus simple hypothesis.

The likelihood of the (previous) samples x and y is

$$\frac{1}{\delta^{m+n}} \cdot L(\frac{x-\lambda}{\delta}\ ,\ \frac{y-\lambda}{\delta})$$

where

$$L(x,y) \ = \ \prod_1^m p(x_i) \cdot \prod_1^n q(y_j)\ .$$

A is then the quasi-linear discriminant region that minimizes

$$\int_A \frac{1}{\delta^{m+n+1}}\, L(\frac{x-\lambda}{\delta}\ ,\ \frac{y-\lambda}{\delta})\, q(\frac{z-\lambda}{\delta})\, dxdydz$$

under the fairness assumption

$$\int_A \frac{1}{\delta^{m+n+1}}\, L(\frac{x-\lambda}{\delta}\ ,\ \frac{y-\lambda}{\delta})\, [p(\frac{z-\lambda}{\delta}) + q(\frac{z-\lambda}{\delta})]\, dxdydz = 1\ ,$$

the misclassification error (independent of λ and δ) being the minimum value of the first integral. As both integrals are independent of λ and δ, we can take $\lambda = 0$, $\delta = 1$.

Owing to the exchangeability, each integral over A is the double of the integral over $A \cap \{x_1 \geq x_2\}$, also a quasi-linear region.

Then

$$\int_{A\cap\{x_1 \geq x_2\}} L(x_1, x_2, x_3, \ldots, x_m, y_1, y_2, \ldots, y_n) q(z) \prod_1^m dx_i \cdot \prod_1^n dy_j \cdot dz$$

by the transformation

$$\begin{aligned} x_1 &= \alpha + \beta \\ x_2 &= \alpha \\ x_i &= \alpha + \beta\xi_i \qquad i = 3, \ldots, m \\ y_i &= \alpha + \beta\eta_j \qquad j = 1, \ldots, n \\ z &= \alpha + \beta\zeta \end{aligned}$$

whose jacobian is β^{m+n-1}, has the value

$$\int_{A\cap\{\beta>0\}} L(\alpha+\beta, \alpha, \alpha+\beta\xi_i, \alpha+\beta\eta_j) q(\alpha+\beta\zeta) \beta^{m+n-1} d\alpha d\beta \prod_1^m d\xi_i \cdot \prod_1^n d\eta_j \cdot d\zeta$$

Denoting, now, by $\overline{A} = \{(\xi_3, \ldots, \xi_m, \eta_1, \ldots, \eta_n, \zeta)\}$ the region of R^{m+n-1} such that $(1, 0, \xi_3, \ldots, \xi_m, \eta_1, \ldots, \eta_n, \zeta) \in A$, the integral is equal to

$$\int_{\overline{A}} \prod_3^m d\xi_i \cdot \prod_1^n d\eta_j \cdot d\zeta \int_{-\infty}^{+\infty} d\alpha \int_0^{+\infty} d\beta\ \beta^{m+n-1} L(\alpha+\beta, \alpha, \alpha+\beta\xi_i,$$

$$\alpha+\beta\eta_j) q(\alpha+\beta\zeta) = \int_{\overline{A}} \prod_3^m d\xi_i \cdot \prod_1^n d\eta_j \cdot d\zeta\ \overline{L}_q(\xi, \eta, \zeta) .$$

The fairness condition can, with an analoguos notation, be written as

$$\int_{\overline{A}} \prod_3^m d\xi_i \cdot \prod_1^n d\eta_j \cdot d\zeta\ [\overline{L}_p(\xi, \eta, \zeta) + \overline{L}_q(\xi, \eta, \zeta)] = 1 .$$

The Neyman-Pearson theorem gives then the region

$$\overline{A} : \frac{\overline{L}_q(\xi,\eta,\zeta)}{\overline{L}_p(\xi,\eta,\zeta) + \overline{L}_q(\xi,\eta,\zeta)} \leq c' \quad (\leq 1), \text{ or } \quad \frac{\overline{L}_p(\xi,\eta,\zeta)}{\overline{L}_q(\xi,\eta,\zeta)} \geq c$$

If in $\overline{L}_q(\xi,\eta,\zeta)$ we make the transformation

$$\alpha = \lambda' + \delta' x_2 \quad , \qquad \beta = \delta'(x_1 - x_2)$$

whose Jacobian is $x_1 - x_2$, and noting that

$$\xi_i = \frac{x_i - x_2}{x_1 - x_2} \qquad \eta_j = \frac{y_j - y_2}{x_1 - x_2} \qquad \zeta = \frac{z - x_2}{x_1 - x_2} \quad ,$$

we get

$$\overline{L}_q(\xi,\eta,\zeta) = \int_{-\infty}^{+\infty} d\lambda' \int_0^{+\infty} d\delta' \; \delta'^{m+n-1} (x_1 - x_2)^{m+n}$$
$$\cdot L(\lambda + \delta' x_i, \; \lambda' + \delta' y_j) q(\lambda' + \delta' z)$$

so that region A (equivalent to $\overline{A}$) is

$$A : \frac{\int_{-\infty}^{+\infty} d\lambda' \int_0^{+\infty} d\delta' \; \delta'^{m+n-1} L(\lambda' + \delta' x_i, \; \lambda' + \delta' y_j) p(\lambda' + \delta' z)}{\int_{-\infty}^{+\infty} d\lambda' \int_0^{+\infty} d\delta' \; \delta'^{m+n-1} L(\lambda' + \delta' x_i, \; \lambda' + \delta' y_j) q(\lambda' + \delta' z)} \geq c$$

c being computed to give equal misclassification errors, i.e., such that

$$\int_A L(x,y)[p(z) + q(z)] dx dy dz = 1 .$$

Evidently if x, y, z, are vectors we have a similar result with different (λ', δ') for each component.

It follows naturally that, as regards the natural exigence of quasi-linearity in discrimination, the basic result is the substitution of the likelihoods in the hypothesis testing scheme by "averaged" likelihoods, with improper independent priors 1 and δ'^{m+n-1} in the form given in the last expression, or (as $\lambda' = -\lambda/\delta$, $\delta' = 1/\delta$) 1 and

δ^{m+n-1} for the form initially given. This way we get misclassification errors independent of the location and dispersion parameters.

4. Non-existence of quasi-linear discrimination with constant errors

Although this result could be expected because two close populations shall give rise to larger misclassification errors than two very distinct populations (in an extreme case, two orthogonal populations), it seems useful to consider, in detail, one example.

Let P and Q be normal populations with expected values μ and ν and equal standard deviations σ. Let us suppose that we have samples $(x_1,\ldots,x_n)$ and $(y_1,\ldots,y_n)$, of equal size, and let z be the observation to discriminate between P and Q.

To get a fair quasi-linear discriminant region, let us follow the usual technique. Supposing μ, ν and σ to be known, a fair test for P and Q (that is, with equal errors of 1st and 2nd kind) is given by:

$$\text{if} \quad (\mu-\nu)\,(z - \tfrac{\mu+\nu}{2}) \geq 0 \quad \text{accept P},$$

$$\text{if} \quad (\mu-\nu)\,(z - \tfrac{\mu+\nu}{2}) \leq 0 \quad \text{accept Q}.$$

Replacing now μ and ν by their sufficient statistics $\overline{x}$ and $\overline{y}$, we get the natural quasi-linear discriminant region for P

$$A : (\overline{x} - \overline{y})\,(z - \tfrac{\overline{x}+\overline{y}}{2}) \geq 0 ,$$

with A^c discriminating for Q. It is immediate that A is quasi-linear.

The misclassification errors, by the transformation

$$\xi = (\overline{x} - \tfrac{\mu+\nu}{2})/\sigma, \quad \eta = (\overline{y} - \tfrac{\mu+\nu}{2})/\sigma, \quad \zeta = (z - \tfrac{\mu+\nu}{2})/\sigma, \quad \Delta = \frac{\mu-\nu}{2\sigma}$$

are

$$\frac{n}{(2\pi)^{3/2}} \int_A e^{-n/2[(\xi-\Delta)^2+(\eta+\Delta)^2]-1/2(\zeta+\Delta)^2} d\xi d\eta d\zeta$$

and

$$1 - \frac{n}{(2\pi)^{3/2}} \int_A e^{-n/2[(\xi-\Delta)^2+(\eta+\Delta)^2]-1/2(\zeta-\Delta)^2} d\xi d\eta d\zeta$$

with

$$A : (\xi-\eta)\ (\zeta - \frac{\xi+\eta}{2}) \geq 0\ .$$

If we interchange ξ and η in the integral of the second misclassification error, we see that this error is equal to the first one, showing then the existence of fair quasi-linear misclassification errors.

Let us now compute the misclassification error, as a function of Δ, an analogue to the power-function.

We have

$$M_n(\Delta) = \frac{n}{(2\pi)^{3/2}} \int_A e^{-n/2[(\xi-\Delta)^2+(\eta+\Delta)^2]-1/2(\xi-\Delta)^2} d\xi d\eta d\zeta$$

$$= M_n(-\Delta)$$

which can be written as

$$M_n(\Delta) = \int_{\xi\geq\eta,\zeta\geq\frac{\xi+\eta}{2}} d\xi d\eta d\zeta \quad \frac{n}{(2\pi)^{3/2}}\, e^{-n/2[(\xi-\Delta)^2+(\eta+\Delta)^2]-1/2(\zeta+\Delta)^2}$$

$$+ \int_{\xi\leq\eta,\zeta\leq\frac{\xi+\eta}{2}} d\xi d\eta d\zeta \quad \frac{n}{(2\pi)^{3/2}}\, e^{-n/2[(\xi-\Delta)^2+(\eta+\Delta)^2]-1/2(\zeta+\Delta)^2}$$

$$= Z_n(\Delta) + Z_n(-\Delta)\ .$$

But

$$Z_n(\Delta) = \int_{\xi\geq\eta,\zeta\geq\frac{\xi+\eta}{2}} d\xi d\eta d\zeta \quad \frac{n}{(2\pi)^{3/2}} e^{-n/2[(\xi-\Delta)^2+(\eta+\Delta)^2]-1/2(\zeta+\Delta)^2}$$

$$= \int_{\eta'-\xi'\leq 2\Delta,\zeta'\geq\frac{\xi'+\eta'}{2}} d\xi' d\eta' d\zeta' \quad \frac{n}{(2\pi)^{3/2}} e^{-n/2(\xi'^2+\eta'^2)-\zeta'^2/2}$$

$$= \int_C da'db'd\zeta' \frac{n}{(2\pi)^{3/2}} e^{-1/2(a'^2+b'^2+\zeta'^2)}$$

with

$$C = \{a' \leq \sqrt{2n}\Delta \ , \quad b' \leq \sqrt{2n}(\zeta'-\Delta)\}$$

$$a' = \sqrt{n/2}(\eta'-\xi') \ , \quad b' = \sqrt{n/2}(\xi'+\eta') \ .$$

Letting $\Phi(x) = \int_{-\infty}^{x} \frac{1}{(2\pi)^{1/2}} e^{-t^2/2}dt$,

we have

$$Z_n(\Delta) = \Phi(\sqrt{2n}\Delta) \int_{b'<\sqrt{2n}(\zeta'-\Delta)} db'd\zeta' \quad \frac{1}{2\pi} e^{-1/2(b'^2+\zeta'^2)}$$

$$= \Phi(\sqrt{2n}\Delta)(1 - \Phi(\frac{\sqrt{2n}}{2n+1}\Delta))$$

so that

$$M_n(\Delta) = 1/2 - 2(\Phi(\sqrt{2n}\Delta)-1/2)(\Phi(\frac{\sqrt{2n}}{2n+1}\Delta)-1/2)$$

which shows that for $\Delta > 0$, $M_n(\Delta)$ is a decreasing function of Δ from $M_n(0) = 1/2$ to $M_n(+\infty) = 0$. Note also that for $\Delta \neq 0$, $M_n(\Delta) \to 1/2$ when $n\to\infty$. Consequently we cannot find a fair quasi-linear discriminant region with constant misclassification error function, smaller than the other ones.

5. *A general set-up for discrimination*

Let the two populations P and Q have densities of the same form $f(\cdot|\omega,\theta)$ with the (vector) parameters ω

and θ. We will suppose that, although unknown, ω is equal for both populations (the δ in previous example) which differ only in the values of θ (the μ and ν in previous example). We dispose of samples $(x_1,\ldots,x_m)$ and $(y_1,\ldots,y_n)$ for P and Q, respectively, and we want to discriminate z between P and Q.

Denoting the likelihood of the samples x and y by

$$L(x,y|\omega,\ \theta_P,\ \theta_Q) = \prod_1^m f(x_i|\omega,\ \theta_P) \cdot \prod_1^m f(y_j|\omega,\ \theta_Q)\ ,$$

the misclassification errors of a region A, discriminating for P, are

$$\int_A L(x,y|\omega,\ \theta_P,\ \theta_Q) f(z|\omega,\ \theta_Q)\,dxdydz$$

and

$$1 - \int_A L(x,y|\omega,\ \theta_P,\ \theta_Q) f(z|\omega,\ \theta_P)\,dxdydz\ ;$$

the fairness assumption imposing the equality of those two probabilities, when possible, we will denote its common value by $M_A(\theta,\ \omega_P,\ \omega_Q)$.

As we have shown in section 4, even in a simple case, we cannot find quasi-linear discriminant regions with constant misclassification errors.

A region can be found, by generalization of the Neyman-Pearson theorem, that minimizes $M_A(\theta^o,\ \omega_P^o,\ \omega_Q^o)$ under the general assumption of fairness. Let us give its proof and, later, discuss its applicability. The generalization is as follows:

Let $f(x|\tau)$ be a family of functions of x, parameterized by τ and let g(x) be another function, all of them integrable over the common domain of definition.

Then among the regions A such that

$$\int_A f(x|\tau)\,dx = c(\tau)$$

the region for which

$$\int_A g(x)\,dx = \min$$

is given by

$$A : g(x) \leq \int f(x|\tau)b(\tau)d\tau$$

where the function $b(\tau)$ is determined to verify the first condition.

In fact, let A' be another region satisfying also the first condition and let $D = A \cap A'$. We have

$$\int_{A-D} g(x)dx \leq \int dx(\int_{A-D} f(x|\tau)d\tau) = \int d\tau b(\tau)(\int_{A-D} f(x|\tau)dx)$$

$$= \int d\tau b(\tau)(\int_{A'-D} f(x|\tau)dx)$$

$$= \int_{A'-D} dx(\int f(x|\tau)b(\tau)d\tau) \leq \int_{A'-D} g(x)dx .$$

This result is immediately generalized when we have a finite set of conditions of the type of the first one.

The application of this result to the problem under discussion is immediate once θ^o, ω_P^o, ω_Q^o are chosen, but it leaves us with a difficult integral equation in $b(\tau)$ to solve. We can, already, remark that the solution obtained is of the type of generalized bayesian solutions, and we can try uniform or beta priors for bounded parameters, exponential, uniform or gamma priors for one-sided unbounded parameters and uniform or normal priors for two-sided (unbounded) parameters and check the result later.

6. *The quasi-linear discrimination problem*

Under the general set-up, given in the preceding section, we will suppose that ω (the common unknown parameters) are all the parameters except the location ones, that is, dispersion, shape, dependence, etc. and that θ are the location parameters.

The densities for P and Q are, then, respectively

$$\frac{1}{\delta} f(\frac{x-\lambda_P}{\delta}|\omega_o) \quad \text{and} \quad \frac{1}{\delta} f(\frac{x-\lambda_Q}{\delta}|\omega_o) ,$$

with a slight change of notation, the likelihood for samples $(x_1,\ldots,x_m)$ and $(y_1,\ldots,y_n)$ being

$$L(x,y|\lambda,\delta,\omega_o) = \frac{1}{\delta^{m+n}} \prod_1^m f(\frac{x_i-\lambda_P}{\delta}|\omega_o) \cdot \prod_1^n f(\frac{y_j-\lambda_Q}{\delta}|\omega_o) .$$

For any fair discriminant region A (for P) we must have

$$\int_A L(x,y|\lambda,\delta,\omega_o)\ \frac{1}{\delta}[f(\frac{z-\lambda_P}{\delta}|\omega_o) + f(\frac{z-\lambda_Q}{\delta}|\omega_o)]dxdydz = 1\ ;$$

we want, yet, A to be quasi-linear and, if possible, to be independent of ω_o.

As

$$M_A(\lambda,\delta,\omega_o) = \int_A L(x,y|\lambda,\delta,\omega_o)\ \frac{1}{\delta}\ f(\frac{z-\lambda_Q}{\delta}|\omega_o)dxdydz\ ,$$

the condition of independence of ω_o is given by $\frac{\partial M_A}{\partial \omega_o} = 0$ under regularity conditions, giving thus a finite set of additional conditions on A. Note that M_A depends only on the difference $\frac{\lambda_P-\lambda_Q}{\delta}$ as it follows from the quasi-linearity assumption, the misclassification error being 1/2 if $\lambda_P = \lambda_Q$.

Evidently we could apply previous techniques, which would amount to using (improper) priors 1 and δ^{m+n+2} for λ_P (say) and δ and $b(\omega_o)$ for ω_o. It does not seem useful to write out formulae owing to effective computational difficulty.

7. *The use of maximum likelihood estimation*

For the moment let us return to the general set-up, given in section 5, with common parameters ω and specific parameters θ_P and θ_Q, and let us suppose that m and n are large.

Under regularity conditions, $L(x,y|\omega,\theta_P,\theta_Q)$ can be approached by the multinormal distribution of the maximum likelihood estimators $\hat{\omega}$, $\hat{\theta}_P$ and $\hat{\theta}_Q$ with expected values ω, θ_P, θ_Q and a covariance matrix $1/n\cdot C + o(n^{-1})$; see Cox (1961) for details.

If we require for $(\omega,\theta_P,\theta_Q)$ -the real values of the parameters - the misclassification error to be minimum, the approximate discriminant region given by the generalization of Neyman-Pearson theorem is

$$f(z|\omega,\theta_Q)\hat{L}(\hat{\omega},\hat{\theta}_P,\hat{\theta}_Q|\omega,\theta_P,\theta_Q) \le \int\hat{L}(\hat{\omega},\hat{\theta}_P,\hat{\theta}_Q|\omega,\theta_P,\theta_Q)$$
$$\cdot[f(z|\omega,\theta_P) + f(z|\omega,\theta_Q)]b(\omega,\theta_P,\theta_Q)d\omega d\theta_P d\theta_Q$$

where $\hat{L}$ denotes the multinormal maximum likelihood approximation of L.

As $(\hat{\omega},\hat{\theta}_P,\hat{\theta}_Q)$ is close to $(\omega,\theta_P,\theta_Q)$ to the order n^{-1} in the variance, this region is approximated by

$$f(z|\omega,\theta_Q) \le [f(z|\hat{\omega},\hat{\theta}_P) + f(z|\hat{\omega},\hat{\theta}_Q)]b(\hat{\omega},\hat{\theta}_P,\hat{\theta}_Q)$$
$$/\hat{L}(\hat{\omega},\hat{\theta}_P,\hat{\theta}_Q|\omega,\theta_P,\theta_Q)$$

and, to the same order of approximation, we can substitute the parameters by their maximum likelihood estimators. We have then, finally, the region of the form

$$A : \frac{f(z|\hat{\omega},\hat{\theta}_P)}{f(z|\hat{\omega},\hat{\theta}_Q)} \ge t(\hat{\omega},\hat{\theta}_P,\hat{\theta}_Q) .$$

Let us look at it in another way and analyse, later, the errors.

Supposing the parameters $(\omega,\theta_P,\theta_Q)$ to be exactly known, the result of section 2 gives as best region:

$$A(\omega,\theta_P,\theta_Q) : f(z|\omega,\theta_P) \ge c(\omega,\theta_P,\theta_Q)f(z|\omega,\theta_Q) ,$$

the misclassification errors verifying the inequalities already given and c being determined by the fairness condition.

The misclassification error is then

$$M(\omega,\theta_P,\theta_Q) = \int_{A(\omega,\theta_P,\theta_Q)} f(z|\omega,\theta_Q)dz = 1 - \int_{A(\omega,\theta_P,\theta_Q)} f(z|\omega,\theta_P)dz$$

We will take now $A(\hat{\omega},\hat{\theta}_P,\hat{\theta}_Q) = \hat{A}$ as the *estimated* discriminant region for P (where we have replaced the parameters by their maximum likelihood estimators). The fairness assumption is not expected to be exactly verified but only approximately.

Let us introduce, for computation sake, the functions

$$N_Q(\omega,\theta_Q;\ \omega',\theta'_P,\theta'_Q) = \int_{A(\omega',\theta'_P,\theta'_Q)} f(z|\omega,\theta_Q)dz$$

and

$$N_P(\omega,\theta_P;\ \omega',\theta'_P,\theta'_Q) = 1 - \int_{A(\omega',\theta'_P,\theta'_Q)} f(z|\omega,\theta_P)dz \ .$$

It is evident that

$$M(\omega,\theta_P,\theta_Q) = N_Q(\omega,\theta_Q;\ \omega,\theta_P,\theta_Q) = N_P(\omega,\theta_P;\ \theta_P,\theta_Q).$$

The misclassification error

$$\hat{M}_Q(\omega,\theta_P,\theta_Q) = \int_{\hat{A}} L(x,y|\omega,\theta_P,\theta_Q)f(z|\omega,\theta_Q)dxdydz$$

$$= \int L(x,y|\omega,\theta_P,\theta_Q)N_Q(\omega,\theta_Q;\ \hat{\omega},\hat{\theta}_P,\hat{\theta}_Q)dxdy \ .$$

Using the series expansion of $N(\hat{\omega},\hat{\theta}_P,\hat{\theta}_Q)$ about $(\omega,\theta_P,\theta_Q)$ we get immediately

$$\hat{M}_Q(\omega,\theta_P,\theta_Q) = M(\omega,\theta_P,\theta_Q) + O(1/\min(m,n))$$

and also

$$\hat{M}_P(\omega,\theta_P,\theta_Q) = M(\omega,\theta_P,\theta_Q) + O(1/(m+n)) \ .$$

The fairness condition is, then, verified to the order $1/\min(m,n)$, the misclassification errors being equal to the best ones, also to the order $1/\min(m,n)$.

8. *Discrimination between two normal populations*

Let P and Q be two multinormal populations with expected values $\mu_P^T = (\mu_{P1},\ldots,\mu_{Pk})$ and $\mu_Q^T = (\mu_{Q1},\ldots,\mu_{Qk})$ and common variance matrix $\Sigma = (\sigma_{ij})$; for more details see Anderson (1958).

The ratio of the two densities for a new observation $z^T = (z_1,\ldots,z_n)$ is

$$\exp\{z^T\Sigma^{-1}(\mu_P-\mu_Q) - 1/2(\mu_P+\mu_Q)^T\Sigma^{-1}(\mu_P-\mu_Q)\}$$

the statistic between brackets being the test statistic.

Let $\mu = (\mu_P-\mu_Q)^T\Sigma^{-1}(\mu_P+\mu_Q)$. As the test statistic is normal with expected value $1/2\ \mu$ and variance $1/2\ \mu$ for population P and normal with expected value $-1/2\ \mu$ and variance $1/2\ \mu$ for population Q the fair region is

$$A : z^T\Sigma^{-1}(\mu_P-\mu_Q) - 1/2(\mu_P+\mu_Q)\Sigma^{-1}(\mu_P-\mu_Q) \geq 0$$

with misclassification error $1 - \Phi(\sqrt{\mu}/2)$.

If $\bar{x}$ and $\bar{y}$ denote the average of observations for P and Q populations and S the empirical matrix of covariances we have as (approximate) discriminant region

$$A : z^T S^{-1}(\overline{x-y}) - 1/2(\overline{x+y})^T S^{-1}(\overline{x-y}) > 0 \ ;$$

this region is approximately fair (to the order $1/\min(m,n)$) and is, evidently, quasi-linearly invariant. The misclassification error is equal to $1-\Phi(\sqrt{\mu}/2)$ to the order $1/\min(m,n)$).

9. *Discrimination for Gumbel populations*

Let $G(x) = \exp(-e^{-x})$ be the standard Gumbel distribution. Suppose that populations P and Q differ in location parameters λ_P and λ_Q and have the same dispersion parameters δ.

A fair region for P in the test of P against Q is

$$e^{-z/\delta}\cdot(e^{\lambda_Q/\delta} - e^{\lambda_P/\delta}) \geq c(\lambda_P,\lambda_Q,\delta) \ .$$

If we have $\lambda_P < \lambda_Q$, we get

$$Z \leq \omega(\lambda_P,\lambda_Q,\delta)$$

where ω is the solution of $G(\frac{\omega-\lambda_P}{\delta}) + G(\frac{\omega-\lambda_Q}{\delta}) = 1$.

Putting $\omega = \lambda_P - \delta \log \tau$, $\lambda_Q = \lambda_P + \delta \log \phi$ we get

$$e^{-\tau} + e^{-\phi\tau} = 1 \qquad (\tau=\tau(\phi),\ \phi \geq 1)$$

and the region is

$$Z \leq \lambda_P - \delta \log \tau(\phi) \ .$$

If we have $\lambda_P \geq \lambda_Q$ the region is

$$Z \geq \omega(\lambda_P,\lambda_Q,\delta) \ ,$$

ω verifying the same equation with, now, $\phi \leq 1$; the region turns out to be

$$Z \geq \lambda_P - \delta \log \tau(\phi) \ .$$

The misclassification probability is $e^{-\phi\tau}$ if $\lambda_P \leq \lambda_Q$ and $1 - e^{-\phi\tau}$ if $\lambda_P \geq \lambda_Q$.

The approximate discriminant region, with, approximately, the same misclassification error, is then the previous one where $(\lambda_P,\lambda_Q,\delta)$ are substituted by their maximum likelihood estimators given by the equations

$$m\left(\frac{\overline{x}}{\hat{\delta}} - 1 - \frac{\Sigma x_i e^{-x_i/\hat{\delta}}}{\Sigma e^{-x_i/\hat{\delta}}}\right) + n\left(\frac{\overline{y}}{\hat{\delta}} - 1 - \frac{\Sigma y_i e^{-y_i/\hat{\delta}}}{\Sigma e^{-y_i/\hat{\delta}}}\right) = 0$$

$$\hat{\lambda}_P = -\hat{\delta} \log \frac{\Sigma\, e^{-x_i/\hat{\delta}}}{m} \ , \qquad \hat{\lambda}_Q = -\hat{\delta} \log \frac{\Sigma\, e^{-y_i/\hat{\delta}}}{n} \ .$$

REFERENCES

Anderson, T. W. (1958), *An Introduction to Multivariate Statistical Analysis*, Wiley: New York.

Cox, D. R. (1961), Tests of separate families of hypothesis, *Fourth Berkeley Symposium on mathematical statistics and probability*, vol.I, Univ. of California

Press.
Gumbel, E. J. (1958), *Statistics of Extremes*, Columbia Univ. Press.
Pitman, E. J. (1939), The estimation of the location and scale parameters of a continuous population of any given form, *Biometrika*, 30, 391-421.
Rao, C. R. (1952), *Advanced Statistical Methods in Biometric Research*, Wiley: New York.
Tiago de Oliveira, J. (1966), Quasi-linearly invariant prediction, *Ann. Math. Statist.*, 37, 1684-1687.

THE DISCRIMINANT FUNCTION IN SYSTEMATIC BIOLOGY

R. A. Reyment
University of Uppsala

1. *Introduction*

The subject of my talk concerns how the discriminant function may be applied in systematic biology, including biological pitfalls. To a fair degree, I think, the topic I have been asked to treat may be regarded as reasonably plain sailing, for, as every statistician knows, the discriminant function was explicitly devised by R. A. Fisher as an answer to perhaps one of the most fundamental of all systematic problems of the taxonomic variety. It is, however, a well known fact among practising quantitative taxonomists that the situation covered by the classical discriminant function is in reality rare, smacking not a little of artificiality, and that the basic discriminant, i.e. identification, model is one that is little employed in dynamic studies of today. There seems to have been a tendency for statisticians to have become petrified around this concept and, to the mind of biologist, at least, to have fastened in a maze of preoccupation with significance problematics and a cult of "misclassification". "Hard words", you may say, even "unjustified attack", but we must keep our minds clear about that which may be accepted as the pabulum of mathematical research, may be biologically trivial, and vice versa.

New ideas do appear now and then. Cacoullos (1962, 1965) was clearly moving in a new, exciting direction in his dissertation on the nearness concept in discriminant analysis, and Burnaby (1966) made doubtlessly one of the most significant contributions to biological discrimination theory of the last 15 years in his paper on growth invariance.

My goal in this paper is, as I see my terms of reference, to put before you problems requiring rigorous solutions, and perhaps to promote thought on familiar topics, presented to you in the manner in which the quantitative biologist looks at them. I am, as it were, a representative of the consumers of the product of the mathematicians, and as such, may be reasonably permitted to lay claims on the appearance of the product being marketed.

Kendall and Stuart (1966, p.314) made clear the three classes of problems involved in differentiating between two or more populations on the grounds of multivariate measurements, to whit: *discrimination* concerns the problem of alloting correctly an individual to the one of k populations from which it derives. This is distinct from *classification*, which is concerned with classifying a sample of individuals into groups, which are to be as distinct as possible, and *dissection*, which involves the arbitrary construction of groups. The concept of classification, as expressed in these terms, is the most commonly occurring problem in numerical taxonomy. It is only recently that professional statisticians have begun to invade this virgin ground. Gower (1966) has made a noteworthy contribution to the theory of the subject and Jardine and Sibson (1971) have recently written a book in which the first clear formalization of some of the ideas involved is given.

In the following pages I review significant biological studies in the field, some recently re-analyzed by Blackith and Reyment (1971).

2. *Studying changes of shape in relation to environment: Solution of a biological systematic problem by a linear discriminant function*

A common pest of cassava, cotton and tobacco plants in West Africa are whiteflies of the genus *Bemisia*, a group in which parthenogenetic reproduction occurs. Several workers have suspected that the cassava, cotton and tobacco whitefly all belong to the same species, but that they grow up to look different according to the host-plant upon which they happened to have been attached at the outset of development.

In an attempt at finding a solution of the problem,

Mound (1963) made a collection of the fourth stage larva drawn from populations on cassava and tobacco plants, all of which larvae were descendants from a single parthenogenetically reproducing female. On the grounds of seven characters (dimensions of body parts), he constructed a discriminant function which could be used to establish that the suspected correlation between host and insect is a reality. The discriminant function also proved useful for charting the nature of the control of shape of the insect by the substrate. The effects of nutritional influences could be relegated to a minor role as overall size changes are not important. In fact, it is the growth pattern, rather than the amount of growth, that is determined by the plant.

3. *Distinctions between morphologically similar organisms: The problem of homeomorphy*

Some species of animals are almost indistinguishable in external appearance. This is a common enough situation for living animals, particularly insects, for example beetles, fruit flies, and it is very often seen among fossils, in which the number of characters preserved is strictly limited. The biological term for this condition is "homeomorphy". Classical palaeontological examples are the horn corals and rudistid pelecypods, many upper Mesozoic and lower Mesozoic ammonites (reflected in the names, such as *Aspidoceras*, *Pseudospidoceras: Ptychites*, *Neoptychites*).

This is a type of systematic problems the treatment of which requires more penetrating treatment than allowed in the conventional determination key.

Lyubischev (1959) has studied a situation of this kind in the genus *Halticus*, of the chrysomalid beetles. He measured 21 characters on two species of *Halticus* from the European USSR. The univariate plots for the observations on each of these characters showed that all of them overlap almost entirely. He selected the four "best" of these characters to produce a discriminant function which was successful in avoiding all overlap for his two samples. He then applied his discriminant function to large samples and found overlap to be only about 3%.

In passing, it is worth noting that some parthenogenetic animals offer good experimental possibilities for testing statistically based hypotheses about growth and environment, for their genetical constitutions are invariant. It is for this reason that Reyment and Brännström (1962) chose to work with the parthenogenetic ostracod species *Cypridopsis vidua*.

4. *Systematic use of quadratic discriminant functions*

This is a field that has yet to receive full evaluation. As far as our knowledge goes at the present time, it would seem to have been demonstrated that a quadratic discriminant function is often markedly superior to its linear counterpart in systematic studies. Burnaby (1966b) discussed the application of distribution-free quadratic discriminant functions in palaeontology in relation to a variate transformation for equalizing the determinants of the dispersion matrices.

5. *Discriminant function and the analysis of shape flexibility*

As an example of this kind of problem I shall consider the study by Stower et al. (1960) on East African locusts selected during breeding from localities with different temperatures. Where locusts breed under highly crowded conditions, they adopt the particular body shape and colour of the phase known technically as "gregaria". In other places, where crowding is not a vital factor, the colouring and morphology of the phase termed "solitaria" develops.

The charting of the generalized distances between each of the samples disclosed the existence of two fundamentally distinct variational categories, the one reflecting variation coupled to sexual dimorphism, and the other bound to "phase" variation, reflecting the populational densities at rearing. These are the main effects, but other facets of variation are reflected in these "axes". Thus, the "phase" axis also carries the effects of temperature differences prevailing during the breeding period. For example, those individuals reared under very hot conditions resemble the locusts of the "solitaria" phase, whatever population density existed during their rearing.

This case history was reviewed by Blackith and Reyment (1971) who noted that inasmuch as it was based on only three morphovariables, the discriminatory power of an otherwise interesting experimental technique was probably limited. Further experimental work on the problem is to be encouraged.

Despite the paucity of variables, Stower et al. (op. cit.) were able to construct discriminant function that could, successfully, determine the origin of populations of the phase "congregans" (parental population isolated, progeny crowded) and those of the phase "dissocians" (parents crowded, progeny isolated), although it did not provide fully satisfactory discrimination between axes of variation.

6. *Discrimination Functions and Time Series of Fossils*

In general, this poses a type of question of particular interest in palaeontology from the point of view of discrimination methodology as it concerns the analysis of phylogenetic series and changes in a species over time. This, one might say, is the unique contribution of palaeontology to morphometrics. The question of ontogenetic series is to a certain extent analogous, and has been treated by several statistical writers, among them C. R. Rao.

We shall now consider two kinds of problems relevant to variation of organisms over time.

The first of these is one of the classical examples of multivariate analysis and we could designate it "The case of the Egyptian Skulls".

The Case of the Egyptian Skulls

A well known textbook favourite is provided by Barnard's (1935) analysis of some stratigraphical data on male skulls from Egypt, to whit:

Late Predynastic skulls
Sixth to twelfth dynasty skulls

Twelfth to thirteenth dynasty skulls
Ptolemaic skulls

The information was extracted from the published writings of three archaeologists. Seven characters were taken for analysis. The aims of the analysis were, firstly, to demonstrate how variation with time of the skull characters may be measured and, secondly, to determine, for the characters showing significant variation with time, some 'compound' able to specify any individual skull, which would show the maximum change with time. As is well known, Barnard's analysis succeeded in establishing a chronocline, not only for four "optimal" individual variables, but also for all four of them, considered simultaneously.

As in geology, archaeologists make use of the concept of stratigraphy or the "law of the superposition of strata" first formulated in 1808 by the "father of English geology", William Smith. In the above study of four samples of skulls, obtained from four distinct time points in the history of ancient Egypt, interest attached to a multivariable, quantitative treatment of the material. Rao (1952, p.266) has also taken a long look at the skull data.

Let us digress for a few moments. The oft-repeated analysis of this material has never been made with its biological nature in mind. The biologist would expect one or more of the following concepts to apply:

1) a genetical change had been occurring in the ancient population (chronologically homogeneous) of Egypt and that the linear discriminant function was constructed between samples from points on a multivariate cline (we note in passing that at least one of the samples is biologically unsatisfactory).

2) That we are dealing with racially different groups in which significant anthropomorphological differences occur and that the morphometric differences seen are a result of immigration of racially unlike populations and mixing.

3) The effects of environmental factors have been responsible for the statistical differences observed. This implies that, at different periods of time, considerable

differences in eating habits, climatic conditions and mode of life existed.

4) The skulls have "behaved" as inanimate objects and the differences are fortuitous.

We now come to the second class of problems, and in so doing, I lead you into the twilight zone of clandestine multivariate analysis.

7. *Allotting a specimen to the nearest population*

Reyment (1963a, 1963b) calculated a set of dispersion matrices, based on carapace dimensions, for chronologically separated occurrences of the Paleocene (Lower Cenozoic) ostracod *Trachyleberis teiskotensis* sampled from a borehole in Western Nigeria. In a way, the problem that had to be solved may be regarded as a variety of the case of the Egyptian skulls. Even visual inspection of the means of the variables for each sampled level showed quite unambiguously that something must have happened during the evolution of the time series of the ostracod species. A chart of discriminant scores for the samples was prepared, the basis for which was a discriminant function, constructed between the oldest and youngest samples. The essential feature of this chart is that it succeeded in bringing out more fully and clearly a definite shift in size and shape with time.

The statistically questionable nature of what was done here is that the discriminant function model was applied in a situation in which it was known that the samples being "checked for identification" did not, and could not, have come from either of the universes providing the samples upon which the discriminant function was based. Owing to the fact that the chronocline turned out to be non-linear (S-shaped), the samples lying nearest the end members of the sequence were, as regards the discriminatory scores, not their nearest neighbours in a morphometric sense, although they were the biologically most closely related evolutionary neighbours.

The palaeontologist is not merely interested in establishing the direction of change in an evolutionary sequence of this kind, but also in obtaining an exact idea of the relative magnitudes of the multivariate changes. Inasmuch as the multivariate size shifts involved at the infraspecific level over the geologically short period of time involved here could shown to be almost entirely related to shifts in the ecological stimuli, an analytical refinement suggests itself. This is to compute the principal components for each sample and to base the study on a reduced number of transformed variables, hopefully relatable to environmental effects. Some of the aspects of this kind of problem were looked at recently by Blackith and Reyment (1971, p.61). It is clearly one that would profit from the attention of professional statisticians. An alternative approach may be made by means of canonical variates, but this is less satisfactory for displaying the nature of the chronocline.

Cacoullos (1962) has examined the question of comparing generalized distances. He treated the problem of deciding to which of k populations an individual is closer, the individual not belonging to either population. This is in part an answer to the question raised in the foregoing paragraph. I have used this method (Reyment, 1963a) in order to study morphometric likeness, or nearness, in *Actinocamax verus*, an Upper Cretaceous belemnite (Class Cephalopoda), from the Russian Platform. This analysis led to a useful conclusion concerning the approximate rate of shift in size of the rostrum; a slow start, followed by an extended period of acceleration.

An important potential field of application for the nearest-neighbour model occurs in biogeography and its palaeontological counterpart, palaeobiogeography. Interest often centres around morphological differences that have developed in various parts of the area of distribution of a species. Well known examples are provided by the frogs *Rana esculenta* and *R. temporaria* in Europe and *R. pipiens* in North America. Inasmuch as a terrestrial species varies continuously over most of its area of distribution, apart from isolated populations (for example, the North African population of *esculenta*), one cannot claim the biogeographical discrimination model to be exactly analogous to the chronological discrimination model. I (Reyment, 1961) constructed a linear discriminant function between the Swedish and North African populations of

Rana esculenta, which was used to demonstrate the existence of a morphological gradient across Europe.

The two concepts of chronological and biogeographical discrimination were combined in a study I made on Paleocene ostracods in northern and western Africa (Reyment, 1966).

8. *Old friends revisited*

The idea of discriminant functions was illustrated by Fisher (1936) at its inception on measurements on four dimensions of three species of *Iris*, *I. versicolor*, *I. setosa*, and *I. virginica*, obtained from the botanist E. Anderson. The variables measured were X_1 = sepal length, X_2 = sepal width, X_3 = petal length, X_4 = petal width. This material has been "done to death" in the textbook literature, but only by statisticians. As far as I am aware, the *Iris* data has never been looked at from a basically biological point of view. Owing to the historical importance of these data, it is well worthwhile giving them a morphometric examination; that is, a statistical analysis made with biological interests to the fore. The kind of taxonomical problem represented by the *Iris* example is, of course, quite out of date.

Let us first take a look at the species *versicolor* and *setosa*.

A quick check of the univariate measures of skewness and kurtosis shows that only one result differs significantly from univariate normality, namely, that for variable 4 of *setosa*.

The dispersion matrices appear to diverge somewhat. A superficial inspection of these is not liable to tell us much about the situation, although we might suspect that the differences could arise from taxonomically significant growth patterns peculiar to each of the species.

This suspicion is further strengthened by the latent roots and vectors of the two matrices (Table 1). The first pair of latent vectors show that for *setosa*, most of the variation resides in the first two elements, corresponding to X_1 and X_2.

I am now going to allow myself a digression on the subject of the transformation of biological data in order to bring about a normalizing effect in cases, where such data show evidence of deviations from normality. I have given some thought to the rationality of this procedure and must confess that I am still uncertain about what really happens. I attempted recently to examine some aspects of this (Reyment, 1971), applying some results of Mardia (1970). As regards transformations of multivariate data, I found that the effect of logarithmically transforming the data is a mixed blessing. Apart from the fact that the investigator is removed one step from the biological relations in his data, the betterment in the multivariate interconnections as regards normality is mostly slight, despite a general tendency to improvement in the univariate values.

Thus as far as 'normalizing' data is concerned, I think I have been able to show that in the multivariate case, caution must be exercized at all times. One might think that only the non-normal variables should be normalized. This is not admissible for much taxometric work, which brings me to my second point.

In studies directed towards unravelling the relationships between morphological variables (usually the case in work proceeding in any way beyond the level of pure pigeon-holing), the variables cannot be treated differently. Apart from whatever mathematical considerations may apply, the fact remains that differential growth is an almost permanent "accompaniment" of characters locked in a growth relationship: one might say, that this is an *a priori* property of organisms, an occupational hazard.

One may think of growth in the context of "growing up", and we are all familiar with what takes place, morphologically, within our own species.

Palaeontology adds the dimension of time to the growth concept; we can also conceive a more abstract condition, that of phylogenetic growth; i.e. phylogenetic shifts in size and shape, transcending the species boundary. In this more special connotation, growth may take a negative direction, and frequently does, although in the case of the mammals, the group best known to the layman, the direction has been positive - we need only think of the phylogenetic lineages of the horses and of the pachyderms to appreciate the significance of this remark.

The concept of allometry, formalized in 1932 by Huxley, was as originally conceived, concerned with differential growth relationships between two variables. Various suggestions for generalizing the allometry concept have been made by practicians of multivariate analysis in biology. The one that seems to lie nearest the truth at the present time is that of Jolicoeur (1963). This stands or falls with the growth and shape interpretation of principal components analysis, being based on the first latent vector of the dispersion matrix of the logarithmically transformed data. Hopkins (1966) has argued along similar lines, but using a plausible factor-analytical model.

A discriminant analysis on systematic biological data must therefore be made with the thought in mind that a transformation might be in order. Although I have considered this question on several occasions, and for a wide range of organisms, I must admit that I am still uncertain about whether a general approach to the question is possible. Here is not the time and place to debate the niceties of the question and this is another field requiring the attentions of a biologically oriented mathematical statistician. However, it is not without interest to see what effects the logarithmic transformation has on a taxonomic discriminant analysis.

And now back to our flowers. I redid the kurtosis and skewness calculations on the logarithmically transformed data for *setosa* and *versicolor* and found now that for *setosa* the skewness situation had been slightly bettered, but the kurtosis position for it had, on the whole, worsened a little. The *versicolor* data were by and large in worse shape than before, significant skewness having been introduced into variable 3, where none had hitherto existed. In summary, the evidence of the *Iris* material is such as to indicate that for mere reasons of 'normalizing' the data, where these are moderately out of line, the logarithmic transformation of them does not seem to serve a very useful purpose. Palaeontological data are not uncommonly quite seriously different from normally distributed, and one would have little choice but to attempt a betterment of the distributions, if it were desired to make use of some multivariate technique, the prerequisite of which might be the requirement of multivariate normality. The bias in palaeontological material derives from postmortal sifting.

A cursory inspection of the dispersion matrices for transformed and untransformed data shows that the variation is appreciably differently apportioned for the two categories. The latent vectors of the two categories of dispersion matrices bear out this difference in a most remarkable manner. For the non-transformed material of *setosa*, the variation lies entirely with x_1 and x_2. For the transformed material, it lies entirely with x_4. The first latent vector of non-transformed *versicolor* is dominated by x_2 and x_3. For the transformed data, it is x_2, x_3 and x_4 that are dominant (Table 1).

Two things seem to arise from the inspection of the latent vectors. Firstly, that the logarithmic transformation results in a certain change in shape of the scatter ellipsoids, a change that reflects divergences in the patterns of variation of the species, as evidenced in their samples. Secondly, the differences in the corresponding elements of the latent vectors for the two species are great enough to cause one to suspect that the scatter ellipsoids have significantly different orientations. The differences in the latent roots suggests that the shapes for the two species are also quite different.

A test for homogeneity of matrices shows indeed that the dispersion matrices are different ($B^2 = 69.9$; $\beta^2 = 0.44$; d.f. = 10). The same test applied to the transformed data showed that even here we have a difference in the homogeneity of the matrices, with an even greater degree of significance ($B^2 = 113.1$; $\beta^2 = 0.44$; d.f. = 10).

Let us sum up the results obtained so far. The species *setosa* and *versicolor* do not deviate seriously from the normal distribution in respect of the four characters viewed singly, but when these are taken collectively, highly significant heterogeneity is found. This is heightened for the logarithmically transformed data. The logarithmic transformation also brings about a remarkably different apportionment of the loadings of the principal components vectors.

Using a rough-and-ready test (Reyment, 1969) based

on a result of Anderson (1963), and first outlined by Jolicoeur (1963), I tested the orientations of the four axes of the ellipsoids of scatter for the two samples. All differ highly significantly. Again, this is accentuated for the transformed observations.

And now the discriminant function. I have, in Table 2a), summarized the results I obtained for these and the corresponding D^2 by the usual calculations, and the Anderson-Bahadur (1962) method for calculating a linear discriminant function for heterogeneous dispersion matrices.

We note that the D^2, using the Anderson-Bahadur method, is somewhat larger for both sets of data and that the logarithmic transformation not only gives larger values of D^2, but also a different emphasis to the weighting coefficients of the discriminant functions.

The coefficients of the discriminant function as found by Kendall and Stuart (1966, p.319) are:

(-3.1, -18.4, 22.2, 31.5), which are, with the coefficient of x_1 as unity, -1.0, -5.9, +7.1, +10.1. The cut-off point for assigning specimens to *setosa*, respectively, *versicolor*, was calculated to be 14.2.

The slight difference between these values and those of Table 2, are probably due to rounding off in the matrix inversion. My calculations were done on a CDC 3600 computer. Rao (1952) obtained (rounded) (-3.1, -18.0, 21.8, 30.8) with $D^2 = 103.2$.

The species *setosa* and *virginica* were analysed in the same way as for *setosa* and *versicolor*. The univariate statistics for the latter show the data to conform with the requirements of the normal distribution. Also, here, the logarithmic transformation produces significant changes in the elements of the latent vectors of the dispersion matrices, and the multivariate test of homogeneity of dispersion matrices gives a highly significant result for heterogeneity. All axes of the dispersion matrices are significantly differently oriented. Here, as before, the biologist would take this as being indicative of different patterns of growth in the four variables.

The results are summarized in Table 2b). Again, the coefficients for the two discriminants are slightly different for the raw data and very different for the transformed data. The generalized distances are in both cases greater for the heterogeneity hypothesis. The fact that the results for the transformed observations deviate so strongly, suggests to me that there are indeed complicated growth patterns in the material, worthy of a detailed quantitative analysis, based preferably on a more comprehensive suite of variables.

Our old friends, the *Iris* species, are thus not a straightforward and uncomplicated as would appear. What may be taken as a rather simple problem from the point of view of a discriminant analysis becomes, when biological considerations are included, far more complex. The analytical approach I have used in disclosing an interesting biological situation is crude, and I am the first to admit it. Here is clearly a field for the activity of the professional mathematician, who would also be able to treat the geometrical aspects of the problem more rigorously and perhaps the discriminatory possibiliiies that would appear to lie with the dispersion matrices.

Finally, we shall take a look at the comparison between *versicolor* and *virginica*. The corresponding elements of the latent vectors of the raw data are close indeed for this kind of material (*versicolor*: 0.7, 0.3, 0.6, 0.2; *virginica*: 0.7, 0.2, 0.6, 0.1), although the subsequent vectors display quite different patterns. The dispersion matrices are not equal, statistically speaking; the first axes do not differ significantly in orientation, but all the others do. The logarithmically transformed data do not differ in the results yielded by them for the latent vectors, except with respect to the first principal axis, which is differently oriented as well. The heterogeneity test for the dispersion matrices does not quite reach significance.

The respective discriminant functions are given in Table 2c). Summarizing these results we find that the first two combinations produced discriminant functions well capable of a complete separation of the two sets of data into discrete heaps. The present combination is less successful in discriminating between the individuals of the two species and there is appreciable overlap and some misidentification. This is to be expected, as the squared generalized distances are much less for this pair than for the others.

In the discussion terminating his paper, Fisher (1936, p.182) used the word 'misclassification', when 'misidentification' was intended. He then went on to apply his method to the concept of allopolyploidy, noting that botanists had shown that *setosa* is diploid, *virginica* is tetraploid and *versicolor* is hexaploid. The problem to be evaluated was whether *versicolor* could be a polyploid hybrid of the two other species. He found that *I. setosa* and *I. versicolor* do not overlap, the values for *virginica* and *versicolor* do, as was seen from my analysis of the data (see also Fisher, 1936, Fig.1). One of the conclusions arising from Fisher's analysis, and one to which one may readily subscribe, is that the four measurements chosen for taxonomic study do not give a very adequate basis for discrimination.

Kendall (1957) illustrated his chapter on discriminant functions by analysing the data on *setosa* and *versicolor*. He obtained the discriminate function

$$Y = x_1 + 5.9037x_2 - 7.1299x_3 - 10.1036x_4 .$$

The *Iris* data were also taken up by Kendall and Stuart (1966, p.317), partly repeating the original presentation of Fisher (1936), and partly demonstrating the size-shape discrimination concept for equally correlated variables (Penrose, 1947). The same data were also used to illustrate a fact not infrequently overlooked in taxonomical studies, namely, that most of the discriminatory power of a discriminant function may lie with one variable alone. For our data on *setosa* and *versicolor*, this happens to lie with petal width. Rao (1952, p.247) also used *setosa* and *versicolor* to illustrate the method of calculating a discriminant function and for testing the differences between the means for significance.

Reyment (1969b) employed the Anderson-Bahadur (1962) procedure for discriminant functions with heterogeneous dispersion matrices for studying the morphometrical manifestations of sexual dimorphism in living vertebrates and invertebrates.

The same concept of shape and size discrimination was illustrated by Kendall (1957) with reference to the well known data on the herring, analysed by Penrose (1947), in addition to the irises. This is clearly often a useful approach, but one that hardly meets the requirements of a full taxonomic treatment of size and shape. But I think this is rather obvious.

9. *Burnaby's Problem: The question of growth invariance*

Burnaby (1966a) treated a problem of central biological importance in biological systematical applications of discriminant functions. The question he studied concerned growth invariance and the bearing of this on discriminant functions and generalized distances. The need for a general approach to the analysis of multivariate growth had been brought to Burnaby's attention by his work on molluscs, which do not have a terminal growth size, but in which growth continues throughout life. Growth is not the only nuisance factor in morphometric work and some environmental components may exert an influence on the shape of the hard parts, as is the case in several groups of bivalved molluscs. Burnaby solved the problem by producing a method of computation of a discriminant function orthogonal to vectors representing variation which is extraneous to the desired comparisons. That the method has not caught on is probably an outcome of the fact that the "nuisance vectors" are in practice very difficult to estimate. Rao (1966) elaborated and generalized on the problem brought to light by Burnaby.

10. *An example: Evolution in Upper Cretaceous echinoids*

To round off my presentation, I wish to give here a brief analysis of an actual example, drawn from the realm of quantitative systematic palaeontology. It is in itself not a particularly remarkable case study, although it certainly does not lack interest, but I think it serves its purpose well, namely, that of providing an insight into the way in which palaeontologists think when dealing with a problem of this kind. The data studied were taken from a publication by Nichols (1959).

Nichols (op. cit.) studied the variation in four major characters of the test of species of the Cretaceous genus of sea urchins, *Micraster*. This species complex yields an excellent example of continuous evolution.

Palaeoecological studies have brought to light that these morphological changes seem to be independent of environmental conditions registered in the sediments. It was suggested by Nichols that the morphological variation in the echinoids was either due to a change in the ecological niche inhabited by the organism, or a bettered adaptive response of the animal to its niche. Comparative laboratory studies made by Nichols on living echinoids provided evidence in support of the first alternative, that is, that there was a gradual adaptation in shape, correlated with a change in niche, with the urchins tending to burrow more deeply into the substrate as time progressed. A sideline of this evolutionary trend produced a form that ploughed through the sediment rather than burrowed downwards in it.

Nichols measured four characters: length, breadth and height of test, and the number of tube feet, as gauged from the ambulacral zones.

This example is useful in another respect. There are no convenient breaks in the sequence to allow the delineation of "species". We are instead faced with a continuously varying complex, with intermediates between any two chosen categories. This brings home the fundamental difference between palaeontological and neontological data in relation to statistical thinking. A neontological sample is drawn from a biological population with only one spatial dimension (in the normal sense), namely that of geographical dispersion. A palaeontological sample, has in addition to this 'dimension', a second one of chronological dispersal. This is not the place to take up the question of the averaging effect of mixed chronological samples and the distortional effect of geological agencies, but I think it is necessary to make clear this difference in the nature of the two kinds of material. It is hoped that the statistical implications of this will eventually attract the interest of professional statisticians, and a start has indeed been made by Burnaby (1966a) and Rao (1966).

For convenience, recent taxonomic revisions (biometrically based) of the micrasters have been grouped around a few arbitrarily chosen points in the evolutionary sequence. The most noticeable average changes in the test in the evolution of the micraster complex are that it becomes relatively broader and the site of maximum broadness tends to shift posteriorly, together with the point of maximum height.

Blackith and Reyment (1971, p.254) re-analyzed Nichol's data, using standard methods of multivariate analysis. A principal coordinates analysis of part of the material indicates a high level of likelihood for evolutionary connections between the species *Micraster cortestudinarium*, *M. coranguinum*, and *M. corbovis*, but that this is in the form of a linkage *corbovis-cortestudinarium-coranguinum*, without evidence for a linkage between *coranguinum* and *corbovis*. The ruling theory about the micrasters postulates the existence of intermediate categories. The canonical variates analysis made by Blackith and Reyment (1971, p.255) did not come up with convincing support for this line of thought and the 'intermediates' are, on the basis of the measurements employed (there is every reason to believe that the characters measured are inadequate), not located where they should be in the plots of the canonical means.

I have here pursued certain avenues of enquiry that were opened up by the previously conducted analyses of the echinoid data. It is of interest to see to what extent the variability of the various taxonomic categories differ from each other on the basis of the characters measured. A logical approach to this question is by means of the discriminant function. The discriminant functions have been calculated between critical pairs of samples. Owing to pronounced curvilinearity in the data, it was considered prudent to work on their logarithmically transformed values, notwithstanding the remarks made earlier on in this text.

The species considered in the present study are from the Upper Cretaceous of the south of England; they are:

Micraster corbovis, *M. cortestudinarium*, and *M. coranguinum*.

Normality of the data: The data of both species *Micraster cortestudinarium* and *M. corbovis* are normally distributed for the four variables. Variable three of *M. coranguinum* deviates from normality both with respect to skewness as well as to kurtosis.

Comparison between M. coranguinum and M. corbovis:

The dispersion matrices differ significantly on the basis of the usual test of homogeneity and the directions of the first two principal axes differ significantly. The lengths of the third and four vectors differ hardly from nought. There is slight overlap in the histograms of the plot of the transformed observational vectors for both the heterogeneous and homogeneous discriminant functions, with the heterogeneous compound giving a slightly cleaner result and a larger value of the generalized distance. These values are given in Table 3.

Comparison between M. coranguinum and M. cortestudinarium:

Here, again, there is pronounced significant heterogeneity in the dispersion matrices, both with respect to the degree of inflation of scatter ellipsoids and their relative orientations. There is strong overlap in the plot of the transformed observational vectors and neither of the discriminant functions (see Table 3) is efficient in keeping the individuals of the two species apart.

Comparison between M. cortestudinarium and M. corbovis:

The tests for homogeneity of the dispersion matrices of these two samples show that the scatter ellipsoids agree in both the degree of inflation and in orientation. The discriminant function is entirely unsuccessful in distinguishing between all but extreme members of the samples (Table 3).

Analytic Remarks: The three species considered here are taken by specialists of the Cretaceous echinoids to be closely related and, as shown by the principal coordinates study of the data (Blackith and Reyment, 1971, p.254), the direction of evolution seems to have been from *corbovis* to *cortestudinarium* to *coranguinum*. It is now in order to enquire whether this is reflected in the discriminant function study.

The analysis indicates that *cortestudinarium* and *corbovis* are morphologically close, so close as to invalidate the use of a discriminant function for the purpose of identification, and the dispersion matrices are homogeneous. *M. cortestudinarium* and *M. coranguinum* have heterogeneous covariance matrices, but the discriminant functions do not succeed in correctly identifying more than the more extreme individuals of both samples. This is, of course, as it should be if the evolutionary model is correct. The acid test is provided by the comparison between the end numbers of the evolutionary series, *Micraster coranguinum* and *M. corbovis*. The palaeobiological theory requires that these two species should be the most unlike of the three; one would expect the greatest generalized distance to occur between them, their dispersion matrices to be the most different in shape and orientation, and for the heterogeneous discriminant function to do a reasonably efficient job of identification. And this is indeed so.

In summary, the discriminant analysis of the echinoids might not be particularly interesting to a statistician, but to the palaeontologist, the result is a huge success. One of the enlightening features of the study is, for the biologist, the fact that it not only reaffirms the indications yielded by other methods of analysis, but also provides a reasonably coherent picture of the relative characteristics of the growth patterns.

REFERENCES

Anderson, T. W. (1963), Asymptotic theory for principal components analysis, *Ann. Math. Statist.*, 34, 122-148.

Anderson, T. W. and Bahadur, R. R. (1962), Two sample comparisons of dispersion matrices for alternatives of immediate specificity, *Ann. Math. Statist.*, 33, 420-431.

Barnard, M. M. (1935), The secular variation of skull characters in four series of Egyptian skulls, *Ann. Eugenics*, 6, 352-371.

Blackith, R. E. and Reyment, R. A. (1971), *Multivariate Morphometrics*, Academic Press: London.

Burnaby, T. P. (1966a), Growth-invariant discriminant functions and generalized distances, *Biometrics*, 22, 96-110.

Burnaby, T. P. (1966b), Distribution-free quadratic discriminant functions in Palaeontology, *Computer Ap-*

plications in the Earth Sciences, *Comp. Contr. Geol. Surv.*, Kansas, 7, 70-76.

Cacoullos, T. (1962), Comparing Mahalanobis distances, Doctoral dissertation, Columbia University, New York.

Cacoullos, T. (1965), Comparing Mahalanobis distances, I: Comparing distances between k known normal populations and another unknown, *Sankhyā* A, 27, 1-22; II: Bayes procedures when the mean vectors are unknown, *Sankhyā* A, 27, 23-32.

Fisher, R. A. (1936), The use of multiple measurements in taxonomic problems, *Ann. Eugenics*, 7, 179-188.

Gower, J. C. (1966), Some distance properties of latent roots and vectors methods used in multivariate analysis, *Biometrika*, 53, 325-338.

Hopkins, J. W. (1966), Some considerations in multivariate allometry, *Biometrics*, 22, 747-760.

Huxley, J. S. (1932), *Problems of Relative Growth*, Methuen and Co.: London.

Jolicoeur, P. (1963), The degree of generality of robustness in *Martes americana*, *Growth*, 27, 1-27.

Kendall, M. G. (1957), *A Course in Multivariate Analysis*, Griffin: London.

Kendall, M. G. and Stuart, A. (1966), *The Advanced Theory of Statistics*, Vol.3: Design and Analysis, and Time Series, Griffin: London.

Lyubischev, A. A. (1959), The application of biometrics in taxonomy, *Vestnik Leningr. Univ.*, 14, 128-136.

Mardia, K. (1970), Measures of multivariate skewness and kurtosis with applications, *Biometrika*, 57, 519-530.

Morrison, D. F. (1967), *Multivariate Statistical Methods*, McGraw Hill Inc.

Mound, L. A. (1963), Host-correlated variation in *Bemisia tabaci* (Gennadius) (Homoptera: Aleyrodidae), *Proc. Roy. Ent. Soc. London* A, 38, 171-180.

Nichols, D. (1959), Changes in the Chalk heart-urchin *Micraster* in relation to living forms, *Phil. Trans. Roy. Soc. London*: Ser. B, 693, 242, 347-437.

Penrose, L. S. (1947), Some notes on discrimination, *Ann. Eugenics*, 13, 228-237.

Rao, C. R. (1952), *Advanced Statistical Methods in Biometrical Research*, Wiley: New York-London.

Rao, C. R. (1966), Discriminant functions between composite hypotheses and related problems, *Biometrika*, 53, 339-345.

Reyment, R. A. (1961), A note on geographical variation in European *Rana*, *Growth*, 25, 219-227.

Reyment, R. A. (1963a), Paleontological applicability of certain recent advances in multivariate statistical analysis, *Geol. Fören. Stockh. Förh.*, 85, 236-265.

Reyment, R. A. (1963b), Studies on Nigerian Upper Cretaceous and Lower Tertiary Ostracoda, II, Danian, Paleocene and Eocene Ostracoda, *Stockh. Contr. Geol.*, 10, 1-286.

Reyment, R. A. (1966), Studies on Nigerian Upper Cretaceous and Lower Tertiary Ostracoda, III, Stratigraphical, Paleoecological and Biometrical Conclusions, *Stockh. Contr. Geol.*, 14, 1-142.

Reyment, R. A. (1969a), A multivariate paleontological growth problem, *Biometrics*, 25, 1-8.

Reyment, R. A. (1969b), Some case studies in the statistical analysis of sexual dimorphism, *Bull. Geol. Instn. Univ. Uppsala NS1*, 4, 97-119.

Reyment, R. A. (1971), Multivariate normality in morphometric analysis, *Math. Geol.*, 3, 4, 357-368.

Reyment, R. A. and Brännström, B. (1962), Certain aspects of the physiology of *Cypridopsis* (Ostracoda: Crustacea), *Stockh. Contr. Geol.*, 9, 207-242.

Reyment, R. A. and Ramdén, H. A. (1970), Fortran IV program for canonical variates analysis for the CDC 3600 computer, *Comp. Contr. Geol. Surv. Kansas*, 47, 1-39.

Stower, W. J., Davies, D. E. and Jones, I. B. (1960), Morphometric studies of the Desert Locust *Schistocerca gregaria* (Forsk.), *J. Anim. Ecol.*, 29, 309-339.

Table 1. Latent vectors for *Iris* species.

setosa						*versicolor*					
NT			T			NT			T		
I	II	III	I	II	III	I	II	III	I	II	III
0.7	-0.6	0.4	0.0	-0.4	-0.2	0.7	0.7	-0.3	0.3	0.3	0.6
0.7	0.6	-0.3	0.1	-0.7	-0.5	0.3	-0.6	-0.7	0.4	-0.8	0.3
0.1	-0.5	-0.8	0.1	-0.6	0.8	0.6	-0.3	0.6	0.5	0.5	0.2
0.1	-0.5	-0.2	1.0	0.1	0.0	0.2	-0.3	0.1	0.7	0.1	-0.7

I, II, III = first, second and third latent vectors

The final latent root is almost zero.

NT = non-transformed; T = transformed.

Table 2. The discriminant functions and D^2 for *Iris* species.

a) For *setosa-versicolor*

	Raw data					Transformed data				
	X_1	X_2	X_3	X_4	D^2	X_1	X_2	X_3	X_4	D^2
Normal method	-3.1	-18.0	21.8	30.8	103.2	-25.7	-125.8	201.7	28.0	124.8
Anderson-Bahadur	-2.3	-16.5	26.3	32.4	108.4	-62.8	-133.3	220.5	35.9	130.4

b) For *setosa-virginica*

Normal method	-15.8	-12.0	36.5	36.8	195.2	-89.3	-93.9	289.6	38.4	199.8
Anderson-Bahadur	-14.6	-10.6	41.9	45.6	209.7	-207.6	-104.9	389.7	62.8	225.8

c) For *versicolor-virginica*

Normal method	-3.6	-5.6	7.0	12.4	14.2	-39.9	-37.8	62.6	47.6	13.07
Anderson-Bahadur	-3.4	-5.7	6.9	12.6	14.2	dispersion matrices homogeneous				

Table 3. Discriminant functions for 3 species of echinoids from southern England

	X_1	X_2	X_3	X_4	D^2	DF
coranguinum-corbovis						
heterogeneous	377.7	-318.6	-24.2	-40.7	19.0	57
homogeneous	322.8	-275.7	-19.2	-26.2	17.7	
cortestudinarium-coranguinum						
heterogeneous	-226.1	153.7	23.4	66.4	8.3	57
homogeneous	-196.6	128.1	21.1	59.8	7.7	
cortestudinarium-corbovis						
homogeneous	131.5	-99.6	-16.2	4.6	3.8	57

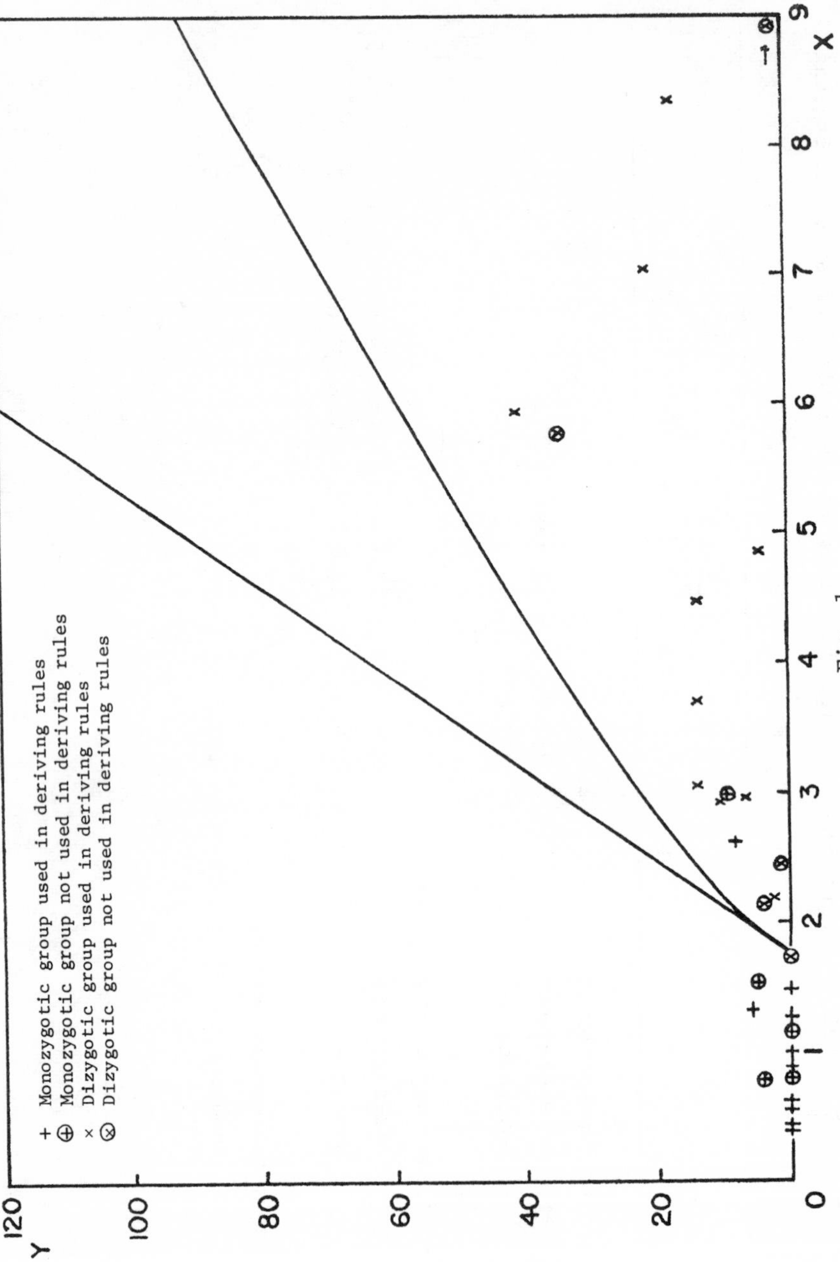
+ Monozygotic group used in deriving rules
⊕ Monozygotic group not used in deriving rules
× Dizygotic group used in deriving rules
⊗ Dizygotic group not used in deriving rules
Y
120
100
80
60
40
20
0
1
2
3
4
5
6
7
8
9
X

Fig. 1

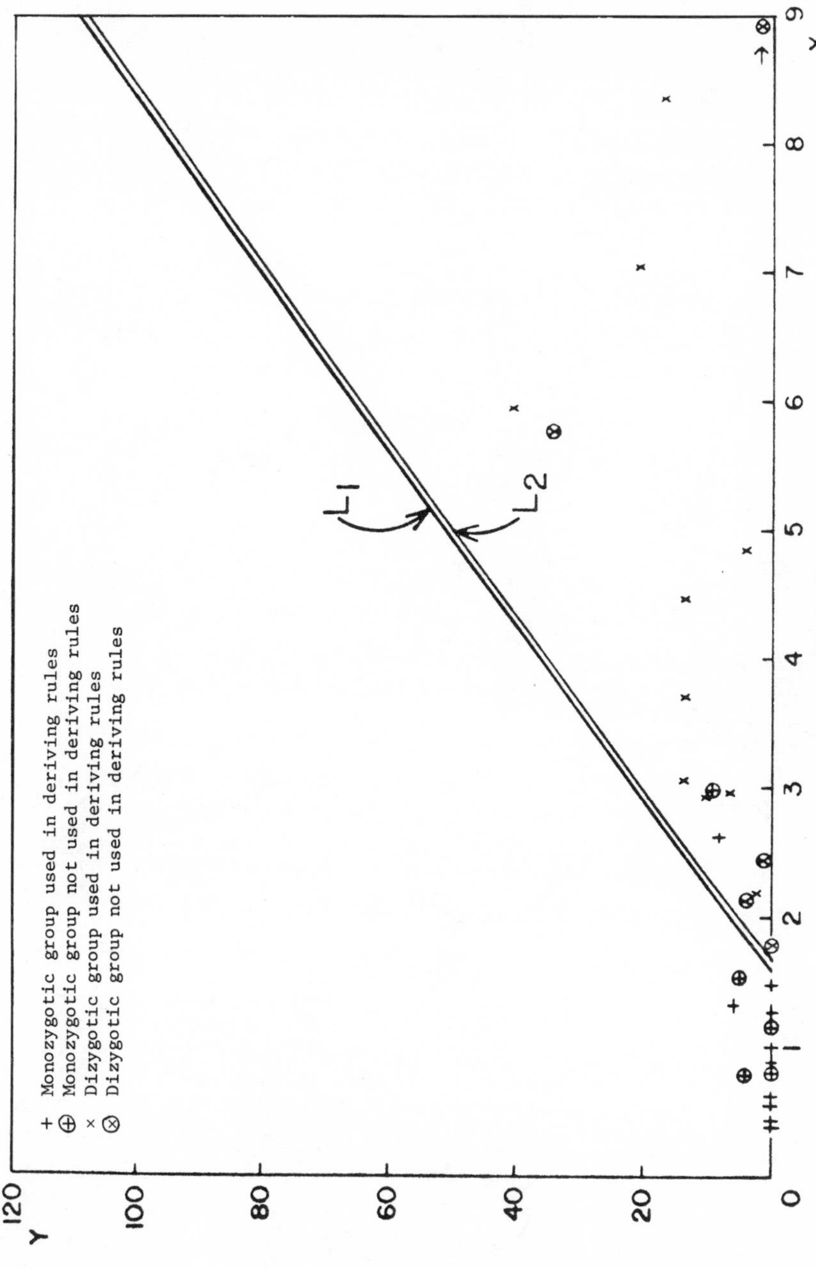

+ Monozygotic group used in deriving rules
⊕ Monozygotic group not used in deriving rules
× Dizygotic group used in deriving rules
⊗ Dizygotic group not used in deriving rules
L1
L2
Y
X

Fig. 2

CLASSIFYING WHEN POPULATIONS ARE ESTIMATED

Willem Schaafsma
University of Groningen

1. *Introduction*

Fisher (1936) introduced the discriminant function for distinguishing between two multivariate normal distributions with identical variance-covariance matrices. His motivation was archaoelogical: he was consulted by Barnard who was trying to classify skeletal remains.

Wald (1944) made an impressive study of the distribution of the discriminant function which is obtained when unknown parameters are replaced for estimators. Subsequent contributions were made by Anderson (1951), Sitgreaves (1952), Rao (1954), John (1961) and many others. Most of these contributions deal with distribution theory for statistics which arise when estimators replace unknown parameters. Anderson derived the generalized likelihood-ratio and proposed this as a suitable statistic.

I became interested in the subject because one of my clients, Dr. G.N. van Vark (1970), was studying prehistoric human skeletal material. He was especially interested in sex-diagnosis for reasons whatsoever. Deviating from van Vark's problems, I became interested in the following problems which may be regarded as ultimate simplifications of the general multivariate assignment problems.

Problem 1. On the basis of the outcome of $X_0, X_1, \ldots, X_n$ an individual has to be classified into (assigned to) one of two populations: a standard population π_1, and a non-standard population π_2. X_0 describes the score of the individual under classification. If this individual is from π_1, then X_0 has the univariate $N(0,1)$ distribution (we assume that the distribution of our measurement in π_1 is known and normal, X_0 is obtained by standardizing). If the individual is from π_2, then X_o has the $N(\mu,1)$-distribution where μ is unknown, $\mu \in (-\infty,+\infty)$. $X_1,\ldots,X_n$ is an independent random sample from $N(\mu,1)$, they describe the scores for n individuals, drawn at random from π_2.

Is it possible to study this problem from the Neyman-Pearson point of view where an error of the first kind is committed when the individual is classified into π_2 whereas it actually comes from π_1? Is the usual estimator $X = (X_1+\ldots+X_n)/n$ for μ the best possible one?

Problem 2. On the basis of the outcome of $X_o,X_1,\ldots,X_{n_1+n_2}$ an individual has to be classified into one of two populations, π_1 and π_2. X_o describes the score of the individual that has to be classified, X_o has the $N(\mu_i,1)$ distribution if the individual is from π_i $(i=1,2)$. $X_1,\ldots,X_{n_1}$ is an independent random sample from $N(\mu_1,1)$ and $X_{n_1+n_1},\ldots,X_{n_1+n_2}$ from $N(\mu_2,1)$. If the two populations are quite dissimilar, then it might be of interest to study this problem also from the Neyman-Pearson point of view (supposing again that π_1 is a standard population and that we wish to control the probability of an error of the first kind). We shall study also some loss-function formulations for Problem 2 and we try to get rid of the assumption $\sigma^2 = 1$ (Section 8).

Remark. It might help the reader to know which results are claimed to be "new" or "important". If the Neyman-Pearson formulation is used, then we shall construct a *uniformly most powerful invariant similar size-*α *test* ϕ^*, both for Problem 1 (Section 3) and for Problem 2 (Section 5, see the sentence before the remark). Apparently ϕ^* appeared for the first time in Schaafsma (1971).

If the symmetric loss-function (6.1) is used, then the problems exhibit so much symmetry that we can construct a *uniformly minimum risk invariant (behavioral) rule* δ^*, both for Problem 1 (Section 6), for Problem 2 (Section 7) and, what is very important, for a modification of Problem 2 which is obtained by letting *the common variance* σ^2 *be unknown* (Section 8). Procedure δ^* corresponds to the classical rule, based on the "plug-in-method", only if $n_1 = n_2$. I was wondering whether δ^* belongs to statistical folklore. Professor Somesh Das Gupta referred me to his paper (1965), the thesis of his student Albert Kinderman (1972) and to Kudô's (1959, 1960). It turned out that, also in the $n_1 \neq n_2$ case, δ^* indeed belongs to statistical folklore though it can still be claimed that this paper is the first one where δ^* is proved to be U.M.R. invariant. The machinery of Section 4 and the rotation (3.3) show "more symmetry" in the problem than what is indicated in these references.

2. *Problem 1 reformulated.*

The outcome of $X_0, X_1, \ldots, X_n$ is observed in order to take an action from $A = \{a_1, a_2\}$ where a_i means that the individual to be classified (that with score x_0) is assigned to population π_i (i=1,2). $\Theta = \{0,1\} \times R$ consists of all $\theta = (\delta, \mu)$ where $\delta = 0$ if our individual is from π_1 and $\theta = 1$ if he comes from π_2. The corresponding probability distribution P_θ is determined by the n+1-dimensional p.d.f.

$$p_\theta(x_0, \ldots, x_n) = (2\pi)^{-\frac{1}{2}(n+1)} \exp[-\frac{1}{2}(x_0 - \delta\mu)^2 + \sum_{i=1}^{n}(x_i - \mu)^2\}]. \tag{2.1}$$

Obviously $(X_0, \sum_{i=1}^{n} X_i)$ is a sufficient statistic (Factorization Criterion). Introducing

$$X = X_0 \ ; \quad Y = \sum_{i=1}^{n} X_i / n^{\frac{1}{2}}$$

we also have that (X,Y) is sufficient; if $\theta = (\delta, \mu)$ then X and Y are independent, X having the $N(\delta\mu, 1)$ and Y the $N(\mu n^{1/2}, 1)$ distribution. We shall construct tests for testing $H : \delta = 0$ against $A : \delta = 1$ on the basis of the outcome of (X,Y).

If (X,Y) were a complete sufficient statistic, then minimality would follow and $\overline{X} = Y/n^{1/2}$ would be a U.M.V. U. estimator for μ.

Unfortunately, (X,Y) is not a complete sufficient

statistic. Nevertheless it is a minimal sufficient or necessary and sufficient statistic (my proof is based on Dynkin, 1951). $\overline{X}$ will not be a U.M.V.U. estimator for μ: the class of unbiased estimators based on (X,Y) does not contain an element which is uniformly best (my proof only works for n=1 and n=3). Nevertheless $\overline{X}$ is admissible and has minimax risk with respect to squared error loss. (Proofs in the report mentioned in the Acknowledgements).

3. *Construction and comparison of some level-α tests for Problem 1.*

We apply reduction by sufficiency and describe Problem 1 in terms of the minimal sufficient statistic (X, Y). In order to make the theory of this section also applicable in a later stage (see Section 5), we reparametrize by taking $\theta = (\theta_1, \theta_2)$ where $\theta_1 = E_\theta(X)$ and $\theta_2 = E_\theta(Y)$. By doing so we obtain the following testing problem.

X and Y are independent, X has the $N(\theta_1,1)$ distribution and Y the $N(\theta_2,1)$ distribution. It is known that $\theta = (\theta_1,\theta_2) \in \Theta = \Theta_o \cup \Theta_1 \cup \Theta_2$ where $\Theta_o = \{(0,0)\}$, $\Theta_1 = \{\theta;\ \theta_1 = 0;\ \theta_2 \neq 0$, $\Theta_2 = \{\theta;\ \theta_1 \cos\psi = \theta_2 \sin\psi \neq 0\}$ where $\psi = \text{arc cotg } n^{1/2}$, n denoting the original sample size. Note that $\theta \in \Theta_1$ indicates that the individual to be classified belongs to π_1. We shall construct level-α tests for testing $H : \theta \in \Theta_o \cup \Theta_1$ (our individual belongs to π_1) against $A : \theta \in \Theta_2$ (it belongs to π_2).

Wald's test. First we derive a level-α test along the lines of Wald (1944) who was dealing with the much more complicated multivariate assignment problem. If μ were known, the Neyman-Pearson Fundamental Lemma shows that for testing the simple hypothesis $\theta = (0, \mu n^{1/2})$ against the simple alternative $\theta = (\mu, \mu n^{1/2})$, one should reject iff $x > u_\alpha$ in case $\mu > 0$, and iff $x < -u_\alpha$ in case $\mu < 0$. Here u_α is such that $P(U > u_\alpha) = \alpha$ if U has the N(0,1)-distribution. Substituting the obvious (see Section 2, Questions 3 and 4) estimator $Y/n^{1/2}$ for μ, we arrive at what will be called Wald's test $\phi_W^{(1)}$ (see Figure 1).

$$\phi_W^{(1)} = I_{\{(x,y)\ ;\ x>u_\alpha,\ y>0\}} + I_{\{(x,y);\ x<-u_\alpha,\ y<0\}}. \quad (3.1)$$

where of course I_A denotes the indicator-function of set $A \subset R^2$. One observes that $\phi_W^{(1)}$ is *similar* size-α:

$$E_\theta\{\phi_W^{(1)}(X,Y)\} = \alpha \qquad \text{for all} \quad \theta \in \Theta_o \cup \Theta_1. \quad (3.2)$$

Anderson's L.R. test. Next we apply the generalized likelihood-ratio principle (Anderson studied the multivariate assignment problem from this point of view). Introduce

$$\begin{pmatrix} T_1 \\ T_2 \end{pmatrix} = \begin{pmatrix} \cos\frac{1}{2}\psi & -\sin\frac{1}{2}\psi \\ \sin\frac{1}{2}\psi & \cos\frac{1}{2}\psi \end{pmatrix} \begin{pmatrix} X \\ Y \end{pmatrix}, \quad (3.3)$$

a rotation over $\frac{1}{2}\psi$. In terms of (T_1,T_2) our testing problem becomes as follows. T_1 and T_2 are independent $N(\tau_1,1)$ and $N(\tau_2,1)$ respectively where $\tau = (\tau_1,\tau_2)$ is restricted to the two lines making angles $\frac{1}{2}\psi$ with the t_2, τ_2-axis. The hypothesis H has to be tested that τ belongs to the line determined by $\tau_1 \cos\frac{1}{2}\psi + \tau_2 \sin\frac{1}{2}\psi = 0$. The alternative A states that τ belongs to the other line which is defined by $\tau_1 \cos\frac{1}{2}\psi = \tau_2 \sin\frac{1}{2}\psi$ $(\neq 0)$. Elementary computations show that the L.R. principle leads to the following test

$$\phi_A = I_{\{(t_1,t_2);\ t_1 t_2 \geqq c_\alpha\}}$$

where c_α is to be determined such that the test becomes of size-α: $\sup_\tau P_\tau\{T_1T_2 \geqq c_\alpha\} = \alpha$ when τ varies over the line determined by H.

Case $\alpha = .50$ *(see Fig. 2).* Now ϕ_A rejects iff

$\{T_1T_2 \geq 0\}$ and it is easily seen that we obtained an *unbiased* size-α test. Remark that $\phi_A \neq \phi_W^{(1)}$ in case $\alpha = .50$.

Case $\alpha < .50$ *(see Fig. 3).* Studying $P_\tau\{T_1T_2 \geq c_\alpha\}$ as a function of τ over the line determined by H, we found (numerically) that, unfortunately, the maximum is not arrived in $\tau = (0,0)$: a lot of computations are necessary in order to find c_α and ϕ_A is not unbiased size-α because the power will be smaller than α in a neighbourhood of (0,0) (I only made computations for cases like $\alpha = .05$ and $\alpha = .01$).

A family of tests ϕ_χ. It is not difficult to propose other tests. In Fig. 4 tests ϕ_χ are introduced. For these tests one can prove easily that

$$E_{(0,\theta_2)}\{\phi_\chi(X,Y)\} , \tag{3.4}$$

considered as a function of θ_2, assumes its maximum for $\theta_2 = 0$. Thus it is pretty easy to make ϕ_χ a size-α test by computing $d_\alpha = d_\alpha(\chi)$. It is obvious that it will depend on ψ whether ϕ_χ is unbiased. If ψ is very small, then this will certainly not be the case.

The uniformly most powerful invariant similar size-α *test* ϕ^*. Until now, no satisfactory optimum theory was presented. Now we give a test ϕ^* which is U.M.P. among all tests ϕ that are *both similar size-*α, i.e.

$$E_\phi\{\phi(X,Y)\} = \alpha \quad \text{for all} \quad \theta \in \Theta_0 \cup \Theta_1 \tag{3.5}$$

and invariant in the sense that

$$\phi(x,y) = (-x,-y) \quad \text{for all} \quad (x,y) \in R^2. \tag{3.6}$$

The test ϕ^* was derived in Schaafsma (1971) and rejects with probability 1 iff

$$\begin{aligned} &\Phi(|x + y \operatorname{cotg} \psi| - y \operatorname{cotg} \psi) \\ &- \Phi(-|x + y \operatorname{cotg} \psi| - y \operatorname{cotg} \psi) \geq 1 - \alpha , \end{aligned} \tag{3.7}$$

where Φ denotes the cumulative distribution function of the N(0,1)-distribution.

In the remaining of this paper we only use the fact that there *exists* such a U.M.P. invariant similar size-α test ϕ^*.

Question 1. Is Wald's test $\phi_W^{(1)}$ admissible? No, because $\phi_W^{(1)}$ satisfies (3.5) and (3.6) and thus $\phi_W^{(1)}$ is uniformly less powerful than ϕ^*. This even holds if $\alpha = .50$. Nevertheless the difference might be small. Indeed, Fig. 6 shows that the maximum shortcoming of $\phi_W^{(1)}$ with respect to ϕ^*, that is,

$$\sup_{\theta\varepsilon\Theta_2} \{\beta_{\phi^*}(\theta) - \beta_{\phi_W^{(1)}}(\theta)\} ,$$

is very small, especially if n is large.

Question 2. Is Anderson's test ϕ_A *admissible*? This test is invariant but not similar. I compared the power properties of ϕ_A and ϕ^* numerically and found for $\alpha = .05$ etc. that ϕ^* is uniformly more powerful than ϕ_A.

It is interesting (and relevant for ϕ_A) to remark that the ordering

$$\phi,\phi' \;\varepsilon\; \Phi_\alpha,\; \phi \leqq \phi' \Leftrightarrow \beta_\phi(\theta) \leqq \beta_{\phi'}(\theta) \quad (\theta \;\varepsilon\; \Theta_2)$$

in the class Φ_α of all level-α tests, is not the only natural one. One might also define $\phi \leqq \phi'$ iff not only $\beta_\phi \leqq \beta_{\phi'}$ but also

$$E_\theta\{\phi(X,Y)\} \geqq E_\theta\{\phi'(X,Y)\} \qquad (\theta \;\varepsilon\; \Theta_0 \cup \Theta_1).$$

In this sense it is not attractive for a test to be similar size-α; unless pairs (ϕ,ϕ') are ordered, ϕ^* is no longer better than ϕ_A. It follows from the theory in Section 6 and 7 that ϕ_A is admissible in the sense of the latter ordering if $\alpha = .50$.

4. Problem 2 reformulated.

The parameter space may be chosen such that $\Theta = \{1,2\} \times R \times R$ consists of all $\theta = (\delta,\mu_1,\mu_2)$ where $\delta = i$ if the individual to be classified (that with score X_0) is from π_i (i=1,2). If $\theta = (\delta,\mu_1,\mu_2)$, then the value $p_\theta(x_0,\ldots,x_{n_1+n_2})$ of the p.d.f. p_θ of $(X_0,\ldots,X_{n_1+n_2})$ in $R^{n_1+n_2+1}$ is determined by

$$(2\pi)^{-\frac{1}{2}(n_1+n_2+1)} \times \exp[-\frac{1}{2}\{(x_0-\mu_\delta)^2+\sum_{i=1}^{n_1}(x_i-\mu_1)^2 + \sum_{i=n_1+1}^{n_1+n_2}(x_i-\mu_2)^2\}] \tag{4.1}$$

which results in that (X_0,U,V) is a necessary and sufficient (not complete-sufficient) statistic (see Section 2) where

$$U = \sum_{i=1}^{n_1} X_i/n_1^{\frac{1}{2}} \quad ; \quad V = \sum_{i=n_1+1}^{n_1+n_2} X_i/n_2^{\frac{1}{2}} .$$

Observe that the sufficient statistic (X_0,U,V) is chosen in such a way that X_0,U and V are independent normal with variance 1. If C is an arbitrary *orthogonal* matrix, then

$$\begin{pmatrix} X \\ Y \\ Z \end{pmatrix} = \begin{pmatrix} c_{11} & c_{12} & c_{13} \\ c_{21} & c_{22} & c_{23} \\ c_{31} & c_{32} & c_{33} \end{pmatrix} \begin{pmatrix} X_0 \\ U \\ V \end{pmatrix} \quad ;$$

$$\begin{pmatrix} X_0 \\ U \\ V \end{pmatrix} = \begin{pmatrix} c_{11} & c_{21} & c_{31} \\ c_{12} & c_{22} & c_{32} \\ c_{13} & c_{23} & c_{33} \end{pmatrix} \begin{pmatrix} X \\ Y \\ Z \end{pmatrix} \tag{4.2}$$

defines a 1:1 transformation such that (X,Y,Z) is equally well sufficient and X, Y and Z are also independent, normal and with variance 1. I shall construct a particular orthogonal matrix C such that our problem obtains a form similar to that of the problem studied in Sections 2 and 3. We have to test the hypothesis $H : \delta = 1$ against $A : \delta = 2$. Next we define the rotation matrix C by

$$
\begin{aligned}
(c_{11}\ c_{12}\ c_{13}) &= (1+n_1)^{-\frac{1}{2}} (n_1^{\frac{1}{2}}, -1, 0) \\
(c_{21}\ c_{22}\ c_{23}) &= \{n_2(1+n_1)(1+n_1+n_2)\}^{-\frac{1}{2}} \\
&\quad \times(-n_2, -n_2\, n_1^{\frac{1}{2}}, (1+n_1)n_2^{\frac{1}{2}}) \\
(c_{31}\ c_{32}\ c_{33}) &= (1+n_1+n_2)^{-\frac{1}{2}} (1, n_1^{\frac{1}{2}}, n_2^{\frac{1}{2}})
\end{aligned}
\tag{4.3}
$$

(the idea was to obtain a problem as represented in Fig. 7). Note that X, Y and Z are independent, normal and with variance 1, because C is a rotation matrix. Also that $\theta = (1,\mu_1,\mu_2)$ (the individual to be classified is from π_1, the expectations in π_1 and π_2 are given by μ_1 and μ_2 respectively) implies that

$$
\begin{pmatrix} E_\theta(X) \\ E_\theta(Y) \\ E_\theta(Z) \end{pmatrix} = (1+n_1+n_2)^{-\frac{1}{2}} \begin{pmatrix} 0 \\ (1+n_1)^{\frac{1}{2}} n^{\frac{1}{2}} (\mu_2-\mu_1) \\ (1+n_1)\mu_1 + n_2\mu_2 \end{pmatrix} ,
$$

while $\theta = (2,\mu_1,\mu_2)$ implies that

$$\begin{pmatrix} E_\theta(X) \\ E_\theta(Y) \\ E_\theta(Z) \end{pmatrix} = \begin{pmatrix} n_1^{\frac{1}{2}}(1+n_1)^{-\frac{1}{2}}(\mu_2-\mu_1) \\ n_1 n_2^{\frac{1}{2}}(1+n_1)^{-\frac{1}{2}}(1+n_1+n_2)^{-\frac{1}{2}}(\mu_2-\mu_1) \\ (1+n_1+n_2)^{-\frac{1}{2}}\{n_1\mu_1 + (1+n_2)\mu_2\} \end{pmatrix} .$$

In terms of X, Y and Z our testing problem obtains the following form if we *reparametrize* by taking $\theta = (\theta_1,\theta_2,\theta_3) = (E_\theta(X), E_\theta(Y), E_\theta(Z))$. X, Y and Z are independent $N(\theta_1,1)$, $N(\theta_2,1)$ and $N(\theta_3,1)$, respectively. It is known that $\theta \in \Theta = \Theta_o \cup \Theta_1 \cup \Theta_2$ where $\Theta_o = \{\theta;\ \theta_1=\theta_2=0\}$ corresponds to the situations $\mu_1 = \mu_2$ where our measurement has the same expectation in both populations, $\Theta_1 = \{\theta;\ \theta_1=0,\ \theta_2\neq 0\}$ consists of all θ's which occur if our individual is from π_1 and $\mu_1 \neq \mu_2$, Θ_2 consists of all θ's such that

$$n_1^{\frac{1}{2}} n_2^{\frac{1}{2}} \theta_1 = (1+n_1+n_2)^{\frac{1}{2}} \theta_2 \neq 0 , \tag{4.4}$$

or in other words of all θ's such that our individual is from π_2 and $\mu_1 \neq \mu_2$. The hypothesis $H: \theta\in\Theta_o \cup \Theta_1$ should be tested against the alternative $A: \theta\in\Theta_2$. Fig. 7 shows that we are essentially back in the situation of Sections 2 and 3 (see Section 5) where now

$$\psi = \text{arc tg } \{n_1^{-\frac{1}{2}} n_2^{-\frac{1}{2}} (1+n_1+n_2)^{\frac{1}{2}}\}. \tag{4.5}$$

Of course we might consider similar questions as at the end of Section 2. We would obtain similar results with respect to the obvious estimators $U/n_1^{1/2}$ and $V/n_2^{1/2}$ for μ_1 and μ_2 respectively.

5. *Construction and comparison of some level-α tests for Problem 2.*

The testing problem (H,A) obtained at the end of

Section 4 is invariant under the group $G = \{g_c;\ c\varepsilon R\}$ of translations $g_c: R^3 \to R^3$ where

$$g_c\{(x,y,z)\} = (x,y,z+c) \tag{5.1}$$

and where the induced transformations $\bar{g}_c : \Theta \to \Theta$ are the restrictions of g_c to $\Theta \subset R^3$. A maximal invariant under G is (X,Y), thus reduction by invariance leads to the problem formulated at the beginning of Section 3 where ψ is determined by (4.5). The tests $\phi_W^{(1)}$, ϕ_A, ϕ_χ and especially ϕ^* of Section 3 might be used and the arguments at the end of Section 3 for comparing these tests still apply.

Remark concerning Wald's test. If we argue along the lines of Wald, we do not arrive at $\phi_W^{(1)}$ but at a test $\phi_W^{(2)}$ of the form ϕ_χ. This is shown as follows. If μ_1 and μ_2 were known, then (N.-P. Fund. Lemma) one should take action a_2 (the individual is assigned to π_2) iff $x_0 > \mu_1 + \mu_\alpha$ in case $\mu_2 > \mu_1$ or $x_0 < \mu_1 - u_\alpha$ in case $\mu_2 < \mu_1$. Using the obvious estimators $U/n_1^{1/2}$ and $V/n_2^{1/2}$ for μ_1 and μ_2, and replacing u_α by a quantity c which later on is to be determined such that we obtain a size-α test, we find that

$$\phi_W^{(2)} = I_{(A_1 \cap B) \cup (A_2 \cap B')}$$

where $A_1 = \{X_0 - U/n_1^{1/2} > c\}$, $B = \{U\ n_2^{1/2} < V\ n_1^{1/2}\}$ and $A_2 = \{X_0 - U/n_1^{1/2} < -c\}$. Using (4.2) and (4.3) we can rewrite these events in terms of X, Y and Z. We find $A_1 = \{X > c(1+n_1^{-1})^{-1/2}\}$, $B = \{X\ n_2^{1/2} + Y\ n_1^{1/2}(1+n_1+n_2)^{1/2} > 0\}$ and $A_2 = \{X > -c(1+n_1^{-1})^{-1/2}\}$. This shows that $\phi_W^{(2)}$ is of the form ϕ_χ, where

$$\chi = \text{arc cotg}\ \{n_1^{\frac{1}{2}}\ n_2^{-1}(1+n_1+n_2)^{\frac{1}{2}}\}\ . \tag{5.2}$$

6. *A very simple loss-function formulation for Problem 1.*

If the two populations π_1 and π_2 play a similar role such that both kinds of error are equally damaging, then one might start from the loss-function

$$L(\theta,a_i)=0 \quad (\theta\varepsilon\Theta_i) \; ; \quad L(\theta,a_i)=1 \quad (\theta\varepsilon\Theta_{3-i}) \quad (i=1,2), \qquad (6.1)$$

instead of from some Neyman-Pearson level-α restriction. Remark that in (6.1) the loss remains undefined for $\theta\varepsilon\Theta_0$. It is clear that this section will have much less applications than the following two sections: if π_1 and π_2 play a similar part, then it seems to be unreasonable to assume that the distribution of our measurement is the known $N(0,1)$ distribution if the individual is from π_1, and an unknown $N(\mu,1)$-distribution if it were from π_2.

Wald's procedure δ_W. If μ were known, then one would take action a_2 iff the event $\{|X_0-0| > |X_0-\mu|\}$ occurs. Substituting the obvious estimator $\overline{X} = Y/n^{1/2}$ for μ, we obtain Wald's procedure δ_W which takes action a_2 iff the event $\{|X_0-0| > |X_0-\overline{X}|\}$ occurs.

Maximum likelihood procedure. It seems plausible to take action a_i iff the event $\{\hat{\theta} \varepsilon \Theta_i\}$ occurs where $\hat{\theta}$ denotes the maximum likelihood estimator of θ. It is well known that $\hat{\theta}$ does not depend on a reduction by sufficiency, nor on the parametrization (apart from trivial transformations). In terms of the original observations $(X_1,X_2,\ldots, X_n)$ one finds that the M.L. procedure takes action a_2 iff the event

$$\{|X_0-0| > (1+n^{-1})^{-\frac{1}{2}}|X_0-\overline{X}|\} \qquad (6.2)$$

occurs. In terms of (T_1,T_2) the M.L. procedure takes action a_2 iff $\{T_1T_2 \wedge 0\}$ occurs: one obtains the likelihood-ratio test ϕ_A in case $\alpha = .50$ (see Fig. 2).

The uniformly minimum risk invariant procedure δ^*. If Problem 1 is reduced by sufficiency and next reformulated in terms of (T_1,T_2) by means of (3.3), then one immediately recognizes that there is so much symmetry, that the M. L. procedure δ^* which takes action a_2 with probability 1 if $\{T_1T_2 > 0\}$ occurs, with probability 1/2 if $\{T_1T_2 = 0\}$ occurs and with probability 0 if $\{T_1T_2 < 0\}$ occurs, is a U.M.R. invariant rule.

Minimax risk procedures. Our decision problem being of type I (see Schaafsma, 1969), the minimax risk procedures can be characterized easily as the class of all unbiased size-1/2 tests. Obviously δ_W is not minimax while δ^* and many other procedures have minimax risk. It is not true that δ^* has U.M.R. among all minimax risk procedures: such a procedure does not exist.

Proving that δ^* is U.M.R. invariant. In terms of (T_1,T_2) (see Fig. 2) our problem is invariant under the commutative group $G = \{g_0,\ldots,g_3\}$ of order 4 where the transformations $g\colon R^2 \to R^2$ are defined by

$$\begin{cases} g_0\{(t_1,t_2)\} = (t_1,t_2);\ g_1\{(t_1,t_2)\} = (-t_1,t_2) \\ g_2\{(t_1,t_2)\} = (t_1,-t_2);\ g_3\{(t_1,t_2)\} = (-t_1,-t_2) \end{cases} \tag{6.3}$$

The induced mappings $\bar{g}\colon \Theta \to \Theta$ are defined by restricting g to $\Theta \subset R^2$. Moreover, if $\mathring{A} = \{a_1,a_2\}$, then the induced mappings $\tilde{g}\colon \mathring{A} \to \mathring{A}$ are defined by

$$\tilde{g}_i(a_h) = a_h \quad (i=0,3); \qquad \tilde{g}_i(a_h) = a_{3-h} \quad (i=1,2). \tag{6.4}$$

Our loss-function (6.1) is indeed such that our decision problem is invariant under G. It is interesting to remark that *there do not exist nonrandomized invariant rules* $d\colon R^2 \to \mathring{A}$ because these should satisfy

$$d[g\{(t_1,t_2)\}] = g[d\{(t_1,t_2)\}] \tag{6.5}$$

for all $(t_1,t_2) \in R^2$ and all $g \in G$. The equality (6.5) can obviously not be satisfied for points (t_1,t_2) with $t_1 = 0$ or $t_2 = 0$. Consequently there do not exist *randomized invariant* rules in the sense of Ferguson (1967), Section 4.2. But there do exist *invariant behavioral rules*.

Our behavioral rule δ can be determined by the test-function $\phi : R^2 \to [0,1]$ (we use the notation $\delta = \delta_\phi$) which expresses that when (t_1,t_2) is observed, then action a_2 is taken with probability $\phi(t_1,t_2)$. The general definition of invariance of a behavioral rule reduces to the obvious requirements

$$\phi(-t_1,t_2) = \phi(t_1,-t_2) = 1-\phi(t_1,t_2) = 1-\phi(-t_1,-t_2) \quad (6.6)$$

We shall show that δ^*, or more precisely $\delta^* = \delta_{\phi^*}$ with ϕ^* determined by (6.6) and

$$\phi^*(t_1,t_2) = 1(t_1,t_2\ 0);\ \phi^*(t_1,t_2) = \tfrac{1}{2}(t_1=0 \textit{ or } t_2=0), \quad (6.7)$$

is U.M.R. among all δ_ϕ's satisfying (6.6). The proof is straightforward: we write $R(\theta,\delta_\phi)$ in terms of the values of ϕ in the positive quadrant $Q_0 = \{(t_1,t_2);t_1,t_2>0\}$. This is possible by using (6.6). Let $O_i = g_i(Q_0)$ (i=1, 2,3) denote the respective quadrants. Let p_θ denote the p.d.f. of the bivariate $N(\theta,I)$ distribution and let $t = (t_1,t_2) \varepsilon R^2$. Recalling $p_\theta(t) = p_{\overline{g}(\theta)}\{g(t)\}$ we obtain for $\theta \varepsilon \Theta_1$

$$R(\theta,\delta_\phi) = \sum_{j=0}^{3} \int_{g_j^{-1}(Q_j)} \phi\{g_j^{-1}(t)\}p_\theta\{g_j^{-1}(t)\}dt$$

$$= \int_{Q_0} \{p_{\overline{g}_1(\theta)}(t)+p_{\overline{g}_2(\theta)}(t)\}dt$$

$$+ \int_{Q_0} \phi(t)\{p_\theta(t)-p_{\overline{g}_1(\theta)}(t)-p_{\overline{g}_2(\theta)}(t)+p_{\overline{g}_3(\theta)}(t)\}dt$$

$$\geq \int_{Q_0} \{p_\theta(t)+p_{\overline{g}_3(\theta)}(t)\}dt \quad = \quad R(\theta,\delta_{\phi^*})$$

with equality iff $\phi = 1$ a.e. in quadrant Q_0, because it can be proved easily that

$$p_\theta(t)-p_{\overline{g}_1(\theta)}(t)-p_{\overline{g}_2(\theta)}(t)+p_{\overline{g}_3(\theta)}(t) \quad < \quad 0$$

holds for all $\theta \in \Theta_1$ and $t \in Q_0$.

7. A very simple loss-function formulation for Problem 2.

The Neyman-Pearson approach of Sections 4 and 5 is only appropriate if both populations play a dissimilar role. Often both populations play a similar part. (I refer to Van Vark who tried to make a sex-diagnosis of prehistoric human skeletal material). In this section we give a theory for loss-function (6.1). This theory will contain the theory of Section 6 as a limiting case ($n_1 \to \infty$, $\overline{X}_1 \to 0$).

Wald's procedure δ_W. If μ_1 and μ_2 were known, then one would take action a_2 iff the event $\{|X_0-\mu_1| > |X_o-\mu_2|\}$ occurs. Substituting the obvious estimators $\overline{X}_1 = U/n^{1/2}$ and $\overline{X}_2 = V/n^{1/2}$ (see Section 4) for μ_1 and μ_2 respectively, we obtain Wald's procedure δ_W which takes action a_2 iff the event $\{|X_0-X_1| > |X_0-\overline{X}_2|\}$ occurs.

Maximum likelihood procedures. Das Gupta (1965) studied the maximum likelihood principle in much more general situations. In terms of the original observations $(X_0,X_1,\ldots,X_{n_1+n_2})$, one easily finds that the M.L. procedure takes action a_2 iff the event

$$\{(1+n_1^{-1})^{-1/2}|X_0-\overline{X}_1| > (1+n_2^{-1})^{-1/2}|X_0-\overline{X}_2|\} \tag{7.1}$$

occurs. This procedure was also studied by Kudô (1959, 1960).

The U.M.R. invariant procedure δ^*. In terms of (X,Y,Z) (see the end of Section 4), Problem 2 is invariant under the translations (5.1). Hence we may delete Z. Next apply (3.3). The M.L. procedure (7.1) is obviously equivalent to the rule which takes action a_2 iff the event $\{T_1T_2 > 0\}$ occurs. It follows from the end of Section 6 that the M.L. rule δ^* which takes action a_2 with probabilities 1, 1/2, and 0 respectively if $\{T_1T_2 > 0\}$, $\{T_1T_2 = 0\}$ or $\{T_1T_2 < 0\}$ occurs, is U.M.R. invariant.

Remark. Comparing δ_W and δ^* we should distinguish between the cases $n_1 < n_2$, $n_1 = n_2$ and $n_1 > n_2$. Obviously (see (7.1)) $\delta_W = \delta^*$ ("almost everywhere") iff $n_1 = n_2$. Optimum properties of $\delta_W = \delta^*$ in this special case have been discussed often. If $n_1 < n_2$, then δ_W assigns more observations to π_2 than δ^*: each observation $(x_0, x_1, \ldots, x_{n_1+n_2})$ which leads to action a_2 when using δ^* also leads to action a_2 when using δ_W. Hence δ_W has a smaller risk than δ^* if $\theta \varepsilon \Theta_2$ and a larger risk than δ^* if $\theta \varepsilon \Theta_1$ (in case $n_1<n_2$): δ^* is not uniformly better than δ_W, δ_W is not invariant. Das Gupta (1965) and Kudô (1959, 1960) proved certain optimum properties of the M.L. rule (7.1); their optimum properties are less compelling than the U.M.R. invariant property established above.

Risk-functions, misclassification probabilities. If δ is a (an "almost everywhere") non-randomized rule, then δ is determined by the set $E \subset R^2$ of all (x,y) or (t_1,t_2) to which action a_2 is assigned. It follows from (6.1) that

$$R(\theta,\delta) = P_\theta(E) \quad (\theta\varepsilon\Theta_1) \; ; \; R(\theta,\delta) = 1-P_\theta(E) \quad (\theta\varepsilon\Theta_2). \quad (7.2)$$

For δ_W we use E_W as a notation for the corresponding set, for δ^* this set is denoted by $E^* = \{T_1T_2 > 0\}$. E^* will always (and E_W only in case $n_1=n_2$) correspond to a right angle. Hence $P_\theta(E^*)$ can always be expressed in terms of the distribution function Φ of the N(0,1)-distribution. In order to express $P_\theta(E_W)$ in the $n_1 \neq n_2$ case, one needs bivariate distribution functions (see John, 1961, Hills, 1966). After some computations we obtained for our U.M.R. invariant rule δ^* that the misclassification probabilities are determined by

$$R(\theta,\delta^*) = \frac{1}{2} - 2\{\Phi(\Delta(\theta)\sin \tfrac{1}{2}\psi) - \tfrac{1}{2}\}\{\Phi(\Delta(\theta)\cos \tfrac{1}{2}\psi) - \tfrac{1}{2}\}, \quad (7.3)$$

where ψ is determined by (4.5) and

$$\Delta(\theta) = (1+n_1+n_2)^{-1/2}(1+n_i)^{1/2} n_{3-i}^{1/2} |\mu_1-\mu_2| \, , \quad (7.4)$$

if $\theta \varepsilon \Theta_i$ (i=1,2) and μ_1, μ_2 are the expectations corresponding to θ. Expression (7.3) seems to be much more explicit than the formulas in Kudô (1959, 1960). For a proof of (7.3) we refer to the report mentioned in the Acknowledgements. John and Hills considered the problem to estimate $P_\theta(E_W)$ on the basis of the observations. We get a more clear-cut estimation problem when dealing with δ^*. Let the parameter $g(\theta)$ be defined by the expression in the right-hand side of (7.3); obviously $g(\theta) = h(|\mu_1-\mu_2|)$ and we could consider the problem to estimate g on the basis of $X_1,\ldots,X_{n_1+n_2}$. It is intuitively clear that the obvious M.L. estimator $h(|X_1-X_2|)$ will underestimate the misclassification probability. This requires further investigation.

Minimax risk procedures. Our decision problem being of type I (see Schaafsma, 1969), minimax risk rules are in 1:1 correspondence with the class of all unbiased size-1/2 tests. If follows immediately (see Fig. 2) that δ^* is a minimax risk procedure and that δ_W will be minimax iff $n_1 = n_2$. The U.M.R. invariant rule δ^* is not a unique minimax risk rule: other minimax procedures arise when E^* is rotated over an angle $\leqq 1/2\psi$ (see Fig. 2). This shows also that δ^* does not have U.M.R. among the minimax procedures: this optimum property cannot be satisfied (cf. Das Gupta, 1965).

8. *Extending Section 7 to the case σ^2 unknown.*

δ^* is defined by the *homogeneous* inequalities (7.1), δ^* is not only invariant under the translations

$$g_c\{(x_0x_1\ldots x_{n_1+n_2})\} = (x_0+c,x_1+c,\ldots,x_{n_1+n_2}+c) \qquad (8.1)$$

but also, surprisingly, under the scale transformations

$$g_d\{(x_0x_1\ldots x_{n_1+n_2})\} = (dx_0,dx_1,\ldots,dx_{n_1+n_2}). \qquad (8.2)$$

This is surprising because our original Problem 2 with $\sigma^2 = 1$ is obviously not invariant under these scale changes. δ^* being invariant under (8.2), we might hope that δ^* *is also optimal for the case σ^2 unknown.*

Making this rigorous, suppose that $(X_0,\ldots,X_{n_1+n_2})$ has the p.d.f.

$$(2\pi\sigma^2)^{-\frac{1}{2}(1+n_1+n_2)} \exp[-\tfrac{1}{2}\sigma^{-2}\{(x_0-\mu_\delta)^2 + \sum_{i=1}^{n_1}(x_i-\mu_1)^2 + \sum_{i=n_1+1}^{n_1+n_2}(x_i-\mu_2)^2\}] \tag{8.3}$$

depending on the unknown parameter $\theta = (\delta,\mu_1,\mu_2,\sigma^2)$ where $\delta = i$ if our individual to be classified is from population π_i $(i=1,2)$. It follows from (8.1) that (X_0,U,V,W) is a (necessary and) sufficient statistic where U and V are defined in Section 4 and

$$W = \sum_{i=1}^{n_1}(X_i-U/n_1^{1/2})^2 + \sum_{i=n_1+1}^{n_1+n_2}(X_i-V/n_2^{1/2})^2 \tag{8.4}$$

is chosen in such a way that X_0,U,V and W are independent r.v.'s whatever θ may be. Of course we may equally well work with (X,Y,Z,W), using (4.2). Invariance under the group G of translations (8.1), or rather

$$g_c\{(x,y,z,w)\} = (x,y,z+c,w) \tag{8.5}$$

shows that we may restrict our attention to (X,Y,W), thus using reduction by invariance. Next applying the rotation (3.3), we might equally well try to solve the decision problem written in terms of (T_1,T_2,W): T_1,T_2 and W are independent, T_i having the $N(\tau_i,\sigma^2)$-distribution $(i=1,2)$ and W having the $\sigma\,\chi^2_{n_1+n_2-2}$-distribution where $(\tau_1,\tau_2,\sigma^2)\ \varepsilon\ \Theta$ and $\Theta = \Theta_0 \cup \Theta_1 \cup \Theta_2$ where Θ_0 is defined by $\tau_1=\tau_2=0$, Θ_1 by $\tau_1 \cos 1/2\psi + \tau_2 \sin 1/2\psi = 0$ $(\tau_1,\tau_2 \neq 0)$ and Θ_2 by $\tau_1 \cos 1/2\psi = \tau_2 \sin 1/2\psi \neq 0$. Now we might proceed in two different manners: (i) our problem is obviously invariant under scale changes

$$g_\alpha(t_1,t_2,w) = (\alpha t_1,\alpha t_2,\alpha^2 w)$$

leaving us with $(T_1W^{-1/2},T_2W^{-1/2})$ as a maximal invariant. Next we might try to generalize the theory at the end of Section 6, (ii) we might try to get rid of W by conditioning w.r. W.

We choose (ii). Our problem in terms of (T_1,T_2,W) is invariant under the group $G = \{g_0,\ldots,g_3\}$ where the transformations $g\colon R^2 \times R^+ \to R^2 \times R^+$ are defined by adding everywhere a third coordinate w to (6.3). We restrict our attention to invariant *behavioral* rules $\delta = \delta_\phi$ where $\phi\colon R^2 \times R^+ \to [0,1]$ should satisfy the following invariance requirements instead of (6.6)

$$\phi(-t_1,t_2,w) = \phi(t_1,-t_2,w) = 1-\phi(t_1,t_2,w) = 1-\phi(-t_1,-t_2,w) \tag{8.6}$$

We show that δ^*, or more precisely $\delta^* = \delta_{\phi^*}$ with δ^* determined by (8.6) and

$$\begin{aligned} \phi^*(t_1,t_2,w) &= 1(t_1,t_2 > 0) \; ; \\ \phi^*(t_1,t_2,w) &= \tfrac{1}{2}(t_1=0 \textit{ or } t_2=0) \end{aligned} \tag{8.7}$$

is U.M.R. among all δ_ϕ's satisfying (8.6). Remark that this procedure δ^* is exactly the same procedure as that derived in Section 6: no attention is paid to the outcome of W. Taking $\theta = (\tau_1,\tau_2,\sigma^2) \in \Theta_1$, fixed, we write

$$R(\theta,\delta_\phi)$$
$$= \int_{R^+} \{\int_{R^2} \phi(t_1,t_2,w) p_{\tau_1,\sigma^2}(t_1) p_{\tau_2,\sigma^2}(t_2) dt_1 dt_2\} p_{\sigma^2}(w) dw$$

and we remark that this integral can be minimized by minimizing the integrand for each value of w. The important observation has to be made that the minimization is performed by (8.7) (see the end of Section 6) and that the procedure obtained obviously does not depend on σ^2 (nor on $\tau_1,\tau_2 \in \Theta_1$).

9. *Other problems.*

Sections 2-5 deal with the Neyman-Pearson formulation: the attention is restricted to the class of (invariant, similar) size-α tests. This may be appropriate if the two populations play a dissimilar part. In the sections 6,...,8 we adopted a decision-theoretic formulation based on the loss-function (6.1). This may be appropriate if the two populations play a similar part and the statistician is forced to take action a_1 or a_2. The loss-function (6.1) produced a nice optimum theory: there exists a U.M.R. invariant procedure. This is caused by the fact that the problem is not only invariant under $G = \{g_0, g_3\}$ but even under $G = \{g_0, g_1, g_2, g_3\}$ (see (6.3) and (6.4)). We shall discuss a few modifications of the problems considered. It will turn out that the corresponding optimum theory will be hard.

Modification 1. In the case of two dissimilar populations (Sections 2-5) we might use the loss-function

$$\begin{cases} L(\theta,a_1) = 0 \quad (\theta\varepsilon\Theta_1) \ ; \quad L(\theta,a_2) = b \quad (\theta\varepsilon\Theta_1) \\ L(\theta,a_1) = a \quad (\theta\varepsilon\Theta_2) \ ; \quad L(\theta,a_2) = 0 \quad (\theta\varepsilon\Theta_2) \end{cases} \tag{9.1}$$

instead of the Neyman-Pearson level-α restriction; (9.1) expresses that an error of the first kind costs b/a times as much as an error of the second kind (we assume $b > a$). Lehmann has shown (see Schaafsma, 1969) that for "problems of type I" like our problem, the class of minimax risk procedures coincides with that of all unbiased size-α tests where $\alpha = a/(a+b)$. The unbiased size-α tests of Section 3 have minimax risk when we regard them as decision procedures for the problem with loss-function (9.1). The risk-function of $\delta_{\phi*}$ (see (3.7)) behaves neatly for $\theta \varepsilon \Theta_2$ but badly for $\theta \varepsilon \Theta_1$: the similarity restriction (3.5) makes $\delta_{\phi*}$ take on its maximum risk $ab/(a+b)$ for all $\theta \varepsilon \Theta_0 \cup \Theta_1$. One should not consider $\delta_{\phi*}$ to be the solution of the problem defined by (9.1).

Modification 2. It might happen that the statistician is not forced to take either action a_1 or action a_2; he may remain undecided and take action a_0: no sufficient evidence for allocating the individual. The following loss-function

$$\begin{cases} L(\theta,a_0) = a \ ; \ L(\theta,a_1) = 0 \ ; \ L(\theta,a_2) = b \quad (\theta\varepsilon\Theta_1) \\ L(\theta,a_0) = a \ ; \ L(\theta,a_1) = b \ ; \ L(\theta,a_2) = 0 \quad (\theta\varepsilon\Theta_2) \end{cases} \tag{9.2}$$

where $b > 2a$, is attractive in our opinion. It is seen easily (see Schaafsma, 1969) that there exists a unique minimax risk procedure (uniqueness a.e. of course). But this is a useless one because it takes action a_0 for all possible outcomes.

Modification 3. When I showed Dr. van Vark the U.M.R. invariant procedure δ^* of Sections 6,7,8 and particularly the consequences of (7.1), he told me that he was not impressed and that he considered δ^* to be only of theoretical interest: if the outcome of X_0 would be very far away from the outcome of $1/2(X_1+X_2)$, then he would not assign the individual to either π_1 or π_2. He would believe that the individual is from a third unknown population. This leads to the following reformulation. The outcome of X_0, $X_1,\ldots,X_{n_1+n_2}$ is observed in order to take one of the following three actions. a_i: the individual is assigned to π_i $(i=1,2)$, a_3: the individual is assigned to a third unknown population. Restricting attention to the case $\sigma^2 = 1$ of Section 7, it is assumed that X_0 has the $N(\mu,1)$-distribution, $X_1,\ldots,X_{n_1}$ is an independent r.s. from $N(\mu_1,1)$ and $X_{n_1+1},\ldots,X_{n_1+n_2}$ from $N(\mu_2,1)$. The parameter space $\Theta = R^3 = \Theta_0 \cup \Theta_1 \cup \Theta_2 \cup \Theta_3$ where $\theta=(\mu,\mu_1,\mu_2)$, $\Theta_0 = \{\theta;\mu=\mu_1=\mu_2\}$, $\Theta_1 = \{\theta;\mu=\mu_1\neq\mu_2\}$, $\Theta_2 = \{\theta;\mu=\mu_2\neq\mu_1\}$ and $\Theta_3 = \{\theta;\mu\neq\mu_1$ and $\mu\neq\mu_1\}$. This is a multiple decision problem of type I because $[\Theta_1] \cap [\Theta_2] \cap [\Theta_3] = \Theta_0 \neq \emptyset$ ([ ] denotes closure with respect to the Euclidean topology in R^3). We are attracted by the following loss-function

$$\begin{cases} L(\theta,a_1) = 0 \;;\; L(\theta,a_2) = 1 \;;\; L(\theta,a_3) = b & (\theta\varepsilon\Theta_1) \\ L(\theta,a_1) = 1 \;;\; L(\theta,a_2) = 0 \;;\; L(\theta,a_3) = b & (\theta\varepsilon\Theta_2) \\ L(\theta,a_1) = a \;;\; L(\theta,a_2) = a \;;\; L(\theta,a_3) = 0 & (\theta\varepsilon\Theta_3) \end{cases} \tag{9.3}$$

where b expresses the loss involved when our individual is assigned to a third unknown population whereas it actually comes from π_1 or π_2; a expresses the loss when the individual is assigned to π_1 or π_2 whereas it actually comes from a third population. We should choose $b \gg a$ but it is a bit more difficult to decide whether we should take $a < 1$, $a = 1$ or $a > 1$. The theory of Schaafsma (1969) can be used to characterize the minimax risk procedures and the procedures which are unbiased in Lehmann's sense. One might attack the problem also by first testing the null-hypothesis $H : \theta \in \Theta_0 \cup \Theta_1 \cup \Theta_2$ that the individual is from π_1 or π_2, against $A : \theta \in \Theta_3$. But such testing problems are very difficult to handle.

Modification 4. In practice one usually will have to classify more than one individual. This destroys the relevancy of the theory developed (see theories of empirical Bayes procedures).

Modification 5. It may also happen that one knows that all the individuals to be classified are from the same population: either π_1 or π_2. Instead of one observation X_0 we have an independent random sample $X_{01},\ldots,X_{0m}$.

ACKNOWLEDGEMENTS

The references of Das Gupta helped a lot to revise sections 6, 7 and 8. A referee asked me to increase the readability by skipping technicalities. I decided to do so. The reader interested in details may ask for the original version of this paper, Report TW113.

REFERENCES

Anderson, J. A. (1969), Constrained Discrimination between k Populations, *Journ. Roy. Stat. Soc.* B, 31, 123-139.

Anderson, T. W. (1951), Classification by multivariate analysis, *Psychometrika*, 16, 31-50.

Anderson, T. W. (1958), *An Introduction to Multivariate Statistical Analysis*, Wiley: New York.

Das Gupta, S. (1965), Optimum classification rules for classification into two multivariate normal populations, *Ann. Math. Statist.*, 36, 1174-1184.

Dynkin, E. B. (1961), Necessary and sufficient statistics for a family of probability distributions, *Selected Transl. in Math. Stat. and Prob.*, 1, 17-40.

Fisher, R. A. (1936), The use of multiple measurements in taxonomic problems, *Ann. Eug.*, 7, 179-188.

Ferguson, T. S. (1967), *Mathematical Statistics*, Academic

Press: New York.

Hills, M. (1966), Allocation rules and their error rates, *Journ. Roy. Stat. Soc.* B, 28, 1-31.

John, S. (1961), Errors in discrimination, *Ann. Math. Statist.*, 32, 1125-1144.

Kinderman, A. (1972), On some problems in classification, *Technical Report*, 178, School of Statistics, Minnesota.

Kudô, A. (1959), The classificatory problem viewed as a two-decision problem, I, *Mem. Fac. of Sc. Kyushu Ser.* A, 13, 96-125.

Kudô. A. (1960), The classificatory problem viewed as a two-decision problem, II, *Mem. Fac. of Sc. Kyushu Ser.* A, 14, 63-83.

Marshall, A. W. and Olkin, I. (1968), A general approach to some screening and classification problems, *Journ. Roy. Stat. Soc.* B, 30, 407-443.

Matusita, K. (1967), Classification based on distance in multivariate Gaussian cases, *Proc. Fifth Berk. Symp.*, 1. 299-304.

Rao, C. R. (1954), On a general theory of discrimination when the information about alternative hypotheses is based on samples, *Ann. Math. Statist.*, 25, 651-670.

Rao, C. R. (1967), *Linear Statistical Inference and its Applications*, Wiley: New York.

Schaafsma, W. (1969), Minimax risk and unbiasedness for multiple decision problems of type I, *Ann. Math. Statist.*, 40, 1684-1720.

Schaafsma, W. (1971), Testing statistical hypotheses concerning the expectations of two independent normals, both with variance one, I and II, *Proc. Kon. Ned. Ak. (Indagationes Math.)*, 33, 86-105.

Sitgreaves, R. (1952), On the distribution of two random matrices used in classification procedures, *Ann. Math. Statist.*, 23, 263-270.

Solomon, H. (1956), Probability and statistics in psychometric research, item analysis and classification techniques, *Proc. Third Berk. Symp.*, 5, 169-184.

Van Vark, G. N. (1970), *Some statistical procedures for the investigation of prehistoric human skeletal material*, Thesis, Groningen University.

Wald, A. (1944), On a statistical problem arising in the classification of an individual into one of two groups, *Ann. Math. Statist.*, 15, 145-162.

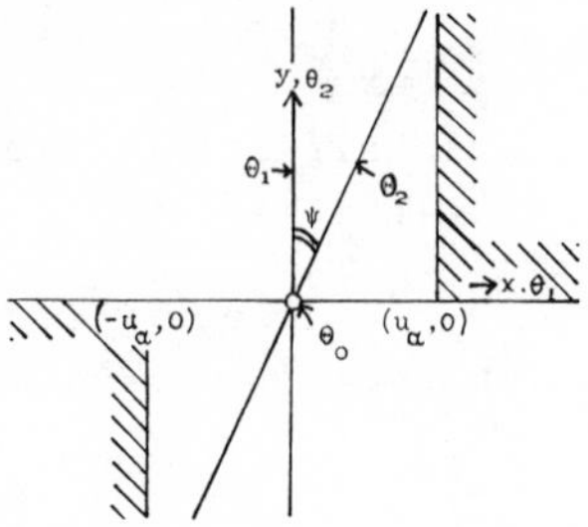

Fig. 1

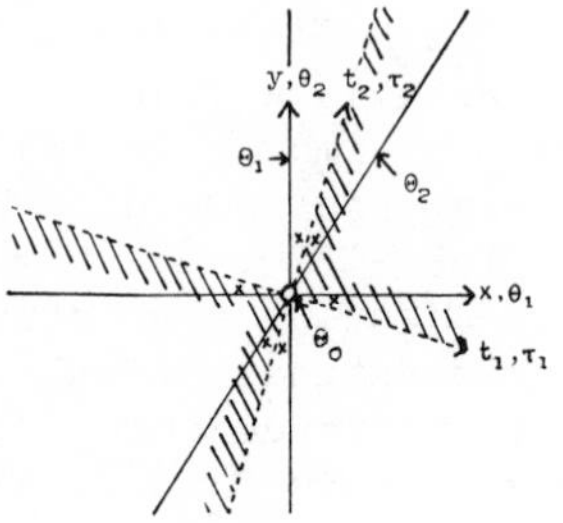

Fig. 2

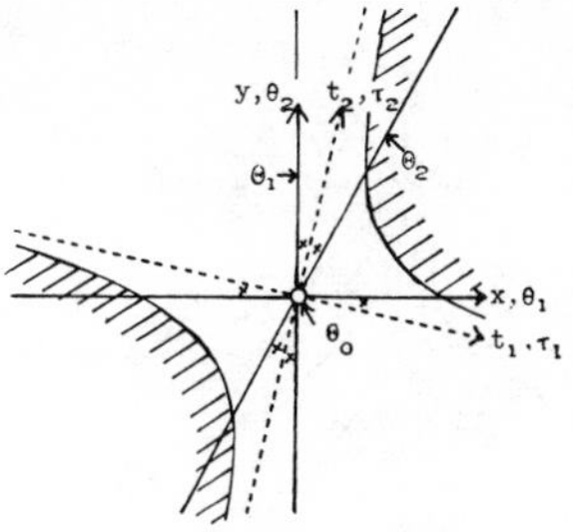

Fig. 3

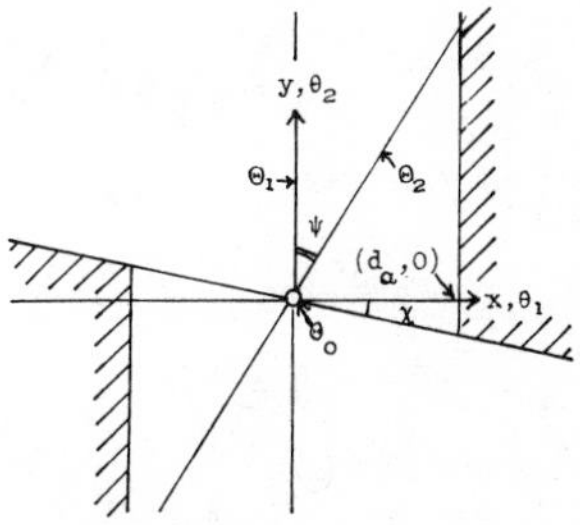

Fig. 4

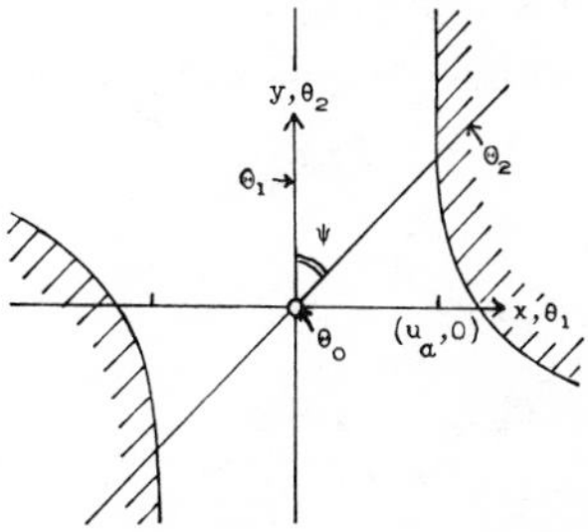

Fig. 5

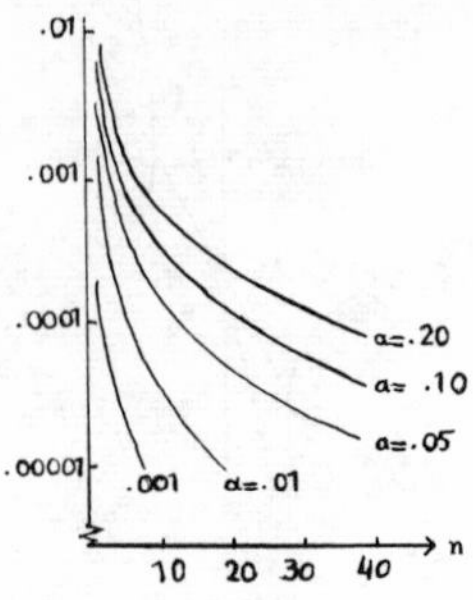
.01
.001
.0001
.00001
.001
α=.01
α=.20
α= .10
α=.05
n
10
20
30
40

Fig. 6

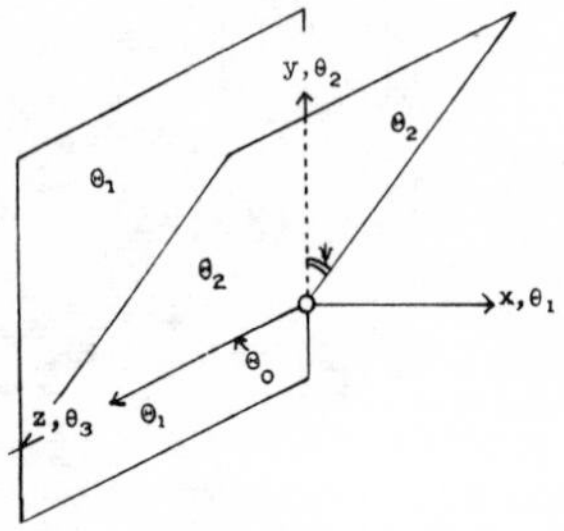
y, θ2
θ1
θ2
θ2
x, θ1
θ0
z, θ3
θ1

Fig. 7

SOME OPERATING CHARACTERISTICS OF LINEAR DISCRIMINANT FUNCTIONS

Rosedith Sitgreaves
California State University, Hayward

Introduction

In many situations we need to classify or assign an individual to one of two groups under conditions of uncertainty. For example, it may be necessary to decide whether a college applicant is or is not a suitable student for the college to which he has applied, or to decide whether a salmon caught in the middle of the Pacific Ocean originated in Japan or in North America. To aid in reaching a decision, it is usual to measure the individual on variables whose values are considered to be affected by the group to which he belongs. These measurements are compared, in some way, with corresponding measurements for individuals known to belong to each of the two possible groups under consideration, and the individual is assigned to the group to which he is "closer".

Among the best known statistical procedures developed to handle classification problems of this kind are Fisher's linear discriminant function and Anderson's classification statistic W, which is a modification of the linear discriminant function. In both cases it is assumed that we have observed a set of p variables for each of $N_1 + N_2 + 1$ individuals. The first N_1 individuals are known to belong to a population Π_1. In the example of college applicants these individuals might be applicants who were admitted and successfully completed the college program. The following N_2 are known to belong to a population Π_2. These individuals might be applicants who were admitted and failed to complete the college program. The last individual is the individual to be classified. That is, a decision is to be made as to whether he belongs to Π_1 or Π_2.

The set of p measurements recorded for each individual, for example, SAT scores plus high school grades, can be written as a p-dimensional vector, say,

$$\underset{\sim}{X}_i \ , \quad i = 1,2,\ldots,N_1+N_2+1 \ .$$

For $i = 1,2,\ldots,N_1$ (i.e., in Π_1),

$$E(\underset{\sim}{X}_i) = \underset{\sim}{\mu}^{(1)}$$

while for $i = N_1+1,\ldots,N_1+N_2$ (i.e., in Π_2),

$$E(\underset{\sim}{X}_i) = \underset{\sim}{\mu}^{(2)} \ .$$

The p×p covariance matrix $\underset{\sim}{\Sigma}$ is assumed to be the same in both populations.

From the $N_1 + N_2$ observations of known populations we compute

$$\overline{\underset{\sim}{X}}^{(1)} = \frac{1}{N_1} \sum_{i=1}^{N_1} \underset{\sim}{X}_i \ , \qquad \overline{\underset{\sim}{X}}^{(2)} = \frac{1}{N_2} \sum_{i=N_1+1}^{N_1+N_2} \underset{\sim}{X}_i$$

and the pooled estimate of the covariance matrix

$$S = \frac{1}{(N_1+N_2-2)} \times \left\{ \sum_{i=1}^{N_1} (\underset{\sim}{X}_i-\overline{\underset{\sim}{X}}^{(1)})(\underset{\sim}{X}_i-\overline{\underset{\sim}{X}}^{(1)})' + \sum_{i=N_1+1}^{N_1+N_2} (\underset{\sim}{X}_i-\overline{\underset{\sim}{X}}^{(2)})(\underset{\sim}{X}_i-\overline{\underset{\sim}{X}}^{(2)})' \right\}.$$

In Fisher's procedure, a vector of coefficients b are calculated, where

$$\underset{\sim}{b} = S^{-1}(\overline{\underset{\sim}{X}}^{(1)} - \overline{\underset{\sim}{X}}^{(2)}) \ .$$

The linear discriminant function is defined as

$$\underset{\sim}{b}'\underset{\sim}{X}_{N_1+N_2+1} = L \ , \text{ say.}$$

The value of L is compared with

$$L_1 = \underset{\sim}{b}'\overline{\underset{\sim}{X}}^{(1)} \quad \text{and} \quad L_2 = \underset{\sim}{b}'\overline{\underset{\sim}{X}}^{(2)} \ .$$

The observation is classified as belonging to Π_1 if

$$|L - L_1| \leq |L - L_2|$$

and to Π_2 otherwise.

In Anderson's procedure the classification statistic W is defined by

$$W = (\bar{\underset{\sim}{X}}^{(1)} - \bar{\underset{\sim}{X}}^{(2)})' S^{-1} (\underset{\sim}{X}_{N_1+N_2+1} - \frac{1}{2}(\bar{\underset{\sim}{X}}^{(1)} + \bar{\underset{\sim}{X}}^{(2)}) = L - \frac{1}{2}(L_1 + L_2).$$

The individual is classified as belonging to Π_1 if the observed value of W is greater than or equal to a preassigned cut-off point w_o and to Π_2 if $W < w_o$. When $w_o = 0$, this procedure is equivalent to Fisher's decision rule given above.

To evaluate the usefulness of any classification procedure we are concerned with calculating the probabilities of misclassification resulting from its use. If $P(i|j)$ denotes the probability of assigning an individual to Π_i when he belongs to Π_j, i,j=1,2, the two probabilities of misclassification using Anderson's procedure are given by

$$P(2|1) = \text{Prob}(W<w_o|\Pi_1) = F_1(w_o) \text{ , say,}$$

and

$$P(1|2) = \text{Prob}(W \geq w_o|\Pi_2) = 1 - F_2(w_o) .$$

If we write*

$$\lambda^2 = \frac{1}{4}(\underset{\sim}{\mu}^{(1)} - \underset{\sim}{\mu}^{(2)})' \Sigma^{-1} (\underset{\sim}{\mu}^{(1)} - \underset{\sim}{\mu}^{(2)}) .$$

It is known that as N_1 and $N_2 \to \infty$, for any given value of p, the asymptotic distribution of W is normal with variance $4\lambda^2$. The expected value of W is $2\lambda^2$ or $-2\lambda^2$, depending upon whether the observation to be classified belongs to Π_1 or Π_2. Thus, when $w_o = 0$, as N_1 and N_2 approach ∞,

---

*Note that $\lambda^2 = \frac{1}{4} D^2$ where D is the usual definition of the "distance" between the two populations.

$$F_1(0) \to \Phi(-\lambda)$$

and

$$1-F_2(0) \to 1-\Phi(\lambda) \quad = \quad \Phi(-\lambda) \ ,$$

where

$$\Phi(\lambda) \quad = \quad \int_{-\infty}^{\lambda} \frac{e^{-1/2x^2}}{\sqrt{2\pi}} \, dx \quad .$$

For finite sample sizes, the distribution of W under either alternative is extremely complicated. Asymptotic expansions for the distribution function have been given by Bowker and Sitgreaves (1961) in the case $N_1 = N_2 = N$, and by Okamoto (1963) in the general case. However, the terms in these expansions are also complicated, and except in a general way, the effect of changing values of p, N_1 and N_2, is not clear.

In an attempt to throw some light on these effects, this paper examines an approximation for the Bowker-Sitgreaves expansion in the case when $w_o = 0$. Each term in the expansion is approximated through expressions involving $\frac{1}{N^2}$ neglecting terms of $0\left(\frac{1}{N^2}\right)$. The approximated value of the probability of misclassification is compared with the probability calculated directly from the expansion for $\lambda = 1$, $N_1 = N_2 = 25$, and for $p = 1,2,3,4,5,10$, and 12.

1. *An approximation for the probability of misclassification when* $N_1 = N_2 = N$ *and* $w_o = 0$.

In the case that $N_1 = N_2 = N$ and $w_o = 0$, the two probabilities of misclassification are equal, that is,

$$P(1|2) \quad = \quad P(2|1) \quad = \quad F_1(0) \ . \tag{1}$$

We simplify the notation in Bowker and Sitgreaves (1961) by writing

$$u = \frac{h_{11}}{\sqrt{h_2}} \quad , \quad v = \frac{1}{\sqrt{h_2}} \quad .$$

When $w_o = 0$,

$$h_{11} = \sqrt{\frac{2N}{2N+1}}\,(2\lambda^2)$$

$$h_2 = \frac{2N-3}{2N-p-2}\,[4\lambda^2(\frac{2N+2}{2N+1}) + \frac{2p}{N}]\ .$$

The asymptotic expansion for $F_1(0)$ can then be written* as

$$F_1(0) = \Phi(-u) + \frac{1}{N}\sum_{r=3}^{4} a_{r1}(-v)^r\,\Phi^{(r)}(-u) + \frac{1}{N^2}\sum_{r=3}^{8} h_{r1}(-v)^r\,\Phi^{(r)}(-u) + 0(\frac{1}{N^2}) \qquad (2)$$

where

$$a_{31} = a_{41} = 4\lambda^2$$

$h_{31} = (2p-3)\lambda^2$

$h_{41} = p + 2(2p-1)\lambda^2 + (p-1)\lambda^4$

$h_{51} = 8\lambda^2$

$h_{61} = 8\lambda^2 + 8\lambda^4$

$h_{71} = 16\lambda^4$

$h_{81} = 8\lambda^4$

$\Phi^{(3)}(-u) = (u^2-1)\phi(-u)$

$\Phi^{(4)}(-u) = (u^3-3u)\phi(-u)$

$\Phi^{(5)}(-u) = (u^4-6u^2+3)\phi(-u)$

$\Phi^{(6)}(-u) = (u^5-10u^3+15u)\phi(-u)$

$\Phi^{(7)}(-u) = (u^6-15u^4+45u^2-15) \times \phi(-u)$

$\Phi^{(8)}(-u) = (u^7-21u^5+105u^3 -105u)\phi(-u)$

and

*Equation (3.34) in Bowker and Sitgreaves (1961) is incorrect in that each term in the two sums should be multiplied by $(-1)^r$.

$$\phi(-u) = \frac{1}{\sqrt{2\pi}} e^{-1/2u^2} .$$

When $\frac{p}{2N} < \lambda^2$ we can approximate u and r as follows:

$$h_{11} = 2\lambda^2(1 + \frac{1}{N})^{-1/2} \sim 2\lambda^2(1 - \frac{1}{4N} + \frac{3}{32N^2})$$

$$h_2 = 4\lambda^2(1 - \frac{3}{2N})(1 - \frac{p+2}{2N})^{-1}((1 + \frac{1}{N})(1 + \frac{1}{2N})^{-1} + \frac{p}{2N\lambda^2})$$

$$(1 - \frac{3}{2N})(1 - \frac{p+2}{2N})^{-1} \sim 1 + \frac{p-1}{2N} + \frac{(p-1)(p+2)}{4N^2}$$

$$((1 + \frac{1}{N})(1 + \frac{1}{2N})^{-1} + \frac{p}{2N\lambda^2}) \sim 1 + \frac{1}{2N}(1 + \frac{p}{\lambda^2}) - \frac{1}{4N^2}$$

so that

$$h_2 \sim 4\lambda^2(1 + \frac{p}{2N}(\frac{\lambda^2+1}{\lambda^2}) + \frac{1}{4N^2}(p^2\frac{(\lambda^2+1)}{\lambda^2} + p\frac{(2\lambda^2-1)}{\lambda^2} - 4)$$

then

$$v = \frac{1}{\sqrt{h_2}} \sim \frac{1}{2\lambda}\left\{1 - \frac{p}{4N}\frac{\lambda^2+1}{\lambda^2} - \frac{1}{32N^2}(p^2\frac{\lambda^4-2\lambda^2-3}{\lambda^4} + 4p\frac{2\lambda^2-1}{\lambda^2} - 16)\right\} \tag{3}$$

and

$$u = \frac{h_{11}}{\sqrt{h_2}} \sim \lambda\left\{1 - \frac{1}{4N}(1 + p\frac{\lambda^2+1}{\lambda^2} - \frac{1}{32N^2}(\frac{p^2(\lambda^4-2\lambda^2-3)}{\lambda^4} + \frac{6p(\lambda^2-1)}{\lambda^2} - 19)\right\} \tag{4}$$

Thus the first term in the asymptotic expansion can be approximated by

$$\Phi(-\lambda + \frac{1}{4N}(\lambda + \frac{p(\lambda^2+1)}{\lambda}) + \frac{1}{32N^2}(\frac{p^2(\lambda^4-2\lambda^2-3)}{\lambda^3} + \frac{6p(\lambda^2-1)}{\lambda} - 19\lambda)) \ . \tag{5}$$

For the sum

$$\frac{1}{N}\{a_{31}(-v)^3 \ \Phi^{(3)}(-u) + a_{41}(-v)^4 \ \Phi^{(4)}(-u)\} \tag{6}$$

the terms within the braces are approximated through terms of order $\frac{1}{N}$. Then

$$\frac{1}{N}\, a_{31}(-v)^3\Phi^{(3)}(-u) \sim \phi(-\lambda)\left\{-\frac{1}{2N}(\frac{\lambda^2-1}{\lambda}) - \frac{1}{8N^2}(\frac{(p+1)\lambda^6-(5p+3)N^4-3p\lambda^2+3p}{\lambda^3})\right\}$$

$$\frac{1}{N}\, a_{41}(-v)^4\Phi^{(4)}(-u) \sim \phi(-\lambda)\left\{\frac{1}{4N}(\frac{\lambda^2-3}{\lambda}) + \frac{1}{16N^2} \times(\frac{(p+1)\lambda^6-3(3p+2)\lambda^4+(5p+3)\lambda^2+15p}{\lambda^3})\right\} .$$

The sum of these two terms is

$$-\frac{1}{4N}\,\frac{(\lambda^2+1)}{\lambda}\,\phi(-\lambda) + \frac{1}{16N^2}\,\frac{-(p+1)\lambda^6+p\lambda^4+(11p+3)\lambda^2+9p}{\lambda^3}\,\phi(-\lambda) \tag{7}$$

For the terms in the sum

$$\sum_{r=3}^{8} h_{r1}(-v)^r \ \Phi^{(r)}(-u) \tag{8}$$

we need only the first term in each product since these terms are multiplied by $\frac{1}{N^2}$. Thus, we obtain

$$\frac{1}{N^2} h_{31}(-v)^3 \Phi^{(3)}(-u) \sim \phi(-\lambda) \left\{ - \frac{1}{8N^2} \frac{(2p-3)(\lambda^2-1)}{\lambda} \right\}$$

$$\frac{1}{N^2} h_{41}(-v)^4 \Phi^{(4)}(-u) \sim \phi(-\lambda) \left\{ \frac{1}{16N^2} \times \left(\frac{(p-1)\lambda^6+(p+1)\lambda^4-(11p-6)\lambda^2-3p}{\lambda^3} \right) \right\}$$

$$\frac{1}{N^2} h_{51}(-v)^5 \Phi^{(5)}(-u) \sim \phi(-\lambda) \left\{ - \frac{1}{4N^2} \frac{(\lambda^4-6\lambda^2+3)}{\lambda^3} \right\}$$

$$\frac{1}{N^2} h_{61}(-v)^6 \Phi^{(6)}(-u) \sim \phi(-\lambda) \left\{ \frac{1}{8N^2} \frac{(\lambda^6-9\lambda^4+5\lambda^2+15)}{\lambda^3} \right\}$$

$$\frac{1}{N^2} h_{71}(-v)^7 \Phi^{(7)}(-u) \sim \phi(-\lambda) \left\{ - \frac{1}{8N^2} \left(\frac{\lambda^6-15\lambda^4+45\lambda^2-15}{\lambda^3} \right) \right\}$$

$$\frac{1}{N^2} h_{81}(-v)^8 \Phi^{(8)}(-u) \sim \phi(-\lambda) \left\{ \frac{1}{32N^2} \frac{(\lambda^6-21\lambda^4+105\lambda^2-105)}{\lambda^3} \right\}$$

Adding them together, we have as an approximation to the sum given in (8)

$$\frac{1}{32N^2} \frac{(2p-1)\lambda^6-3(2p-3)\lambda^4-7(2p+1)\lambda^2-3(2p+3)}{\lambda^3} \phi(-\lambda) . \qquad (9)$$

From (5), (7), and (9) we obtain the result that for $N_1 = N_2 = N$ and $w_0 = 0$, the common probability of misclassification is approximated by

$$\Phi\ -\lambda + \frac{1}{4N}(\lambda + p\ \frac{(\lambda^2+1)}{\lambda}) + \frac{1}{32N^2}(p^2\ \frac{(\lambda^4-2\lambda^2-3)}{\lambda^3} + 6p\ \frac{(\lambda^2-1)}{\lambda}$$

$$- 19\lambda) \quad - \frac{1}{4N}\ \frac{(\lambda^2+1)}{\lambda}\ \phi(-\lambda) \tag{10}$$

$$+ \frac{1}{32N^2}\ \frac{-3\lambda^6-(4p-9)\lambda^4+(8p-1)\lambda^2+(12p-9)}{\lambda^3}\ \phi(-\lambda)$$

2. *Comparison of the approximation with some exact values*

We will examine the usefulness of the approximation by comparing some approximate values of the probability of misclassification with some values computed directly from (2) in the case when $\lambda^2 = 1$ and $N_1 = N_2 = 25$ for $p = 1,2,3,4,5,10,12$. We look first at the effect of the approximation on the values of u.

TABLE 1

Approximate and calculated values of u

$\lambda = 1$, $N_1 = N_2 = 25$

p	Calculated value	Approximate value
1	.97110	.97115
2	.95150	.95175
3	.93245	.93275
4	.91360	.91415
5	.89506	.89595
10	.80618	.81095
12	.77212	.77975

Except in the case that $p = 12$, the approximation for u is correct to two decimal places for the moderate value of $\lambda = 1$ and relatively small sample sizes. The approximation clearly improves as N becomes larger.

The approximate and calculated probabilities of misclassification in this case are as follows.

TABLE 2

Approximate and calculated values of the probability of misclassification

$\lambda = 1$, $N_1 = N_2 = 25$, $\Phi(-1) = .159$

p	Calculated value	Approximate value
1	.161	.161
2	.166	.167
3	.171	.172
4	.177	.177
5	.182	.180
10	.208	.206
12	.218	.215

The approximation seems reasonably good in this case.

REFERENCES

Bowlsen, A. H. and Sitgreaves, R. (1961), An asymptotic expansion for the distribution function of the W-classification statistic, *Studies in Item Analysis and Prediction*, (H. Solomon, Ed.), Chapter 19, Stanford University Press.

Okamoto, M. (1963), An asymptotic expansion for the distribution of the linear discriminant function, *Ann. Math. Stat.*, 34, 1286-1301.

A BIBLIOGRAPHY OF DISCRIMINANT ANALYSIS

Theophilos Cacoullos and *George P. H. Styan*
University of Athens and McGill University

1. INTRODUCTION

A natural companion to the collection of papers in these *Proceedings* is an exhaustive bibliography of discriminant analysis. In this bibliography, which we call *Aboda*, we list 26 books in Section 2 and 547 research papers in Section 4, which were published up to and including 1972 and 1971, respectively. Technical reports and similar unpublished papers are not included except for the bibliographic works of Hodges (1950) and Posten (1962), which appear in Section 2. Section 3 identifies 38 of the 173 journals and collections in which the research papers listed were published. The remaining 135 journals and collections are identified in Chapter 2 of *A Bibliography of Multivariate Statistical Analysis* by T. W. Anderson, Somesh Das Gupta and George P. H. Styan (Oliver & Boyd, Edinburgh, 1972). This book, which we will refer to as *Abomsa*, provided many references. Additions to Section 4, as well as some references, which we were unable to verify, are given in the *Addenda* (Section 5).

Section 2 is intended to include all books which deal comprehensively with the theory and practice of discriminant analysis, or which concentrate on a related special topic such as pattern recognition. Several books which contain at least one full chapter on discriminant analysis are also included.

Section 4 purports to include all research papers dealing with the theory, methodology and applications of discriminant analysis. Papers in the related areas of clustering or classification (in the sense of taxonomy) and pattern recognition were included only if they had some bearing or conceptual similarity with discrimination procedures. It was difficult to draw the line in many cases and naturally we had to be selective.

The preparation of this bibliography began with a core of over 200 references published through 1966 extracted from *Abomsa*. Many additional references were then found by searching two unpublished lists: A Selected Bibliography on Classification and Discrimination (Preliminary version) by Somesh Das Gupta (17 pp.) and References by Peter A. Lachenbruch (19 pp.). We are grateful to these authors for making these unpublished materials available to us. We then searched the lists of references in the papers of these *Proceedings*, and the leading statistical journals for 1967-1971, as well as *Mathematical Reviews* (through Volume 45(1973), No.2) and *Statistical Theory and Method Abstracts* (through Volume 13, 1972). References to *Mathematical Reviews*, with the prefix MR, are given for 6 books and 237 papers.

The journals and years in which the 547 research papers were published may be broken down as follows:

Journals and Collections Arranged by Number of Papers	
Ann. Math. Statist.	55
Sankhyā and *Sankhyā Ser. A*	39
Biometrika	29
Biometrics	27
Ann. Inst. Statist. Math. Tokyo	24
J. Amer. Statist. Assoc.	18
Psychometrika	16
IEEE Trans. Information Theory	15
J. Roy. Statist. Soc. Ser. B	15
Ann. Eugenics	14
Technometrics	13
	265
161 other journals & collections	282
Total	547

Number of Papers per Year

Pre -1940	44	1965	34
1940-1949	59	1966	43
1950-1954	59	1967	32
1955-1959	59	1968	38
		1969	20
1960	14		
1961	22	1970	29
1962	17	1971	22
1963	28		
1964	27	Total	547

Our bibliographic procedures and style follow those of *Abomsa*. Cross-referencing by all authors of multiple-authored entries is, however, not included in *Aboda*. Moreover, the entries are not classified by subject. A bibliography of about 250 books and research papers, arranged by subject, appears in the paper by Somesh Das Gupta "Theories and methods of classification: A review", a chapter in these *Proceedings*.

We owe a tremendous debt of gratitude to Leslie Ann Hathaway, who patiently and diligently checked references and prepared the card file from which *Aboda* was typed. Her carefulness and enthusiasm added greatly to the final collection of references. We also thank Maria Lam for so efficiently and cheerfully typing the manuscript and Heather Benson for preparing the card file extracted from *Abomsa*. This research was supported in part by the National Research Council of Canada at McGill University.

2. BOOKS

Anderson, T. W. (1958). *An Introduction to Multivariate Statistical Analysis*. John Wiley & Sons, Inc., New York. xii + 374 pp. *Chapter 6*. (MR19, p.992)

Anderson, T. W.; Das Gupta, Somesh & Styan, George P. H. (1972). *A Bibliography of Multivariate Statistical Analysis*. Oliver & Boyd, Edinburgh. ix + 642 pp. *Subject-matter codes 6.1 through 6.5*.

Blackith, R. & Reyment, R. A. (1971). *Multivariate Morphometrics*. Academic Press, Ltd., London. 435 pp.

Bose, P. K. & Chaudhuri, S. B. (1966). *On some Problems associated with* D^2*-statistics and* p*-statistics*. Asia Publishing House, London. xi + 58 pp.

Clyde, Dean J., Cramer, Elliott M. & Sherin, Richard J. (1966). *Multivariate Statistical Programs*. Biometrics Laboratory, University of Miami, Coral Gables, Florida. 61 pp. *Chapter 3*.

Cooley, William W. & Lohnes, Paul R. (1962). *Multivariate Procedures for the Behavioral Sciences*. John Wiley & Sons, Inc., New York. x + 211 pp. *Chapters 6-7*.

Cooley, William W. & Lohnes, Paul R. (1971). *Multivariate Data Analysis*. John Wiley & Sons, Inc., New York. x + 364 pp. *Chapters 9, 10, 12*.

Dempster, A. P. (1969). *Elements of Continuous Multivariate Analysis*. Addison-Wesley Publishing Company, Inc., Reading, Mass. xii + 388 pp. *Chapter 10*.

Dixon, W. J. *editor* (1967). *BMD: Biomedical Computer Programs*. Second Edition. University of California Press, Berkeley & Los Angeles. x + 600 pp. [University of California Publications in Automatic Computation No.2. First published 1965. Third Printing of the Second Edition, revised, 1970] *Programs BMDO4M, BMDO5M, BMDO7M*.

Fu, K. S. (1968). *Sequential Methods in Pattern Recognition and Machine Learning*. Academic Press, Inc., New York. xi + 227 pp.

Hodges, Joseph L., Jr. (1950). Discriminatory Analysis. I. Survey of Discriminatory Analysis. USAF School of Aviation Medicine, Randolph Field, Texas. 115 pp. [Report Number 1, Project Number 21-49-004 under Contract No.AF41(128)-8. Distribution No.1514. Mimeographed]

Hope, Keith (1968). *Methods of Multivariate Analysis, with Handbook of Multivariate Methods Programmed in Atlas Autocode.* University of London Press, Ltd., London. 288 pp. [Paperback edition *Methods of Multivariate Analysis.* Unibooks, London. 165 pp.] *Chapter 7.*

I.B.M. (1970). *System/360 Scientific Subroutine Package, Version III. Programmer's Manual.* Fifth Edition. International Business Machines Corporation, White Plains, New York. 454 pp. First published 1966. [Manual No.GH20-0205-4] *Programs DISCR, MDISC.*

Kendall, M. G. (1957). *A Course in Multivariate Analysis.* Charles Griffin & Company, Ltd., London. 185 pp. [Number 2 of Griffin's Statistical Monographs & Courses. Copyright 1968] *Chapter 9.* (MR19, p. 1093)

Kendall, Maurice G. & Stuart, Alan (1968). *The Advanced Theory of Statistics. Volume 3, Design and Analysis, and Time-Series.* Second Edition. Charles Griffin & Company, Ltd., London. x + 557 pp. [First published 1966] *Chapter 44.*

Kshirsagar, Anant M. (1972). *Multivariate Analysis.* Marcel Dekker, Inc., New York. xiv + 534 pp. *Chapter 6.*

Kullback, Solomon (1968). *Information Theory and Statistics.* Dover Publications, Inc., New York. xv + 399 pp. [First published 1959, John Wiley & Sons, Inc., New York. xvii + 395 pp.] *Chapters 11-13.* (MR21-2325)

Mosteller, Frederick & Wallace, David L. (1964). *Inference and Disputed Authorship: The Federalist.* Addison-Wesley Publishing Company, Inc., Reading, Mass. xvi + 287 pp.

Overall, John E. & Klett, James C. (1972). *Applied Multivariate Analysis.* McGraw-Hill Book Company, New York. xxii + 500 pp. *Chapters 9-16.*

Posten, Harry O. (1962). Bibliography on Classification, Discrimination, Generalized Distance and Related Topics. International Business Machines Corporation, Thomas J. Watson Research Center, Yorktown Heights, New York. 20 pp. [Research Report RC-743. Updated by: "Addenda", 5 pp. "Second Addenda", 5 pp. Both undated. All mimeographed]

Rao, C. Radhakrishna (1952). *Advanced Statistical Methods in Biometric Research.* John Wiley & Sons, Inc., New York. xvii + 390 pp. [Reprinted with corrections, 1970, Hafner Publishing Company, Darien, Conn.] *Chapters 8, 9.* (MR14, p.388)

Rulon, Philip J.; Tiedeman, David V.; Tatsuoka, Maurice M., & Langmuir, Charles R. (1967). *Multivariate Statistics for Personnel Classification.* John Wiley & Sons, Inc., New York. xi + 406 pp.

Sebestyen, George S. (1962). *Decision-making Processes in Pattern Recognition.* The Macmillan Company, New York. viii + 162 pp. (MR27-6620)

Tatsuoka, Maurice M. (1970). *Discriminant Analysis; the Study of Group Differences.* Institute for Personality and Ability Testing, 1602-04 Coronado Drive, Champaign, Ill. iv + 57 pp. [Number 6 of Selected Topics in Advanced Statistics; an Elementary Approach]

Tatsuoka, Maurice M. (1971). *Multivariate Analysis: Techniques for Educational and Psychological Research.* John Wiley & Sons, Inc., New York. xiii + 310 pp. *Chapter 6.*

Terouanne, Eric (1969/70). *L'Analyse en Composantes Canoniques et son Utilisation en Discrimination.* D. E.A. de Mathématiques, Université de Montpellier, Montpellier. ii + 19 pp. [Secrétariat des Mathématiques de la Faculté des Sciences de Montpellier, Publication No.62] (MR41-2840)

3. JOURNALS AND COLLECTIONS

Adaptive, Learning and Pattern Recognition Systems (Mendel & Fu). Adaptive, Learning and Pattern Recognition Systems, Theory and Applications. Edited by J. M. Mendel and K. S. Fu. [Academic Press, Inc., New York, 1970. Volume 66, Mathematics and Science in Engineering Series] (MR42-3968)

Amer. J. Human Genet. American Journal of Human Genetics. (Baltimore)

Amer. J. Psychotherapy. American Journal of Psychotherapy. (Lancaster, Pa.)

Amer. Med. Assoc. Arch. Internal Med. American Medical Association Archives of Internal Medicine. (Chicago)

Amer. Rev. Tuber. Pulm. Diseases. American Review of Tuberculosis and Pulmonary Diseases. (Baltimore) [*Became* American Review of Respiratory Disease]

An. Univ. Bucureşti Mat.-Mec. Analele Universităţii Bucuresti. Matematică-Mecanică. (Bucureşti)

Bull. Canad. Psychol. Assoc. Bulletin of the Canadian Psychological Association. (Toronto)

Bull. Math. Soc. Sci. Math. R. S. Roumanie (N.S.). Bulletin Mathématique de la Societé des Sciences Mathématiques de la République Socialiste de Roumanie. (Bucureşti)

Essays Prob. Statist. (Bose et al.). Essays in Probability and Statistics. Edited by R. C. Bose, I. M. Chakravarti, P. C. Mahalanobis, C. R. Rao & K. J. C. Smith. Dedicated to the memory of Samarendra Nath Roy. [Statistical Publishing Society, Calcutta, 1969. The University of North Carolina Monograph Series in Probability and Statistics, No. 3, The University of North Carolina Press, Chapel Hill, N.C., 1970] (MR41-9310)

Hiroshima Math. J. Hiroshima Mathematical Journal. (Hiroshima) [*Was* Journal of Science of the Hiroshima University. Series A-I (Mathematics)]

Indian J. Genet. Plant Breeding. Indian Journal of Genetics and Plant Breeding. (New Delhi)

Information Sci. Information Scientist. (London)

J. Consult. Psychol. Journal of Consulting Psychology. (Washington, D.C.)

J. Mental Sci. Journal of Mental Science. (London) [*Became* British Journal of Psychiatry]

Lancet. Lancet. (London)

Mat. Zametki. Matematičeskie Zametki. Akademija Nauk Sojuza SSR. (Moscow) [*Translated as* Mathematical Notes of the Academy of Sciences of the USSR]

Math. Biosciences. Mathematical Biosciences. (New York)

Mem. Ist. Ital. Idrobiol. Memorie dell'Istituto Italiano di idrobiologia dott. Marco di Marchi. (Milano)

New England J. Med. New England Journal of Medicine. (Boston)

Proc. EJCC. Proceedings of the Eastern Joint Computer Conference. [9th (Boston), 1959. Pub. Spartan Books, Washington, D.C. *Became* Proceedings of the Fall Joint Computer Conference]

Proc. Fifth Annual Allerton Conf. Circuit System Theory. Proceedings of the Fifth Annual Allerton Conference on Circuit and System Theory, September 1967, Monticello, Ill. [University of Illinois, Urbana, Ill., 1967]

Proc. First Canad. Conf. Appl. Statist. Proceedings of the First Canadian Conference in Applied Statistics, "Statistics '71 Canada". Edited by C. S. Carter, T. D. Dwivedi, I. P. Fellegi, D. A. S. Fraser, J. R. McGregor & D. A. Sprott. [Issued by Sir George Williams University, 1972]

Proc. Fourth Annual Allerton Conf. Circuit System Theory. Proceedings of the Fourth Annual Allerton Conference on Circuit and System Theory, September 1966, Monticello, Ill. [University of Illinois, Urbana, Ill., 1966]

Proc. Hawaii Internat. Conf. Systems Sci. Proceedings of the Hawaii International Conference on Systems Sciences, January 29-31, 1968. (University of Hawaii, Honolulu) [University of Hawaii Press, Honolulu, 1968]

Proc. IEEE. Proceedings of the IEEE. (New York) [Institute of Electrical and Electronics Engineers]

Proc. IRE. Proceedings of the IRE. (New York) [Institute of Radio Engineers]

Proc. Soil Sci. Soc. Amer. Proceedings of the Soil Science Society of America. (Ann Arbor, Mich.)

Punjab Univ. J. Math. Punjab University Journal of Mathematics. (Lahore)

Rev. Belge Statist. Rech. Opérat. Revue Belge de Statistique et de Recherche Opérationnelle. (Bruxelles)

S. African Statist. J. South African Statistical Journal. (Pretoria)

Statist. Ecol. Statistical Ecology. Volume I, Spatial Patterns and Statistical Distributions. Edited by G. P. Patil, E. C. Pielon & W. E. Waters. [Pennsylvania State University Press, University Park, 1971]

Teor. Verojatnost. i Mat. Statist. Teorija Verojatnosteĭ i Matematičeskaja Statistika. (Kiev)

Theory of Optimal Solutions. Theory of Optimal Solutions. (Kiev) [Proceedings of a seminar, 1968 (Kiev). Pub. Akademija Nauk Ukrainskoĭ SSR, Kiev, 1968. In Russian]

Trans. Fourth Prague Conf. Information Theory, Statist. Decision Functions, Random Processes. Transactions of the Fourth Prague Conference on Information Theory, Statistical Decision Functions, Random Processes, August 31 - September 11, 1965. [Academia, Praha, 1967]

Trudy Turk. Nauč. Obšč. Sred. Aziat. Gos. Univ. Tashkent. Trudy Turkestanskoe Naučnoe Obščestvo. Sredne-Aziatskiĭ (Turkestanskiĭ) Gosudarstvennyĭ Universitet. (Tashkent)

Use Comput. Anthropol. (Hymes). The Use of Computers in Anthropology. Edited by Dell Hymes. [Mouton & Co., 's-Gravenhage, 1965]

Z. Pflanzen. Zeitschrift für Pflanzenzüchtung. (Berlin)

Zastos. Mat. Zastosowania Matematyki. Polska Akademia Nauk. Instytut Matematyczny. (Warszawa) [= *Polska Akad. Nauk Zastos Mat.*]

4. RESEARCH PAPERS

Adam, J. & Enke, H. (1970). Application of factor analysis as a separating procedure. (In German) *Biom. Z.*, 12, 395-411.

Adhikari, Bishwanath Prosad (1957). Analyse discriminante des mesures de probabilité sur un espace abstrait. *C.R. Acad. Sci. Paris*, 244, 845-846. (MR18, p.773)

Adhikari, Bishwanath Prosad & Joshi, Devi Datt (1956). Distance, discrimination et résumé exhaustif. *Publ. Inst. Statist. Univ. Paris*, 5, 57-74. (MR19, p.329)

Adke, S.R. (1958). A note on distance between two populations. *Sankhyā*, 19, 195-200. [Amended by *Sankhyā*, 20(1959), 108]

Afifi, A.; Sacks, S.; Liu, V.Y.; Weil, M.H. & Shubin, H. (1971). Accumulative prognostic index for patients with barbituate, gluetethemide and meprobamate intoxication. *New England J. Med.*, 285, 1485-1502.

Aizerman, M.A.; Braverman, E.M. & Rozonoer, L.I. (1964). Probability problem of pattern recognition learning and potential functions method. (In Russian, with summary in English) *Avtomat. i Telemeh.*, 25, 1307-1323. [Translated in *Automat. Remote Control*, 25 (1964), 1175-1190]

Albert, Arthur E. (1963). A mathematical theory of pattern recognition. *Ann. Math. Statist.*, 34, 284-299. (MR26-7461)

Alexakos, C.E. (1966). Predictive efficiency of two multivariate selection techniques in comparison with clinical predictions. *J. Educ. Res.*, 57, 297-306.

Alf, Edward F., Jr., & Dorfman, Donald D. (1967). The classification of individuals into two criterion groups on the basis of a discontinuous payoff function. *Psychometrika*, 32, 115-123.

Allais, D.C. (1966). The problem of too many measurements in pattern recognition and prediction. *IEEE Internat. Conv. Rec.*, 14(7), 124-130.

Ali, S.M. & Silvey, Samuel D. (1966). A general class of coefficients of divergence of one distribution from another. *J. Roy. Statist. Soc. Ser. B*, 28, 131-142. (MR33-4963)

Anderson, Gary J.; Walberg, Herbert J. & Welch, Wayne W. (1969). Curriculum effects on the social climate of learning: a new representation of discriminant functions. *Amer. Educ. Res. J.*, 6, 315-328.

Anderson, J.A. (1969). Constrained discrimination between k populations. *J. Roy. Statist. Soc. Ser. B*, 31, 123-139. (MR42-6996)

Anderson, Marshall W. & Benning, Roger D. (1970). A distribution-free discrimination procedure based on clustering. *IEEE Trans. Information Theory*, IT-16, 541-548. (MR42-8628)

Anderson, Oskar (1926). Ueber die Anwendung der Differenzenmethode ("variate difference method") bei Reihenausgleichungen, Stabilitätsuntersuchungen und Korrelationsmessungen. *Biometrika*, 18, 293-320. [Continued in *Biometrika*, 19(1927), 53-86]

Anderson, T.W. (1951). Classification by multivariate analysis. *Psychometrika*, 16, 31-50. (MR12, p.842)

Anderson, T.W. (1964). On Bayes procedures for a problem with choice of observations. *Ann. Math. Statist.*, 35, 1128-1135.

Anderson, T.W. (1966). Some nonparametric multivariate procedures based on statistically equivalent blocks. *Multivariate Anal. Proc. Internat. Symp. Dayton (Krishnaiah)*, 5-27. (MR35-5101)

Anderson, T.W. & Bahadur, Raghu Raj (1962). Classification into two multivariate normal distributions with different covariance matrices. *Ann. Math. Statist.*, 33, 420-431. (MR25-4609)

Aoyama, Hirojiro (1950). A note on the classification of observation data. *Ann. Inst. Statist. Math. Tokyo*, 2, 17-19. (MR12, p.511)

Aoyama, Hirojiro (1959). On the evaluation of the risk index of the railroad crossing. *Ann. Inst. Statist. Math. Tokyo*, 10, 163-180.

Aoyama, Hirojiro (1965). Dummy variable and its applications to the quantification method. (In Japanese, with summary in English) *Proc. Inst. Statist. Math. Tokyo*, 13, 1-12. [Amended by *Proc. Inst. Statist. Math. Tokyo*, 13(1965), 135-137]

Armitage, P. (1950). Sequential analysis with more than two alternative hypotheses and its relation to discriminant function analysis. *J. Roy. Statist. Soc. Ser. B*, 12, 137-144. (MR12, p.429)

Ashton, E.H.; Healy, M.J.R. & Lipton, S. (1957). The descriptive use of discriminant functions in physical anthropology. *Proc. Roy. Soc. London Ser. B*, 146, 552-572.

Atkinson, A.C. (1970). A method for discriminating between models. (With discussion) *J. Roy. Statist. Soc. Ser. B*, 32, 323-353.

Azorín Poch, Francisco (1962). Notas sobre taxonomía y estadística. *Trabajos Estadíst.*, 13, 249-263.

Bahadur, Raghu Raj (1961). On classification based on response to *n* dichotomous items. *Stud. Item Anal. Predict. (Solomon)*, 169-176. (MR22-12621c)

Baitsch, Helmut & Bauer, Rainald K. (1956). Zum Problem der Merkmalsauswahl für Trennverfahren (Barnardproblem). *Allgemein. Statist. Arch.*, 40, 160-167. (MR18, p.345)

Balakrishnan, V. & Sanghvi, L.D. (1968). Distance between populations on the basis of attribute data. *Biometrics*, 24, 859-865.

Banerjee, Kalishankar & Marcus, L.F. (1965). Bounds in a minimax classification procedure. *Biometrika*, 52, 653-654. (MR34-5236)

Bargmann, Rolf E. (1969). Interpretation and use of a generalized discriminant function. *Essays Prob. Statist. (Bose et al.)*, 35-60.

Barnard, M.M. (1935). The secular variation of skull characters in four series of Egyptian skulls. *Ann. Eugenics*, 6, 352-371.

Baron, D.N. & Fraser, P.M. (1965). The digital computer in the classification and diagnosis of diseases. *Lancet*, 2, 1066-1069.

Bartlett, Maurice S. (1939). The standard errors of discriminant function coefficients. *J. Roy. Statist. Soc. Supp.*, 6, 169-173. (MR1, p.248)

Bartlett, Maurice S. (1939). A note on tests of significance in multivariate analysis. *Proc. Cambridge Philos. Soc.*, 35, 180-185.

Bartlett, Maurice S. (1951). The goodness of fit of a single hypothetical discriminant function in the case of several groups. *Ann. Eugenics*, 16, 199-214. (MR13, p.666)

Bartlett, Maurice S. (1951). An inverse matrix adjustment arising in discriminant analysis. *Ann. Math. Statist.*, 22, 107-111. (MR12, p.639)

Bartlett, Maurice S. & Please, N.W. (1963). Discrimination in the case of zero mean differences. *Biometrika*, 50, 17-21. (MR27-6365)

Bartoszyński, Robert (1971). A note on subjective classifications. *Rev. Inst. Internat. Statist.*, 39, 39-45. (MR44-2290)

Bartoszyński, Robert (1971). On the construction and evaluation of subjective classifications. *Zastos. Mat.*, 12, 1-21. (MR43-8184)

Bašarinov, A.E. (1965) Asymptotically extremal procedures for the parametric recognition of forms. (In Russian) *Radiotehnika*, 10, 812-816. (MR31-4646)

Baten, William Dowell (1944). The discriminant function applied to spore measurements. *Papers Michigan Acad. Sci. Arts Lett.*, 29, 3-7.

Baten, William Dowell (1945). The use of discriminant functions in comparing judges' scores concerning potatoes. *J. Amer. Statist. Assoc.*, 40, 223-228.

Baten, William Dowell & Dewitt, C.C. (1944). Use of the discriminant function in the comparison of proximate coal analyses. *J. Indust. Engrg. Chem.*, 16, 32-34.

Baten, William Dowell & Hatcher, Hazel M. (1944). Distinguishing method differences by use of discriminant functions. *J. Exper. Educ.*, 12, 184-186.

Baten, William Dowell; Tack, P.I. & Baeder, Helen A. (1958). Testing for differences between methods of preparing fish by use of a discriminant function. *Indust. Qual. Control*, 14(7), 7-10.

Bauer, Rainald K. (1954). Diskriminanzanalyse. *Allgemein. Statist. Arch.*, 38, 205-216.

Bauer, Thomas (1957). The practical calculation of Hotelling's T^2. *Indust. Qual. Control*, 14(1), 7-10.

Beall, Geoffrey (1945). Approximate methods in calculating discriminant functions. *Psychometrika*, 10, 205-217.

Behrens, W.U. (1959). Beitrag zur Diskriminanzanalyse. *Biom. Z.*, 1, 3-14.

Bhattacharyya, A. (1943). On a measure of divergence between two statistical populations defined by their probability distributions. *Bull. Calcutta Math. Soc.*, 35, 99-109.

Bhattacharyya, A. (1946). On a measure of divergence between two multinomial populations. *Sankhyā*, 7, 401-406. (MR8, p.282)

Bhattacharyya, B.C. (1941). On alternative method of the distribution of Mahalanobis's D^2-statistic. *Bull. Calcutta Math. Soc.*, 33, 87-92. (MR4, p.23)

Bhattacharyya, D.P. & Narayan, Ram Deva (1941). Moments of the D^2-statistic for populations with unequal dispersions. *Sankhyā*, 5, 401-412. (MR4, p.105)

Binet, F.E. & Watson, G.S. (1956). Algebraic theory of the computing routine for tests of significance on the dimensionality of normal multivariate systems. *J. Roy. Statist. Soc. Ser. B*, 18, 70-78. (MR18, p.243)

Birnbaum, Allan & Maxwell, A.E. (1960). Classification procedures based on Bayes' formula. *Appl. Statist.*, 9, 152-169. (MR22-8619)

Blackith, R.E. (1960). A synthesis of multivariate techniques to distinguish patterns of growth in grasshoppers. *Biometrics*, 16, 28-40.

Blackith, R.E. (1965). Morphometrics. *Theor. Math. Biol. (Waterman & Morowitz)*, 225-249.

Bol'sev, L.N. (1955). A nomogram connecting the parameters of a normal distribution with the probabilities for classification into three groups. (In Russian) *Inžen. Sb. Akad. Nauk SSSR*, 21, 212-214. (MR17, p.53) [Reprinted in *Teor. Verojatnost. i Primenen.*, 2(1957), 124-126; translated in *Theory Prob. Appl.*, 2(1957), 120-122]

Bood, D.M. & Baker, C.B. (1958). Some problems in linear discrimination. *J. Farm Econ.*, 40, 674-683.

Bose, Purnendu Kumar (1947). On recursion formulae, tables and Bessel function populations associated with the distribution of classical D^2-statistic. *Sankhyā*, 8, 235-248. (MR9, p.620)

Bose, Purnendu Kumar (1949). Incomplete probability integral tables connected with Studentised D^2-statistic. *Calcutta Statist. Assoc. Bull.*, 2, 131-137. (MR11, p.527) [Amended by *Sankhyā*, 11(1951), 96]

Bose, R.C. (1935). On the exact distribution and moment-coefficients of the D^2-statistics. *Sci. and Cult.*, 1, 205-206.

Bose, R.C. (1936). On the exact distribution and moment-coefficients of the D^2-statistic. *Sankhyā*, 2, 143-154.

Bose, R.C. (1936). A note on the distribution of differences in mean values of two samples drawn from two multivariate normally distributed populations, and the definition of the D^2-statistic. *Sankhyā*, 2, 379-384.

Bose, R.C. & Roy, S.N. (1935). On the evaluation of the probability integral of the D^2-statistics. *Sci. and Cult.*, 1, 436-437.

Bose, R.C. & Roy, S.N. (1937). On the distribution of Fisher's taxonomic co-efficient and studentized D^2-statistic. *Sci. and Cult.*, 3, 335.

Bose, R.C. & Roy, S.N. (1938). The distribution of the studentised D^2-statistic. (With discussion) *Sankhyā*, 4, 19-38.

Bose, Satyendra Nath (1936). On the complete moment-coefficients of the D^2-statistic. *Sankhyā*, 2, 385-396.

Bose, Satyendra Nath (1937). On the moment-coefficients of the D^2-statistic and certain integral and differential equations connected with the multivariate normal population. *Sankhyā*, 3, 105-124.

Bowker, Albert H. (1960). A representation of Hotelling's T^2 and Anderson's classification statistic W in terms of simple statistics. *Contrib. Prob. Statist. (Hotelling Volume)*, 142-149. (MR22-B1145) [Reprinted in *Stud. Item Anal. Predict. (Solomon)*, 285-292]

Bowker, Albert H. & Sitgreaves, Rosedith (1961). An asymptotic expansion for the distribution function of the W-classification statistic. *Stud. Item Anal. Predict. (Solomon)*, 293-310. (MR23-A4210)

Box, G.E.P. & Hill, W.J. (1967). Discrimination among mechanistic models. *Technometrics*, 9, 57-71.

Breiman, Leo & Wurtele, Zivia S. (1964). Convergence properties of a learning algorithm. *Ann. Math. Statist.*, 35, 1819-1822.

Brier, Glenn W.; Schoot, R.G. & Simmons, V.L. (1940). The discriminant function applied to quality rating in sheep. *Proc. Amer. Soc. Animal Prod.*, 33, 153-160.

Brinegar, Claude S. (1963). Mark Twain and the Quintus Curtius Snodgrass letters: a statistical approach to authorship. *J. Amer. Statist. Assoc.*, 58, 85-96.

Bordman, Keeve (1960). Diagnostic decisions by machine. *IRE Trans. Med. Electronics*, ME-7, 216-219.

Brogden, Hubert E. (1955). Least squares estimates and optimal classification. *Psychometrika*, 20, 249-252.

Brogden, Hubert E. (1964). Simplified regression patterns for classification. *Psychometrika*, 29, 393-396.

Brown, George W. (1947). Discriminant functions. *Ann. Math. Statist.*, 18, 514-528. (MR9, p.195)

Brown, George W. (1950). Basic principles for construction and application of discriminators. (With discussion by John W. Tukey and John C. Flanagan) *J. Clin. Psychol.*, 6, 58-76.

Brown, John (1965). Multiple response evaluation of discrimination. *British J. Math. Statist. Psychol.*, 18, 125-137. (MR39-2284)

Brown, Mark (1971). Discrimination of Poisson processes. *Ann. Math. Statist.*, 42, 773-776.

Bryan, Joseph G. (1951). The generalized discriminant function: mathematical foundation and computational routine. *Harvard Educ. Rev.*, 21(2), 90-95.

Bryson, Marion R. (1965). Errors of classification in a binomial population. *J. Amer. Statist. Assoc.*, 60, 217-224.

Bunke, Olaf (1964). Über optimale Verfahren der Diskriminanzanalyse. *Abhandl. Deutschen Akad. Wiss. Berlin Kl. Math. Phys. Tech.*, 4, 35-41. (MR32-6624)

Bunke, Olaf (1966). Nichtparametrische Klassifikationsverfahren für qualitative und quantitative Beobachtungen. *Wiss. Z. Humboldt-Univ. Berlin Math.-Naturwiss. Reihe*, 15, 15-18. (MR36-1031)

Bunke, Olaf (1967). Stabilität statistischer Entscheidungsprobleme und Anwendungen in der Diskriminanzanalyse. *Z. Wahrschein. Verw. Gebiete*, 7, 131-146. (MR35-6267)

Bunke, Olaf (1970). Non-parametric decision functions with asymptotic minimax optimality. *Theory Prob. Appl.*, 15, 148-152.

Burnaby, T.P. (1966). Growth-invariant discriminant functions and generalized distances. *Biometrics*, 22, 96-110. (MR33-1926)

Burt, Cyril (1950). Appendix: On the discrimination between members of two groups. *British J. Psychol. Statist. Sect.*, 3, 104.

Bush, Kenneth A. & Olkin, Ingram (1959). Extrema of quadratic forms with applications to statistics. *Biometrika*, 46, 483-486. (MR22-8588) [Amended by *Biometrika*, 48(1961), 474-475]

Cacoullos, Theophilos (1965). Comparing Mahalanobis distances. I: Comparing distances between *k* known normal populations and another unknown. *Sankhyā Ser. A*, 27, 1-22. (MR32-4787)

Cacoullos, Theophilos (1965). Comparing Mahalanobis distances. II: Bayes procedures when the mean vectors are unknown. *Sankhyā Ser. A*, 27, 23-32. (MR32-4788)

Cacoullos, Theophilos (1966). On a class of admissible partitions. *Ann. Math. Statist.*, 37, 189-195. (MR33-845)

Camp, Burton H. (1946). The effect on a distribution function of small changes in the population function. *Ann. Math. Statist.*, 17, 226-231. (MR8, p.44)

Cassie, R. Morrison (1963). Multivariate analysis in the interpretation of numerical plankton data. *New Zealand J. Sci.*, 6, 36-59.

Cavalli, Luigi L. (1945). Alcuni problemi della analisi biometrica di popolazioni naturali. *Mem. Ist. Ital. Idrobiol.*, 2, 301-323.

Chaddha, R.L. & Marcus, L.F. (1968). An empirical comparison of distance statistics for populations with unequal covariance matrices. *Biometrics*, 24, 683-694.

Chambers, Elizabeth A. & Cox, D.R. (1967). Discrimination between alternative binary response models. *Biometrika*, 54, 573-578. (MR36-6107)

Charbonnier, A.; Cyffers, B.; Schwartz, D. & Vessereau, A. (1955). Application of discriminatory analysis to medical diagnostic. *Biometrics*, 11, 553-555.

Charbonnier, A.; Cyffers, B.; Schwartz, D. & Vessereau, A. (1957). Discrimination entre ictères médicaux et chirurgicaux à partir des résultats de l'analyse électrophorétique des protéines du sérum. *Bull. Inst. Internat. Statist.*, 35(2), 303-320.

Chattopadhyay, K.P. (1941). Application of statistical methods to anthropological research. *Sankhyā*, 5, 99-104.

Chaudhuri, S.B. (1954). The most powerful unbiased critical regions and the shortest unbiased confidence intervals associated with the distribution of classical D^2-statistic. *Sankhyā*, 14, 71-80. (MR16, p.383)

Chernoff, Herman (1952). A measure of asymptotic efficiency for tests of a hypothesis based on the sum of observations. *Ann. Math. Statist.*, 23, 493-507. (MR15, p.241)

Chino, Sadako (1963). The relation between correlation ratio and success rate in the classification by the quantification method. (In Japanese, with summary in English) *Proc. Inst. Statist. Math. Tokyo*, 11, 7-24.

Choi, Keewhan (1969). Estimators for the parameters of a finite mixture of distributions. *Ann. Inst. Statist. Math. Tokyo*, 21, 107-116. (MR39-6435)

Choi, Keewhan (1969). Empirical Bayes procedure for (pattern) classification with stochastic learning. *Ann. Inst. Statist. Math. Tokyo*, 21, 117-125.

Christensen, C.M. (1953). Multivariate statistical analysis of differences between pre-professional groups of college students. *J. Exper. Educ.*, 21, 221-232.

Chung, C.S. & Morton, N.E. (1959). Discrimination of genetic entities in muscular dystrophy. *Amer. J. Human Genet.*, 11, 339-359.

Clark, Philip J. (1952). An extension of the coefficient of divergence for use with multiple characters. *Copeia*, 61-64.

Clunies-Ross, Charles W. & Riffenburgh, Robert H. (1960). Geometry and linear discrimination. *Biometrika*, 47, 185-189. (MR22-1962)

Cochran, William G. (1962). On the performance of the linear discriminant function. *Bull. Inst. Internat. Statist.*, 39(2), 435-447. (MR29-1690) [Reprinted in *Technometrics*, 6(1964), 179-190]

Cochran, William G. (1964). Comparison of two methods of handling covariates in discriminatory analysis. *Ann. Inst. Statist. Math. Tokyo*, 16, 43-53. (MR30-4338)

Cochran, William G. (1966). Analyse des classifications d'ordre. *Rev. Statist. Appl.*, 14, 5-17.

Cochran, William G. (1968). Commentary on estimation of error rates in discriminant analysis. *Technometrics*, 10, 204-205.

Cochran, William G. & Bliss, C.I. (1948). Discriminant functions with covariance. *Ann. Math. Statist.*, 19, 151-176. (MR10, p.50)

Cochran, William G. & Hopkins, Carl E. (1961). Some classification problems with multivariate qualitative data. *Biometrics*, 17, 10-32. (MR22-12636)

Consul, Prem Chandra (1964). Distribution of the determinant of the sum of products matrix in the non-central linear case. *Math. Nachr.*, 28, 169-179. (MR34-6935)

Consul, Prem Chandra (1966). On the distribution of the ratio of the measures of divergence between two multivariate populations. *Math. Nachr.*, 32, 149-155.

Cooper, Paul W. (1962). The hyperplane in pattern recognition. *Cybernetica*, 5, 215-238.

Cooper, Paul W. (1962). The hypersphere in pattern recognition. *Information and Control*, 5, 324-346.

Cooper, Paul W. (1963). Statistical classification with quadratic forms. *Biometrika*, 50, 439-448. (MR29-1702)

Cooper, Paul W. (1963). The multiple category Bayes decision procedure. *IEEE Trans. Electronic Comput.*, EC-12, 18.

Cooper, Paul W. (1964). Hyperplanes, hyperspheres and hyperquadratics as decision boundaries. *Comput. Information Sci. (Tou & Wilcox)*, 111-138.

Cooper, Paul W. (1965). Quadratic discriminant functions in pattern recognition. *IEEE Trans. Information Theory*, IT-11, 313-315.

Cormack, R.M. (1971). A review of classification. (With discussion) *J. Roy. Statist. Soc. Ser. A*, 134, 321-367.

Cornfield, Jerome (1962). Joint dependence of risk of coronary heart disease on serum cholesterol and systolic blood pressure: a discriminant function analysis. *Proc. Fed. Amer. Soc. Exper. Biol.*, 21 (2), 58-61.

Cornfield, Jerome (1967). Discriminant functions. *Rev. Inst. Internat. Statist.*, 35, 142-153. (MR35-5071)

Couto, Anne (1971). Use of discriminant analysis for selecting students for ninth grade algebra or general mathematics. *Proc. First Canad. Conf. Appl. Statist.*, 229-235.

Cover, Thomas M. (1968). Estimation by the nearest neighbor rule. *IEEE Trans. Information Theory*, IT-14, 50-55.

Cover, Thomas M. (1968). Rates of convergence for nearest neighbor procedures. *Proc. Hawaii Internat. Conf. Systems Sci.*, 413-415.

Cover, Thomas M. & Hart, P.E. (1967). Nearest neighbor pattern classification. *IEEE Trans. Information Theory*, IT-13, 21-27.

Cox, D.R. (1961). Tests of separate families of hypotheses. *Proc. Fourth Berkeley Symp. Math. Statist. Prob.*, 1, 105-123.

Cox, D.R. (1966). Some procedures connected with the logistic qualitative response curve. *Res. Papers Statist. Festschr. Neyman (David)*, 55-71. (MR35-2409)

Cox, D.R. & Brandwood, L. (1959). On a discriminatory problem connected with the works of Plato. *J. Roy. Statist. Soc. Ser. B*, 21, 195-200. (MR21-7814)

Cox, Gertrude M. & Martin, W.P. (1939). Use of a discriminant function for differentiating soils with different azotobacter populations. *Iowa State Coll. J. Sci.*, 11, 323-332.

Cramer, Elliot M. (1967). Equivalence of two methods of computing discriminant function coefficients. *Biometrics*, 23, 153-157.

Crumb, C.B., Jr., & Rupe, C.E. (1959). The automatic digital computer as an aid in medical diagnosis. *Proc. EJCC*, 16, 174-179.

Csiszár, I. & Fischer, János (1962). Informationsentfernungen im Raum der Wahrscheinlichkeitsverteilungen. *Magyar Tud. Akad. Mat. Kutató Int. Közl.*, 7, 159-180. (MR32-9136)

Curnow, R.N. (1970). A classification problem involving human chromosomes. *Biometrics*, 26, 559-566.

Cyffers, B. (1965). Analyse discriminatoire. *Rev. Statist. Appl.*, 13(2), 29-46.

Cyffers, B. (1965). Analyse discriminatoire II. *Rev. Statist. Appl.*, 13(3), 39-65.

Dagnelie, P. (1966). On different methods of numerical classification. *Rev. Statist. Appl.*, 14(3), 55-75.

Das, A.C. (1948). A note on the D^2-statistic when the variances and co-variances are known. *Sankhyā*, 8, 372-374. (MR10, p.134)

Das Gupta, Somesh (1964). Non-parametric classification rules. *Sankhyā Ser. A*, 26, 25-30.

Das Gupta, Somesh (1965). Optimum classification rules for classification into two multivariate normal populations. *Ann. Math. Statist.*, 36, 1174-1184.

Das Gupta, Somesh (1968). Some aspects of discrimination function coefficients. *Sankhyā Ser. A*, 30, 387-400. (MR42-5388)

Das Gupta, Somesh & Bhattacharya, P.K. (1964). Classification between exponential populations. *Sankhyā Ser. A*, 26, 17-24.

Davies, M.G. (1970). The performance of the linear discriminant function in two variables. *British J. Math. Statist. Psychol.*, 23, 165-176.

Davis, A.W. (1968). A system of linear differential equations for the distribution of Hotelling's generalized T_0^2. *Ann. Math. Statist.*, 39, 815-832. (MR37-1000)

Davisson, L.D.; Feustel, E.A. & Modestino, J.W. (1970). The effects of dependence on nonparametric detection. *IEEE Trans. Information Theory*, IT-16, 32-41.

Day, Besse B. & Sandomire, Marion M. (1942). Use of the discriminant function for more than two groups. *J. Amer. Statist. Assoc.*, 37, 461-472. (MR4, p.104)

Day, N.E. (1969). Linear and quadratic discrimination in pattern regognition. *IEEE Trans. Information Theory*, IT-15, 419-420. (MR39-5006)

Day, N.E. & Kerridge, D.F. (1967). A general maximum likelihood discriminant. *Biometrics*, 23, 313-323.

Deev, A.D. (1970). Representation of statistics of discriminant analysis, and asymptotic expansion when space dimensions are comparable with sample size. (In Russian) *Dokl. Akad. Nauk SSSR*, 195, 759-762. (MR43-2797)

Defrise-Gussenhoven, E. (1966). A masculinity-femininity scale based on a discriminant function. *Acta Genet. Statist. Med.*, 16, 198-208.

DeGroot, Morris H. (1966). Optimal allocation of observations. *Ann. Inst. Statist. Math. Tokyo*, 18, 13-28. (MR34-6951)

DeGroot, Morris H.; Feder, Paul I. & Goel, Prem K. (1971). Matchmaking. *Ann. Math. Statist.*, 42, 578-593. (MR44-4841)

Dempster, A.P. (1964). Tests for the equality of two covariance matrices in relation to a best linear discriminator analysis. *Ann. Math. Statist.*, 35, 190-199. (MR28-4626)

Díaz Ungría, A.; Camacho, A. & Ríos, S. (1955). Análisis discriminante de dos muestras de indios venezolanos. *Trabajos Estadíst.*, 6, 237-242.

Dickey, James M. (1968). Estimation of disease probabilities conditioned on symptom variables. *Math. Biosciences*, 3, 249-265. (MR39-2281)

Doktorov, B.Z. (1970). Two problems in the theory of discriminant analysis. (In Russian, with summary in English) *Teor. Verojatnost. i Mat. Statist.*, 3, 50-55. (MR44-3430)

Dolby, James L. (1970). Some statistical aspects of character recognition. *Technometrics*, 12, 231-245.

Duda, R.O. & Fossum, H. (1966). Pattern classification by iteratively determined linear and piecewise linear discriminant functions. *IEEE Trans. Electronic Comput.*, EC-15, 220-232.

Dunn, Olive Jean (1971). Some expected values for probabilities of correct classification in discriminant analysis. *Technometrics*, 13, 345-353.

Dunn, Olive Jean & Varady, Paul V. (1966). Probabilities of correct classification in discriminant analysis. *Biometrics*, 22, 908-924.

Dunsmore, I.R. (1966). A Bayesian approach to classification. *J. Roy. Statist. Soc. Ser. B*, 28, 568-577. (MR35-5068)

Dvoretzky, Aryeh; Kiefer, J. & Wolfowitz, J. (1953). Sequential decision problems for processes with continuous time parameters. I. Testing hypotheses. *Ann. Math. Statist.*, 24, 254-264. (MR14, p.997, p.1279)

Dwyer, Paul S. (1954). Solution of the personnel classification problem with the method of optimal regions. *Psychometrika*, 19, 11-26.

East, D.A. & Ochinsky, L. (1958). A comparison of serological and somatometrical methods used in differentiating between certain East African racial groups, with special reference to D^2-analysis. *Sankhyā*, 20, 31-68.

Eaton, M.L. & Bradley, Efron (1970). Hotelling's T^2 test under symmetry conditions. *J. Amer. Statist. Assoc.*, 65, 702-711. (MR42-3918)

Ebel, Robert L. (1947). The frequency of errors in the classification of individuals on the basis of fallible test scores. *Educ. Psychol. Meas.*, 7, 725-734.

Elashoff, Janet D.; Elashoff, Robert M. & Goldman, G.E. (1967). On the choice of variables in classification problems with dichotomous variables. *Biometrika*, 54, 668-670. (MR36-4713)

Elfving, Gustav (1961). An expansion principle for distribution functions with applications to Student's statistic and the one-dimensional classification statistic. *Stud. Item Anal. Predict. (Solomon)*, 276-284. (MR23-A4209)

Ellison, Bob E. (1962). A classification problem in which information about alternative distributions is based on samples. *Ann. Math. Statist.*, 33, 213-223. (MR25-689)

Ellison, Bob E. (1965). Multivariate-normal classification with covariances known. *Ann. Math. Statist.*, 36, 1787-1793. (MR32-3212)

Enis, P. & Geisser, Seymour (1970). Sample discriminants which minimize posterior squared loss. *S. African Statist. J.*, 4, 85-93. (MR43-1320)

Eysenck, Hans J. (1955). Psychiatric diagnosis as a psychological and statistical problem. *Psychol. Rep.*, 1, 3-17.

Eysenck, S.B. (1956). Neurosis and psychosis: an experimental analysis. *J. Mental Sci.*, 102, 517-529.

Feldman, Jacob (1958). Equivalence and perpendicularity of Gaussian processes. *Pacific J. Math.*, 8, 699-708.

Feldman, Sydney; Klein, Donald F. & Honigfeld, Gilbert (1969). A comparison of successive screening and discriminant function techniques in medical taxonomy. *Biometrics*, 25, 725-734.

Fischer, G.R. (1962). A discriminant analysis of reporting errors in health interviews. *Appl. Statist.*, 11, 148-163.

Fischer, Otto F. (1949). Diskriminační analýsa a hodnocení zkoušek schopností. (Discriminatory analysis and the weighting of the results of psychological measurements) *Statist. Obzor*, 29, 106-129.

Fisher, Ronald A. (1936). The use of multiple measurements in taxonomic problems. *Ann. Eugenics*, 7, 179-188.

Fisher, Ronald A. (1936). "The coefficient of racial likeness" and the future of craniometry. *J. Roy. Anthrop. Inst.*, 66, 57-63.

Fisher, Ronald A. (1938). The statistical utilization of multiple measurements. *Ann. Eugenics*, 8, 376-386. (MR12, p.427) [Reprinted in *Contrib. Math. Statist. (Fisher Reprints)*, 33(1950), 375A-386]

Fisher, Ronald A. (1940). The precision of discriminant functions. *Ann. Eugenics*, 10, 422-429. (MR2, p.235) [Reprinted in *Contrib. Math. Statist. (Fisher Reprints)*, 34(1950), 421A-429]

Frank, Ronald E.; Massy, William F. & Morrison, Donald G. (1965). Bias in multiple discriminant analysis. *J. Market. Res.*, 2, 250-258.

Fréchet, Maurice (1929). Sur la distance de deux variables aléatoires. *C.R. Acad. Sci. Paris*, 188, 368-370.

Fréchet, Maurice (1931). Nouvelles expressions de la "distance" de deux variables aléatoires et de la "distance" de deux fonctions mesurables. *Ann. Soc. Polon. Math.*, 9, 45-48.

Fréchet, Maurice (1957). Sur la distance de deux lois de probabilité. *C.R. Acad. Sci. Paris*, 244, 689-692. (MR18, p.679)

Fréchet, Maurice (1959). Les définitions de la somme et du produit scalaires en termes de distance dans un espace abstrait. (Avec supplément) *Calcutta Math. Soc. Golden Jubilee Commem. Volume*, 1, 151-157, 159-160. [*Sic*]

Freedman, David A. (1967). A remark on sequential discrimination. *Ann. Math. Statist.*, 38, 1666-1670. (MR36-2271) [Amended by *Ann. Math. Statist.*, 39 (1968), 2161]

Friedman, Henry D. (1965). On the expected error in the probability of misclassification. *Proc. IEEE*, 53, 658-659.

Fu, K.S. (1970). Statistical pattern recognition. *Adaptive, Learning and Pattern Recognition Systems (Mendel & Fu)*, 35-79.

Gales, Kathleen (1957). Discriminant functions of socio-economic class. *Appl. Statist.*, 6, 123-132.

Gardiner, J. (1959). The use of profiles for differential classification. *Educ. Psychol. Meas.*, 19, 191-205.

Garrett, Henry E. (1943). The discriminant function and its use in psychology. *Psychometrika*, 8, 65-79.

Geisser, Seymour (1964). Posterior odds for multivariate normal classifications. *J. Roy. Statist. Soc. Ser. B*, 26, 69-76. (MR30-4340)

Geisser, Seymour (1966). Predictive discrimination. *Multivariate Anal. Proc. Internat. Symp. Dayton (Krishnaiah)*, 149-163. (MR35-2417)

Geisser, Seymour (1967). Estimation associated with linear discriminants. *Ann. Math. Statist.*, 38, 807-817. (MR35-1135)

Geisser, Seymour & Desu, M.M. (1968). Predictive zero-mean uniform discrimination. *Biometrika*, 55, 519-524. (MR39-1044)

Georgescu, Horia (1968). Sur quelques problèmes de classification mathématique. *Bull. Math. Soc. Sci. Math. R.S. Roumanie (N.S.)*, 12(60), 1968, no.3, 31-37(1969). (MR40-2193)

Gilbert, Ethel S. (1968). On discrimination using qualitative variables. *J. Amer. Statist. Assoc.*, 63, 1399-1412.

Gilbert, Ethel S. (1969). The effect of unequal variance-covariance matrices on Fisher's linear discriminant function. *Biometrics*, 25, 505-515. (MR39-7742)

Giri, N.C. (1964). On the likelihood ratio test of a normal multivariate testing problem. *Ann. Math. Statist.*, 35, 181-189. (MR31-6318) [Amended by *Ann. Math. Statist.*, 35(1964), 1388]

Giri, N.C. (1965). On the complex analogues of T^2- and R^2-tests. *Ann. Math. Statist.*, 36, 664-670. (MR31-6312)

Giri, N.C. (1965). On the likelihood ratio of a normal multivariate testing problem. II. *Ann. Math. Statist.*, 36, 1061-1065. (MR33-815)

Giri, N.C.; Kiefer, J. & Stein, Charles M. (1963). Minimax character of Hotelling's T^2-test in the simplest case. *Ann. Math. Statist.*, 34, 1524-1535. (MR27-6331)

Glahn, Harry R. (1965). Objective weather forecasting by statistical methods. *Statistician*, 15, 111-142.

Gnanadesikan, R. & Wilk, M.B. (1969). Data analytic methods in multivariate statistical analysis. *Multivariate Anal. II, Proc. Second Internat. Symp. Dayton (Krishnaiah)*, 593-638.

Good, I.J. (1965). Categorization of classification. (With discussion) *Math. Comput. Sci. Biol. Med.*, 115-128.

Goodall, D.W. (1964). A probabilistic similarity index. *Nature*, 203, 1098.

Goodwin, C.N. & Morant, G.M. (1939). The human remains of the Iron Age and other periods from Maiden Castle, Dorset. *Biometrika*, 31, 295-312.

Gower, J.C. (1966). Some distance properties of latent root and vector methods used in multivariate analysis. *Biometrika*, 53, 325-338. (MR35-5075)

Gower, J.C. (1966). A Q-technique for the calculation of canonical variates. *Biometrika*, 53, 588-590. (MR35-7484)

Gower, J.C. (1970). Classification and geology. *Rev. Inst. Internat. Statist.*, 38, 35-41.

Gower, J.C. & Barnett, J.A. (1971). Selecting tests in diagnostic keys with unknown responses. *Nature*, 232, 491-493.

Griffin, John S., Jr.; King, J.H., Jr., & Tunis, C.J. (1963). A pattern identification system using linear decision functions. *IBM Systems J.*, 2, 248-267.

Gupta, Shanti S. (1965). On some selection and ranking procedures for multivariate normal populations using distance functions. *Multivariate Anal. Proc. Internat. Symp. Dayton (Krishnaiah)*, 457-475. (MR38-2882)

Halfina, N.M. (1967). The minimax properties of the complex analogue of the T^2-test. (In Russian) *Mat. Zametki*, 2, 635-644. (MR36-7251)

Han, Chien-Pai (1968). A note on discrimination in the case of unequal covariance matrices. *Biometrika*, 55, 586-587.

Han, Chien-Pai (1969). Distribution of discriminant function when covariance matrices are proportional. *Ann. Math. Statist.*, 40, 979-985. (MR39-3648)

Han, Chien-Pai (1970). Distribution of discriminant function in circular models. *Ann. Inst. Statist. Math. Tokyo*, 22, 117-125. (MR42-3926)

Harper, A. Edwin, Jr. (1950). Discrimination between matched schizophrenics and normals by the Wechsler-Bellevue scale. *J. Consult. Psychol.*, 14, 351-357.

Harter, H. Leon (1951). On the distribution of Wald's classification statistic. *Ann. Math. Statist.*, 22, 58-67. (MR12, p.620)

Hartley, H.O. & Rao, J.N.K. (1968). Classification and estimation in analysis of variance problems. *Rev. Inst. Internat. Statist.*, 36, 141-147. (MR38-830)

Hayashi, Chikio (1950). On the quantification of qualitative data from the mathematico-statistical point of view. *Ann. Inst. Statist. Math. Tokyo*, 2, 35-47. (MR12, p.511)

Hayashi, Chikio (1952). On the prediction of phenomena from qualitative data and the quantification of qualitative data from the mathematico-statistical point of view. *Ann. Inst. Statist. Math. Tokyo*, 3, 69-98.

Hayashi, Chikio (1954). Multidimensional quantification--with the applications to analysis of social phenomena. *Ann. Inst. Statist. Math. Tokyo*, 5, 121-143.

Healy, M.J.R. (1965). Computing a discriminant function from within-sample dispersions. *Biometrics*, 21, 1011-1012.

Healy, M.J.R. (1965). Descriptive uses of discriminant functions. *Math. Comput. Sci. Biol. Med.*, 93-102.

Hellman, Martin E. & Raviv, Josef (1970). Probability of error, equivocation, and the Chernoff bound. *IEEE Trans. Information Theory*, IT-16, 368-372.

Hicks, C. (1955). Some applications of Hotelling's T^2. *Indust. Qual. Control*, 11(9), 23-26.

Higgins, Gerald F. (1970). A discriminant analysis of employment in defense and nondefense industries. *J. Amer. Statist. Assoc.*, 65, 613-622.

Highleyman, W.H. (1962). Linear decision functions, with applications to pattern recognition. *Proc. IRE*, 50, 1501-1514. (MR25-1032)

Hill, William J.; Hunter, William G. & Wichern, Dean W. (1968). A joint design criterion for the dual problem of model discrimination and parameter estimation. *Technometrics*, 10, 145-160. (MR36-4732)

Hills, M. (1966). Allocation rules and their error rates. *J. Roy. Statist. Soc. Ser. B*, 28, 1-31. (MR33-5063)

Hills, M. (1967). Discrimination and allocation with discrete data. *Appl. Statist.*, 16, 237-250.

Hoeffding, Wassily & Wolfowitz, J. (1958). Distinguishability of sets of distributions. (The case of independent and identically distributed chance variables) *Ann. Math. Statist.*, 29, 700-718. (MR20-2057)

Hoel, Paul G. (1944). On statistical coefficients of likeness. *Univ. California Publ. Math.*, 2(1), 1-8. (MR6, p.6)

Hoel, Paul G. & Peterson, Raymond P. (1949). A solution to the problem of optimum classification. *Ann. Math. Statist.*, 20, 433-438. (MR11, p.191)

Hollingsworth, T.H. (1959). Using an electronic computer in a problem of medical diagnosis. *J. Roy. Statist. Soc. Ser. A*, 122, 221-231.

Holloway, Lois Nelson & Dunn, Olive Jean (1967). The robustness of Hotelling's T^2. *J. Amer. Statist. Assoc.*, 62, 124-136. (MR35-3778)

Hopkins, J.W. & Clay, P.P.F. (1963). Some empirical distributions of bivariate T^2 and homoscedasticity criterion m under unequal variance and leptokurtosis. *J. Amer. Statist. Assoc.*, 58, 1048-1053. (MR27-6344)

Horst, Paul (1956). Multiple classification by the method of least squares. *J. Clin. Psychol.*, 12, 3-16.

Horst, Paul (1956). Least square multiple classification for unequal subgroups. *J. Clin. Psychol.*, 12, 309-315.

Horst, Paul & Smith, Stevenson (1950). The discrimination of two racial samples. *Psychometrika*, 15, 271-290.

Horton, I.F.; Russell, J.S. & Moore, A.W. (1968). Multivariate-covariance and canonical analysis: a method for selecting the most effective discriminators in a multivariate situation. *Biometrics*, 24, 845-858.

Hotelling, Harold (1931). The generalization of Student's ratio. *Ann. Math. Statist.*, 2, 360-378.

Hsu, P.L. (1938). Notes on Hotelling's generalized T. *Ann. Math. Statist.*, 9, 231-243.

Hsu, P.L. (1945). On the power functions for the E^2-test and the T^2-test. *Ann. Math. Statist.*, 16, 278-286. (MR7, p.212)

Hudimoto, Hirosi (1956). On the distribution-free classification of an individual into one of two groups. *Ann. Inst. Statist. Math. Tokyo*, 8, 105-118. (MR19, p.472)

Hudimoto, Hirosi (1957). A note on the probability of the correct classification when the distributions are not specified. *Ann. Inst. Statist. Math. Tokyo*, 9, 31-36. (MR19, p.1094)

Hudimoto, Hirosi (1963). On the classification-I. The case of two populations. (In Japanese, with summary in English) *Proc. Inst. Statist. Math. Tokyo*, 11, 31-38.

Hudimoto, Hirosi (1964). On a distribution-free two-way classification. *Ann. Inst. Statist. Math. Tokyo*, 16, 247-253.

Hudimoto, Hirosi (1965). On the classification-II. (In Japanese, with summary in English) *Proc. Inst. Statist. Math. Tokyo*, 12, 274-276. (MR32-3220)

Hudimoto, Hirosi (1968). On the empirical Bayes procedure (1). *Ann. Inst. Statist. Math. Tokyo*, 20, 169-185. (MR38-2878)

Hughes, R.E. & Lindley, D.V. (1955). Application of biometric methods to problems of classification in ecology. *Nature*, 175, 806-807.

Ihm, Peter (1954). Anwendung von Hotellings verallgemeinerten *T*-test zur Prüfung der Differenz zweier Mittelwertpaare. *Z. Indukt. Abstam. Vererbungsl.*, 86, 143-156.

Ihm, Peter (1965). Automatic classification in anthropology. *Use Comput. Anthropol. (Hymes)*, 357-376.

Ikeda, Sadao (1962). Necessary conditions for the convergence of Kullback-Leibler's mean information. *Ann. Inst. Statist. Math. Tokyo*, 14, 107-118.

Isaacson, Stanley L. (1954). Problems in classifying populations. *Statist. Math. Biol. (Kempthorne et al.)*, 107-117.

Ivanović, Branislav V. (1954). Sur la discrimination des ensembles statistiques. *Publ. Inst. Statist. Univ. Paris*, 3, 207-269. (MR16, p.1038)

Ivanović, Branislav V. (1957). Nov način odre-divanja otstojanja izmedu višedimenzionalnih statističkih skupova sa primenom u problemu klasifikacije srezova FNRJ prema stepenu ekonomske razvijenosti. (Nouvelle méthode de détermination de la distance entre les ensembles statistiques à plusieurs dimensions avec l'application dans le problème de la classification des départements de la RFPY d'après leur degré du développement économique) *Statist. Rev. Beograd*, 7, 125-154.

Jackson, Esther C. (1968). Missing values in linear multiple discriminant analysis. *Biometrics*, 24, 835-844.

Jackson, Robert W.R. (1950). The selection of students for freshman chemistry by means of a discriminant function. *J. Exper. Educ.*, 18, 209-214.

Jeffers, J.N.R. (1967). The study of variation in taxonomic research. *Statistician*, 17, 29-43.

Jennison, R.F.; Penfold, J.B. & Roberts, J.A. Fraser (1948). An application to a laboratory problem of discriminant function analysis involving more than two groups. *British J. Soc. Med.*, 2, 139-148.

John, S. (1959). The distribution of Wald's classification statistic when the dispersion matrix is known. *Sankhyā*, 21, 371-376. (MR22-1958)

John, S. (1960). On some classification problems. I, II. *Sankhyā*, 22, 301-308, 309-316. (MR23-A4212) [Amended by *Sankhyā Ser. A*, 23(1961), 308]

John, S. (1961). Errors in discrimination. *Ann. Math. Statist.*, 32, 1125-1144. (MR24-A2487)

John, S. (1963). On classification by the statistics R and Z. *Ann. Inst. Statist. Math. Tokyo*, 14, 237-246. (MR28-670) [Amended by *Ann. Inst. Statist. Math. Tokyo*, 17(1965), 113]

John, S. (1964). Further results on classification by W. *Sankhyā Ser. A*, 26, 39-46. (MR34-894)

John, S. (1966). On the evaluation of probabilities of convex polyhedra under multivariate normal and t distributions. *J. Roy. Statist. Soc. Ser. B*, 28, 366-369. (MR35-6237)

John, S. (1970). On identifying the population of origin of each observation in a mixture of observations from two normal populations. *Technometrics*, 12, 353-363.

John, S. (1970). On identifying the population of origin of each observation in a mixture of observations from two gamma populations. *Technometrics*, 12, 365-368.

Johns, Milton V., Jr. (1961). An empirical Bayes approach to non-parametric two-way classification. *Stud. Item Anal. Predict. (Solomon)*, 221-232. (MR22-12648)

Johnson, M.C. (1955). Classification by multivariate analysis with objectives of minimizing risk, minimizing maximum risk, and minimizing probability of misclassification. *J. Exper. Educ.*, 23, 259-264.

Johnson, Palmer O. (1950). The quantification of qualitative data in discriminant analysis. *J. Amer. Statist. Assoc.*, 45, 65-76.

Jutzi, E. (1964). Anwendung der Diskriminanzanalyse in der Medizin. *Abhandl. Deutschen Akad. Wiss. Berlin Kl. Math. Phys. Tech.*, 4, 73-74.

Kabe, D.G. (1963). Some results on the distribution of two random matrices used in classification procedures. *Ann. Math. Statist.*, 34, 181-185. (MR26-833) [Amended by *Ann. Math. Statist.*, 35(1964), 924]

Kabe, D.G. (1968). On the distributions of direction and collinearity factors in discriminant anaylisis. *Ann. Math. Statist.*, 39, 855-858. (MR37-4911)

Kakutani, Shizuo (1948). On equivalence of infinite product measures. *Ann. Math. Ser. 2*, 49, 214-224.

Kalmus, H. & Maynard-Smith, Sheila (1965). The antimode and lines of optimal separation in a genetically determined bimodal distribution, with particular reference to phenylthiocarbamide sensitivity. *Ann. Human Genet.*, 29, 127-138.

Kamensky, L.A. & Liu, C.N. (1964). A theoretical and experimental study of a model for pattern recognition. *Comput. Information Sc. (Tou & Wilcox)*, 194-218.

Keehn, Daniel G. (1965). A note on learning for Gaussian properties. *IEEE Trans. Information Theory*, IT-11, 126-132. (MR32-5441)

Kendall, Maurice G. (1966). Discrimination and classification. *Multivariate Anal. Proc. Internat. Symp. Dayton (Krishnaiah)*, 165-185.

Kiefer, J. & Schwartz, R. (1965). Admissible Bayes character of T^2-, R^2- and other fully invariant tests for classical multivariate normal problems. *Ann. Math. Statist.*, 36, 747-770. (MR30-5430)

King, William R. (1967). Structural analysis and descriptive discriminant functions. *J. Adv. Res.*, 7, 39-43.

Knopov, P.S. (1968). Some problems on discrimination of hypotheses. (In Russian) *Theory of Optimal Solutions*, no.2, 91-94.

Kossack, Carl F. (1945). On the mechanics of classification. *Ann. Math. Statist.*, 16, 95-98.

Kossack, Carl F. (1949). Some techniques for simple classification. *Proc. Berkeley Symp. Math. Statist. Prob.*, 345-352.

Kossack, Carl F. (1963). Statistical classification techniques. *IBM Systems J.*, 2, 136-151.

Krishnaswami, P. & Nath, Rajeshwar (1968). Bias in multinomial classification. *J. Amer. Statist. Assoc.*, 63, 298-303. (MR37-2376)

Kronmal, R. & Tarter, M. (1968). The estimation of probability densities and cumulatives by Fourier series methods. *J. Amer. Statist. Assoc.*, 63, 925-952. (MR37-7023)

Krzanowski, W.J. (1971). A comparison of some distance measures applicable to multinomial data, using a rotational fit technique. *Biometrics*, 27, 1062-1068.

Kshirsagar, A.M. (1963). Confidence intervals for discriminant function coefficients. *J. Indian Statist. Assoc.*, 1, 1-7. (MR29-1701)

Kshirsagar, A.M. (1964). Distribution of the direction and collinearity factors in discriminant analysis. *Proc. Cambridge Philos. Soc.*, 60, 217-225. (MR29-2903)

Kshirsagar, A.M. (1969). Distributions associated with the factors of Wilks' Λ in discriminant analysis. *J. Austral. Math. Soc.*, 10, 269-277. (MR41-2838)

Kshirsagar, A.M. (1971). Goodness of fit of a discriminant function from the vector space of dummy variables. *J. Roy. Statist. Soc. Ser. B*, 33, 111-116.

Kudo, Akiô (1959). The classificatory problem viewed as a two-decision problem. *Mem. Fac. Sci. Kyūshū Univ. Ser. A*, 13, 96-125. (MR22-11495)

Kudo, Akiô (1960). The classificatory problem viewed as a two-decision problem, II. *Mem. Fac. Sci. Kyūshū Univ. Ser. A*, 14, 63-83. (MR22-11496)

Kudo, Akiô (1960). The symmetric multiple decision problems. *Mem. Fac. Sci. Kyūshū Univ. Ser. A*, 14, 179-206. (MR23-A741)

Kudo, Akiô (1961). Some problems of symmetric multiple decisions in multivariate analysis. *Bull. Inst. Internat. Statist.*, 38(4), 165-171. (MR27-2041)

Kullback, Solomon (1952). An application of information theory to multivariate analysis, I. *Ann. Math. Statist.*, 23, 88-102. (MR13, p.854)

Kurczynski, T.W. (1970). Generalized distance and discrete variables. *Biometrics*, 26, 525-534.

Lachenbruch, Peter A. (1966). Discriminant analysis when the initial samples are misclassified. *Technometrics*, 8, 657-662. (MR34-3719)

Lachenbruch, Peter A. (1967). An almost unbiased method of obtaining confidence intervals for the probability of misclassification in discriminant analysis. *Biometrics*, 23, 639-645. (MR36-3459)

Lachenbruch, Peter A. (1968). On expected probabilities of misclassification in discriminant analysis, necessary sample size, and a relation with the multiple correlation coefficient. *Biometrics*, 24, 823-834.

Lachenbruch, Peter A. (1968). Letter to the editor. *Technometrics*, 10, 431-432. (MR38-2888)

Lachenbruch, Peter A. & Mickey, M. Ray (1968). Estimation of error rates in discriminant analysis. *Technometrics*, 10, 1-11. (MR36-6065)

Ladd, George W. (1966). Linear probability functions and discriminant functions. *Econometrica*, 34, 873-885.

Lbov, G.S. (1964). Errors in the classification of patterns for unequal covariance matrices. (In Russian) *Akad. Nauk SSSR Sibirsk. Otdel. Inst. Mat. Vyč. Sistemy Sb. Trudov*, 14, 31-38. (MR31-5724)

Ledley, Robert S. & Lusted, L.B. (1959). The use of electronic computers to aid in medical diagnosis. *Proc. IRE*, 47, 1970-1977.

Ledley, Robert S. & Lusted, L.B. (1959). Reasoning foundations of medical diagnosis. *Science*, 130, 9-21.

Lehmann, E.L. (1957). A theory of some multiple decision problems. I. *Ann. Math. Statist.*, 28, 1-25. (MR18, p.955)

Lehmann, E.L. (1957). A theory of some multiple decision problems. II. *Ann. Math. Statist.*, 28, 547-572. (MR20-2822)

le Roux, A.A. (1967). A method of detecting errors of classification by respondents to postal enquiries. *Appl. Statist.*, 16, 64-69.

Lewis, Philip M., II (1962). The characteristic selection problem in recognition systems. *IRE Trans. Information Theory*, IT-8, 171-178.

Lieberman, G.J.; Miller, R.G., Jr. & Hamilton, M.A. (1967). Unlimited simultaneous discrimination intervals in regression. *Biometrika*, 54, 133-145. (MR36-1042)

Li Hoang Tu (1968). A remark on the theory of Hotelling's test. (In Russian, with summary in English) *Vestnik Leningrad. Univ.*, 23(1), 154-155. (MR36-7257)

Linder, Arthur (1963). Trennverfahren bei qualitätiven Merkmalen. *Metrika*, 6, 76-83. (MR28-4629)

Linhart, H. (1959). Techniques for discriminant analysis with discrete variables. *Metrika*, 2, 138-149. (MR21-6067)

Linhart, H. (1961). Zur Wahl von Variablen in der Trennanalyse. *Metrika*, 4, 126-139. (MR23-A3029)

Linnik, Yu. V. & Pliss, V.A. (1966). On the theory of the Hotelling test. (In Russian) *Dokl. Akad. Nauk SSSR*, 168, 743-746. (MR33-3403)

Loginov, V.I. (1968). Probability estimates of the quality of a pattern recognition decision rule. (In Russian) *Izv. Akad. Nauk SSSR Teh. Kibernet.*, 1968 (6), 129-135. (MR42-6987) [Translated (1969) in *Engrg. Cybernetics*, 1968(6), 112-117]

Lohnes, Paul R. (1961). Test space and discriminant space classification models and related significance tests. *Educ. Psychol. Meas.*, 21, 559-574.

Lord, Frederic M. (1952). Notes on a problem of multiple classification. *Psychometrika*, 17, 297-304.

Lorge, Irving (1940). Two-group comparisons by multivariate analysis. *Off. Rep. 1940 Meet. Amer. Educ. Res. Assoc.*, 117-121.

Lu, K.H. (1968). An information and discriminant analysis of fingerprint patterns pertaining to identification of mongolism and mental retardation. *Amer. J. Human Genet.*, 20, 24-43.

Lubin, Ardie (1950). Linear and non-linear discriminating functions. *British J. Psychol. Statist. Sect.*, 3, 90-103.

Lubischew, Alexander A. (1962). On the use of discriminant functions in taxonomy. *Biometrics*, 18, 455-477.

MacNaughton-Smith, P. (1963). The classification of individuals by the possession of attributes associated with a criterion. *Biometrics*, 19, 364-366.

MacQueen, J. (1967). Some methods for classification and analysis of multivariate observations. *Proc. Fifth Berkeley Symp. Math. Statist. Prob.*, 1, 281-297. (MR35-5078)

Mahalanobis, P.C. (1927). Analysis of race mixture in Bengal. *J. and Proc. Asiat. Soc. Bengal*, 23, 301-333.

Mahalanobis, P.C. (1930). A statistical study of certain anthropometric measurements from Sweden. *Biometrika*, 22, 94-108.

Mahalanobis, P.C. (1930). On tests and measures of group divergence. Part I. Theoretical formulae. *J. and Proc. Asiat. Soc. Bengal*, 26, 541-588.

Mahalanobis, P.C. (1936). On the generalized distance in statistics. *Proc. Nat. Inst. Sci. India*, 2, 49-55.

Mahalanobis, P.C. (1949). Historical note on the D^2-statistic. *Sankhyā*, 9, 237-240.

Mahalanobis, P.C.; Majumdar, D.N. & Rao, C. Radhakrishna (1949). Anthropometric survey of the United Provinces, 1941: a statistical study. *Sankhyā*, 9, 89-324.

Mallows, C.L. (1953). Sequential discrimination. *Sankhyā*, 12, 321-338. (MR15, p.453)

Marshall, Albert W. & Olkin, Ingram (1968). A general approach to some screening and classification problems. (With discussion) *J. Roy. Statist. Soc. Ser. B*, 30, 407-443. (MR39-7743)

Martin, E.S. (1936). A study of the Egyptian series of mandibles with special reference to mathematical methods of sexing. *Biometrika*, 28, 149-178.

Massy, William F. (1965). Discrimination analysis of audience characteristics. *J. Adv. Res.*, 5(1), 39-48.

Masuyama, Motosaburo (1952). The misclassification in the sampling inspection. *Rep. Statist. Appl. Res. Un. Japan. Sci. Engrs.*, 1(4), 7-9. (MR14, p.665)

Matusita, Kameo (1956). Decision rule, based on the distance, for the classification problem. *Ann. Inst. Statist. Math. Tokyo*, 8, 67-77. (MR19, p.186)

Matusita, Kameo (1964). Distance and decision rule. *Ann. Inst. Statist. Math. Tokyo*, 16, 305-315. (MR30-2638)

Matusita, Kameo (1966). A distance and related statistics in multivariate analysis. *Multivariate Anal. Proc. Internat. Symp. Dayton (Krishnaiah)*, 187-200. (MR35-1139)

Matusita, Kameo (1967). Classification based on distance in multivariate Gaussian cases. *Proc. Fifth Berkeley Symp. Math. Statist. Prob.*, 1, 299-304. (MR36-2270)

Maung, Khint (1941). Discriminant analysis for Tocher's eye-colour data for Scottish school children. *Ann. Eugenics*, 11, 64-76.

McQuitty, Louis L. (1956). Agreement analysis: classifying persons by predominant patterns of responses. *British J. Statist. Psychol.*, 9, 5-16.

Meilijson, Isaac (1969). A note on sequential multiple decision procedures. *Ann. Math. Statist.*, 40, 653-657. (MR39-1050)

Meisel, William S. (1968). Least-square methods in abstract pattern recognition. *Information Sci.*, 1, 43-54.

Meisel, William S. (1971). On nonparametric feature selection. *IEEE Trans. Information Theory*, IT-17, 105-106.

Melton, Richard S. (1963). Some remarks on failure to meet assumptions in discriminant analyses. *Psychometrika*, 28, 49-53.

Memon, Ahmed Zogo (1970). Distribution of the classification statistic Z when covariance matrix is known. *Punjab Univ. J. Math.*, 3, 59-67. (MR43-4182)

Memon, Ahmed Zogo & Okamoto, Masashi (1970). The classification statistic W^* in covariate discriminant analysis. *Ann. Math. Statist.*, 41, 1491-1499. (MR42-1280)

Memon, Ahmed Zogo & Okamoto, Masashi (1971). Asymptotic expansion of the distribution of the Z statistic in discriminant analysis. *J. Multivariate Anal.*, 1, 294-307.

Merrett, J.D.; Wells, R.S.; Kerr, C.B. & Barr, A. (1967). Discriminant function analysis of phenotype variates in ichthyosis. *Amer. J. Human Genet.*, 19, 575-585.

Metakides, Theocharis A. (1953). Calculation and testing of discriminant functions. *Trabajos Estadíst.*, 4, 339-368. (MR15, p.728)

Meulepas, E. (1967). Discriminant-analysis as a screening test for patients with Turner's syndrome. (In Dutch) *Rev. Belge Statist. Rech. Opérat.*, 7 (3,4), 58-64.

Meyer, H. Arthur & Deming, W. Edwards (1935). On the influence of classification on the determination of a measurable characteristic. *J. Amer. Statist. Assoc.*, 30, 671-677.

Mikiewicz, J. (1963). On levels of confidence in Wroclaw taxonomy. (In Polish) *Polska Akad. Nauk Zastos. Mat.*, 7, 1-40.

Mikiewicz, J. (1969/70). Confidence regions for dendrites. (In Polish, with summaries in Russian & English) *Zastos. Mat.*, 11, 391-421. (MR43-7007)

Miller, Robert G. (1962). Statistical prediction by discriminant analysis. *Meteorol. Monog.*, 4(25), 1-54.

Morant, G.M. (1926). A first study of crainology of England and Scotland from neolithic to early historic times, with special reference to Anglo-Saxon skulls in London museums. *Biometrika*, 18, 56-98.

Morant, G.M. (1928). A preliminary classification of European races based on cranial measurements. *Biometrika*, 20(B), 301-375.

Morant, G.M. (1939). The use of statistical methods in the investigation of problems of classification in anthropology. I. The general nature of the material and the forms of intraracial distributions of metrical characters. *Biometrika*, 31, 72-98.

Mosteller, Frederick (1971). The jackknife. *Rev. Inst. Internat. Statist.*, 39, 363-368.

Mosteller, Frederick & Wallace, David L. (1963). Inference in an authorship problem. *J. Amer. Statist. Assoc.*, 58, 275-309.

Mourier, Édith (1946). Étude du choix entre deux lois de probabilité. *C.R. Acad. Sci. Paris*, 223, 712-714. (MR8, p.162)

Mourier, Édith (1951). Tests du choix entre diverses lois de probabilité. *Trabajos Estadíst.*, 2, 233-260. (MR14, p.191)

Mukherjee, Ramkrishna & Bandyopadhyah, Suraj (1964). Social research and Mahalanobis's D^2. *Contrib. Statist. (Mahalanobis Volume)*, 259-282.

Myers, James H. & Forgy, Edward W. (1963). The development of numerical credit evaluation systems. *J. Amer. Statist. Assoc.*, 58, 799-806.

Nadkarni, U.G. (1968). Discriminatory techniques in classing of fleeces. *J. Indian Soc. Agric. Statist.*, 20, 22-28.

Nagy, George (1968). State of the art in pattern recognition. *Proc. IEEE*, 56, 836-862.

Nair, K. Raghavan (1952). Use of measurements of more than one biometric character regarding size of body-parts for discriminating between phases in the case of six-eye-striped specimens of the desert locust. (Part of longer paper by S.D. Misra, K.R. Nair and M.L. Roonwal, Studies in intraspecific variation, Part IV, p.95-152) *Indian J. Ent.*, 14, 126-136.

Nair, K. Raghavan & Mukerji, H.K. (1960). Classification of natural and plantation teck (*tectona grandis*) grown at different localities of India and Burma with respect to physical and mechanical properties. *Sankhyā*, 22, 1-20.

Namkoong, Gene (1966). Statistical analysis of introgression. *Biometrics*, 22, 488-502.

Nanda, D.N. (1949). Efficiency of the application of discriminant function in plant-selection. *J. Indian Soc. Agric. Statist.*, 2, 8-19. (MR11, p.674)

Nanda, D.N. (1949). The standard errors of discriminant function coefficients in plant-breeding experiments. *J. Roy. Statist. Soc. Ser. B*, 11, 283-290. (MR11, p.674)

Nandi, Harikinkar K. (1946). On the power function of Studentized D^2-statistic. *Bull. Calcutta Math. Soc.*, 38, 79-84. (MR8, p.394)

Narayan, Ram Deva (1949). Some results on discriminant functions. *J. Indian Soc. Agric. Statist.*, 2, 44-59. (MR12, p.192)

Nathanson, James A. (1971). An application of multivariate analysis in astronomy. *Appl. Statist.*, 20, 239-249.

Neĭmark, Ju. I.; Breĭdo, M.D. & Durnovo, A.N. (1970). A linear minimax classification algorithm. (In Russian) *Izv. Akad. Nauk SSSR Teh. Kibernet.*, 1970 (2), 144-152. (MR42-8619) [Translated (1971) in *Engrg. Cybernetics*, 1970(2), 328-336]

Neyman, Jerzy & Pearson, Egon Sharpe (1933). On the problem of the most efficient tests of statistical hypotheses. *Philos. Trans. Roy. Soc. London Ser. A*, 231, 289-337.

Neyman, Jerzy & Pearson, Egon Sharpe (1933). On the testing of statistical hypotheses in relation to probabilities *a priori*. *Proc. Cambridge Philos. Soc.*, 29, 492-510.

Neyman, Jerzy & Pearson, Egon Sharpe (1936). Contributions to the theory of testing statistical hypotheses. *Statist. Res. Mem. London*, 1, 1-37.

Nishida, N. (1971). A note on the admissible tests and classifications in multivariate analysis. *Hiroshima Math. J.*, 1, 427-434.

Ogawa, Junjiro (1960). A remark on Wald's paper: "On a statistical problem arising in the classification of an individual into one of two groups". *Collected Papers 70th Anniv. Nihon Univ. Natur. Sci.*, 3, 11-20.

Okamoto, Masashi (1961). Discrimination for variance matrices. *Osaka Math. J.*, 13, 1-39. (MR25-1615)

Okamoto, Masashi (1963). An asymptotic expansion for the distribution of the linear discriminant function. *Ann. Math. Statist.*, 34, 1286-1301. (MR27-6342) [Amended by *Ann. Math. Statist.*, 39(1968), 1358-1359]

Overall, John E. & Gorham, D.R. (1963). A pattern probability model for the classification of psychiatric patients. *Behav. Sci.*, 8, 108-116.

Overall, John E. & Hollister, L.E. (1964). Computer procedures for psychiatric classification. *J. Amer. Med. Assoc.*, 187, 115-120.

Oyama, T. & Tatsuoka, M.M. (1956). Prediction of relapse in pulmonary tuberculosis: an application of discriminant analysis. *Amer. Rev. Tuber. Pulm. Diseases*, 73, 472-484.

Panse, V.G. (1946). An application of the discriminant function for selection in poultry. *J. Genet.*, 47, 242-248.

Patrick, Edward A. & Fisher, Frederic P., II (1969). Nonparametric feature selection. *IEEE Trans. Information Theory*, IT-15, 577-584.

Pearson, Karl (1926). On the coefficient of racial likeness. *Biometrika*, 18, 105-117.

Pearson, Karl (1928). The application of the coefficient of racial likeliness to test the character of samples. *Biometrika*, 20(B), 294-300.

Pearson, Karl (1928). Note on standardisation of method of using the coefficient of racial likeness. *Biometrika*, 20(B), 376-378.

Pelto, Chester R. (1969). Adaptive nonparametric classification. *Technometrics*, 11, 775-792.

Penrose, L.S. (1945). Discrimination between normal and psychotic subjects by revised examination *M*. *Bull. Canad. Psychol. Assoc.*, 5, 37-40.

Penrose, L.S. (1947). Some notes on discrimination. *Ann. Eugenics*, 13, 228-237. (MR8, p.592)

Penrose, L.S. (1954). Distance, size and shape. *Ann. Eugenics*, 18, 337-343.

Perkal, J. (1966/68). On plane discrimination. (In Polish, with summaries in Russian and English) *Zastos. Mat.*, 9, 315-324. (MR38-5339)

Peterson, D.W. (1970). Some convergence properties of a nearest neighbor rule. *IEEE Trans. Information Theory*, IT-16, 26-31.

Peterson, D.W. & Mattson, R.L. (1966). A method of finding linear discriminant functions for a class of performance criteria. *IEEE Trans. Information Theory*, IT-12, 380-387. (MR34-895)

Phatarfod, R.M. (1965). Sequential analysis of dependent observations. I. *Biometrika*, 52, 157-165. (MR34-8586)

Pickrel, E.W. (1958). Classification theory and techniques. *Educ. Psychol. Meas.*, 18, 37-46.

Porebski, Olgierd R. (1966). On the interrelated nature of the multivariate statistics used in discriminatory analysis. *British J. Math. Statist. Psychol.*, 19, 197-214. (MR39-2267)

Porebski, Olgierd R. (1966). Discriminatory and canonical analysis of technical college data. *British J. Math. Statist. Psychol.*, 19, 215-236.

Predetti, Aldo (1960). In tema di analisi discriminatoria. *Giorn. Econ. Ann. Economia*, 19, 223-258.

Press, Sheldon James (1968). Estimating from misclassified data. *J. Amer. Statist. Assoc.*, 63, 123-133. (MR36-7243)

Quenouille, Maurice H. (1949). Note on the elimination of insignificant variates in discriminatory analysis. *Ann. Eugenics*, 14, 305-308. (MR11, p.259)

Quenouille, Maurice H. (1950). A further note on discriminatory analysis. *Ann. Eugenics*, 15, 11-14. (MR11, p.743)

Quenouille, Maurice H. (1968). The distributions of certain factors occurring in discriminant analysis. *Proc. Cambridge Philos. Soc.*, 64, 731-740.

Quesenberry, C.P. & Gessaman, M.P. (1968). Nonparametric distribution using tolerance regions. *Ann. Math. Statist.*, 39, 664-673. (MR37-3694)

Radcliffe, J. (1966). Factorizations of the residual likelihood criterion in discriminant analysis. *Proc. Cambridge Philos. Soc.*, 62, 743-752. (MR36-1032)

Radcliffe, J. (1967). A note on an approximate factorization in discriminant analysis. *Biometrika*, 54, 665-668. (MR36-4714)

Radcliffe, J. (1968). A note on the construction of confidence intervals for the coefficients of a second canonical variable. *Proc. Cambridge Philos. Soc.*, 64, 471-475.

Radcliffe, J. (1968). The distributions of certain factors occurring in discriminant analysis. *Proc. Cambridge Philos. Soc.*, 64, 731-740. (MR37-1025)

Radhakrishna. S. (1964). Discrimination analysis in medicine. *Statistician*, 14, 147-167.

Rahman, N.A. (1963). On the sampling distribution of the studentized Penrose measure of distance. *Ann. Human Genet.*, 26, 97-106.

Raiffa, Howard (1961). Statistical decision theory approach to item selection for dichotomous test and criterion variables. *Stud. Item Anal. Predict. (Solomon)*, 187-200. (MR22-12661)

Rao, C. Radhakrishna (1946). Tests with discriminant functions in multivariate analysis. *Sankhyā*, 7, 407-414. (MR8, p.162)

Rao, C. Radhakrishna (1947). On the significance of the additional information obtained by the inclusion of some extra variables in the discrimination of populations. *Current Sci.*, 16, 216-217.

Rao, C. Radhakrishna (1947). The problem of classification and distance between two populations. *Nature*, 159, 30-31.

Rao, C. Radhakrishna (1947). A statistical criterion to determine the group to which an individual belongs. *Nature*, 160, 835-836.

Rao, C. Radhakrishna (1948). The utilization of multiple measurements in problems of biological classification. (With discussion) *J. Roy. Statist. Soc. Ser. B*, 10, 159-203. (MR11, p.191)

Rao, C. Radhakrishna (1949). On the distance between two populations. *Sankhyā*, 9, 246-248. (MR11, p.191)

Rao, C. Radhakrishna (1949). On some problems arising out of discrimination with multiple characters. *Sankhyā*, 9, 343-366. (MR11, p.448)

Rao, C. Radhakrishna (1950). Statistical inference applied to classificatory problems. *Sankhyā*, 10, 229-256. (MR12, p.511)

Rao, C. Radhakrishna (1950). A note on the distribution of $D^2_{p+q} - D^2_p$ and some computational aspects of D^2 statistic and discriminant function. *Sankhyā*, 10, 257-268. (MR12, p.428)

Rao, C. Radhakrishna (1951). Statistical inference applied to classificatory problems. II. The problem of selecting individuals for various duties in a specified ratio. *Sankhyā*, 11, 107-116. (MR13, p.480)

Rao, C. Radhakrishna (1951). Statistical inference applied to classificatory problems. III. The discriminant function approach in the classification of time series. *Sankhyā*, 11, 257-272.

Rao, C. Radhakrishna (1953). Discriminant functions for genetic differentiation and selection. (Part IV of Statistical inference applied to classificatory problems) *Sankhyā*, 12, 229-246. (MR15, p.543)

Rao, C. Radhakrishna (1954). A general theory of discrimination when the information about alternative population distributions is based on samples. *Ann. Math. Statist.*, 25, 651-670. (MR16, p.380)

Rao, C. Radhakrishna (1954). On the use and interpretation of distance functions in statistics. *Bull. Inst. Internat. Statist.*, 34(2), 90-97. (MR16, p.1037)

Rao, C. Radhakrishna (1962). Some observations in anthropometric surveys. *Indian Anthrop. Essays Mem. Majumdar*, 135-149.

Rao, C. Radhakrishna (1962). Use of discriminant and allied functions in multivariate analysis. *Sankhyā Ser. A*, 24, 149-154. (MR26-1967)

Rao, C. Radhakrishna (1966). Discriminant function between composite hypotheses and related problems. *Biometrika*, 53, 339-345. (MR35-2424)

Rao, C. Radhakrishna (1966). Covariance adjustment and related problems in multivariate analysis. *Multivariate Anal. Proc. Internat. Symp. Dayton (Krishnaiah)*, 87-103.

Rao, C. Radhakrishna (1969). Inference on discriminant function coefficients. *Essays Prob. Statist. (Bose et al.)*, 587-602. (MR44-3431)

Rao, C. Radhakrishna (1969). Recent advances in discriminatory analysis. *J. Indian Soc. Agric. Statist.*, 21, 3-15.

Rao, C. Radhakrishna & Shaw, D.C. (1948). On a formula for the prediction of cranial capacity. *Biometrics*, 4, 247-253.

Rao, C. Radhakrishna & Varadarajan, V.S. (1963). Discrimination of Gaussian processes. *Sankhyā Ser. A*, 25, 303-330. (MR32-572) [Reprinted (1964) in *Contrib. Statist. (Mahalanobis Volume)*, 363-390] (MR35-7541)

Rao, K. Sambasiva (1951). On the mutual independence of a set of Hotelling's T^2 derivable from a sample of size n from a k-variate normal population. *Bull. Inst. Internat. Statist.*, 33(2), 171-176. (MR17, p.278)

Rao, M.M. (1963). Discriminant analysis. *Ann. Inst. Statist. Math. Tokyo*, 15, 11-24. (MR30-1592)

Riffenburgh, Robert H. (1966). A method of sociometric identification on the basis of multiple measurements. *Sociometry*, 29, 280-290.

Riffenburgh, Robert H. & Clunies-Ross, Charles W. (1960). Linear discriminant analysis. *Pacific Sci.*, 14, 251-256.

Roberts, Richard A. & Mullis, Clifford T. (1970). A Bayes sequential test of m hypotheses. *IEEE Trans. Information Theory*, IT-16, 91-94.

Rohlf, F.J. (1965). Multivariate methods in taxonomy. *Proc. IBM Sci. Comput. Symp. Statist.*, 3-14.

Romanovskiĭ, Vsevolod I. (1925). On the statistical criteria that a given specimen belongs to one of allied species. (In Russian, with summary in English) *Trudy Turk. Nauč. Obšč. Sred. Aziat. Gos. Univ. Tashkent*, 2, 173-184.

Romanovskiĭ, Vsevolod I. (1928). On the criteria that two given samples belong to the same normal population (on the different coefficients of racial likeness). *Metron*, 7(3), 3-46.

Rose, M.J. (1964). Classification of a set of elements. *Comput. J.*, 7, 208-211. (MR31-4738)

Rouvier, R. (1966). L'analyse en composantes principales: son utilisation en génétique et ses rapports avec l'analyse discriminatoire. *Biometrics*, 22, 343-357. (MR33-6770)

Roy, S.N. (1939). A note on the distribution of the Studentized D^2-statistic. *Sankhyā*, 4, 373-380.

Roy, S.N. & Bose, Purnendu Kumar (1940). On the reduction formulae for the incomplete probability integral of the Studentized D^2-statistic. *Sci. and Cult.*, 5, 773-775.

Roy, S.N. & Bose, R.C. (1940). The use and distribution of the Studentized D^2-statistic when the variances and covariances are based on k samples. *Sankhyā*, 4, 535-542. (MR4, p.105)

Rulon, Phillip J. (1951). Distinctions between discriminant and regression analysis and a geometric interpretation of the discriminant function. *Harvard Educ. Rev.*, 21, 80-90.

Rutovitz, Denis (1966). Pattern recognition. (With discussion) *J. Roy. Statist. Soc. Ser. A*, 129, 504-530.

Sakaguchi, Minoru (1952). Notes on statistical applications of the information theory. *Rep. Statist. Appl. Res. Un. Japan. Sci. Engrs.*, 1(4), 27-31. (MR14, p.996)

Sakino, Sigeki & Kōno, Goro (1955). On the forecasting of prognosis in pediatrics by a quantifying method. *Ann. Inst. Statist. Math. Tokyo*, 6, 173-178.

Šalaevskiĭ, O.V. (1968). The minimax character of Hotelling's T^2 test. (In Russian) *Dokl. Akad. Nauk SSSR*, 180, 1048-1050. (MR38-2879)

Salvemini, Tommaso (1959). Transvariazione e analisi discriminatoria. *Mem. Met. Statist. Univ. Roma*, 2, 731-742.

Salvemini, Tommaso (1961). Sulla discriminazione tra due variabili statistiche semplici e multiple. *Statistica, Bologna*, 21, 121-144.

Samuel, Ester (1963). Note on a sequential classification problem. *Ann. Math. Statist.*, 34, 1095-1097. (MR30-3554)

Samuel, Ester & Bachi, Roberto (1964). Measure of distances of distribution functions and some applications. *Metron*, 23, 83-122. (MR33-1871)

Sardar, P.K.; Bidwell, O.W. & Marcus, L.F. (1966). Selection of characteristics for numerical classification of soils. *Proc. Soil Sci. Soc. Amer.*, 30, 269-272.

Särndal, Carl-Erik (1967). On deciding cases of disputed authorship. *Appl. Statist.*, 16, 251-268.

Saxena, Ashok K. (1966). On the complex analogue of T^2 for two populations. *J. Indian Statist. Assoc.*, 4, 99-102.

Saxena, Ashok K. (1967). A note on classification. *Ann. Math. Statist.*, 38, 1592-1593. (MR35-7500)

Schmid, John, Jr. (1950). A comparison of two procedures for calculating discriminant function coefficients. *Psychometrika*, 15, 431-434.

Schwartz, S.C. (1967). Convergence of risk in adaptive pattern recognition procedures. *Proc. Fifth Annual Allerton Conf. Circuit System Theory*, 800-807. (MR41-7804)

Sebestyen, George S. (1961). Recognition of membership in classes. *IRE Trans. Information Theory*, IT-7, 44-50.

Sedransk, N. & Okamoto, Masashi (1971). Estimation of the probabilities of misclassification for a linear discriminant function in the univariate normal case. *Ann. Inst. Statist. Math. Tokyo*, 23, 419-436.

Seltzer, Carl S. (1965). On a new multivariate distribution and its properties. *Amer. J. Phys. Anthrop.*, 23, 101-109.

Shrivastava, M.P. (1941). On the D^2-statistic. *Bull. Calcutta Math. Soc.*, 33, 71-86. (MR4, p.83)

Simons, Gordon (1967). Lower bounds for average sample number of sequential multi-hypothesis tests. *Ann. Math. Statist.*, 38, 1343-1364. (MR36-1037)

Siotani, Minoru (1956). On the distributions of the Hotelling's T^2-statistics. (In Japanese, with summary in English) *Proc. Inst. Statist. Math. Tokyo*, 4(1), 33-42. (MR18, p.243) [Translated in *Ann. Inst. Statist. Math. Tokyo*, 8(1956), 1-14]

Siotani, Minoru (1961). The extreme value of generalised distances and its applications. *Bull. Inst. Internat. Statist.*, 38(4), 591-599.

Sitgreaves, Rosedith (1952). On the distribution of two random matrices used in classification procedures. *Ann. Math. Statist.*, 23, 263-270. (MR15, p.239)

Sitgreaves, Rosedith (1961). Some results on the distribution of the *W*-classification statistic. *Stud. Item Anal. Predict. (Solomon)*, 241-251. (MR23-A2998)

Smith, Cedric A.B. (1947). Some examples of discrimination. *Ann. Eugenics*, 13, 272-282. (MR8, p.593)

Smith, Cedric A.B. (1953). The linear function maximizing intraclass correlation. *Ann. Eugenics*, 17, 286-292.

Smith, H. Fairfield (1936). A discriminant function for plant selection. *Ann. Eugenics*, 7, 240-250. [Amended by *Ann. Eugenics*, 7(1936), 240A]

Smith, J.E. Keith & Klem, Laura (1961). Vowel recognition using a multiple discriminant function. *J. Acoust. Soc. Amer.*, 33, 358.

Smith, M.S. (1954). Discrimination between electroencephalograph recordings of normal females and normal males. *Ann. Eugenics*, 18, 344-350.

Socivko, V.P. (1966). Pattern recognition by means of computing machines. (In Russian) *Theory Prob. Math. Statist. (Izdat. Nauka Uzbek. SSR)*, 55-99. (MR36-1232)

Sokal, Robert R. (1961). Distance as a measure of taxonomic similarity. *Systematic Zool.*, 10, 70-79.

Solomon, Herbert (1956). Probability and statistics in psychometric research: item analysis and classification techniques. *Proc. Third Berkeley Symp. Math. Statist. Prob.*, 5, 169-184. (MR18, p.955)

Solomon, Herbert (1960). Classification procedures based on dichotomous response vectors. *Contrib. Prob. Statist. (Hotelling Volume)*, 414-423. (MR22-11466) [Reprinted (1961) in *Stud. Item Anal. Predict. (Solomon)*, 177-186. (MR22-11472)]

Sorum, M.J. (1971). Estimating the conditional probability of misclassification. *Technometrics*, 13, 333-343.

Specht, Donald F. (1967). Generation of polynomial discriminant functions for pattern recognition. *IEEE Trans. Electronic Comput.*, EC-16, 308-319.

Specht, Donald F. (1971). Series estimation of a probability density function. *Technometrics*, 13, 409-424.

Srivastava, M.S. (1967). Classification into multivariate normal populations when the population means are linearly restricted. *Ann. Inst. Statist. Math. Tokyo*, 19, 473-478. (MR36-6066)

Srivastava, M.S. (1967). Comparing distances between multivariate populations, the problem of minimum distances. *Ann. Math. Statist.*, 38, 550-556. (MR34-8566)

Stein, Charles M. (1956). The admissibility of Hotelling's T^2-test. *Ann. Math. Statist.*, 27, 616-623. (MR18, p.243)

Stevens, W.L. (1945). Análise discriminante. *Questões Met. Inst. Antrop. Coimbra*, 7, 5-54.

Stoller, David S. (1954). Univariate two-population distribution-free discrimination. *J. Amer. Statist. Assoc.*, 49, 770-777. (MR16, p.604)

Stromberg, John L. (1969). Distinguishing among multinomial populations: a decision theoretic approach. *J. Roy. Statist. Soc. Ser. B*, 31, 376-387.

Sutcliffe, J.P. (1965). A probability model for errors of classification. I. General considerations. *Psychometrika*, 30, 73-96. (MR33-838)

Sutcliffe, J.P. (1965). A probability model for errors of classification. II. Particular case. *Psychometrika*, 30, 129-155.

Switzer, Paul (1971). Mapping a geographically correlated environment. (With discussion) *Statist. Ecol.*, 1, 235-269.

Takakura, Setsuko (1962). Some statistical methods of classification by the theory of quantification. (In Japanese, with summary in English) *Proc. Inst. Statist. Math. Tokyo*, 9, 81-105. (MR27-3063)

Tallis, G.M. (1970). Some extension of discriminant function analysis. *Metrika*, 15, 86-91.

Tanaka, K. (1970). On the pattern classification problems by learning I, II. *Bull. Math. Statist.*, 14, 31-50, 61-74.

Tatsuoka, M.M. & Tiedeman, David V. (1954). Discriminant analysis. *Rev. Educ. Res.*, 24, 402-420.

Teichroew, Daniel & Sitgreaves, Rosedith (1961). Computation of an empirical sampling distribution for the *W*-classification statistic. *Stud. Item Anal. Predict. (Solomon)*, 252-275. (MR23-A2994)

Tenebein, Aaron (1970). A double sampling scheme for estimating from binomial data with misclassifications. *J. Amer. Statist. Assoc.*, 65, 1350-1361.

Tiedeman, David V. (1951). The utility of the discriminant function in psychological and guidance investigations. *Harvard Educ. Rev.*, 21(4), 71-80.

Tildesley, M.L. (1921). A first study of the Burmese skull. *Biometrika*, 13, 247-251.

Travers, R.M.W. (1939). The use of a discriminant function in the treatment of psychological group differences. *Psychometrika*, 4, 25-32.

Tyler, Fred T. (1952). Some examples of multivariate analysis in educational and psychological research. *Psychometrika*, 17, 289-296.

Uematu, Tosio (1959). Note on the numerical computation in the discrimination problem. *Ann. Inst. Statist. Math. Tokyo*, 10, 131-135. (MR21-4506)

Uematu, Tosio (1964). On a multidimensional linear discriminant function. *Ann. Inst. Statist. Math. Tokyo*, 16, 431-437. (MR31-842)

Urbakh, V. Yu. (1971). Linear discriminant analysis: loss of discriminatory power when a variate is omitted. *Biometrics*, 27, 531-534.

Van Ryzin, J.R. (1966). Bayes risk consistency of classification procedures using density estimation. *Sankhyā Ser. A*, 28, 261-270. (MR35-1158)

Van Ryzin, J.R. (1967). Non-parametric Bayesian decision procedures for (pattern) classification with stochastic learning. *Trans. Fourth Prague Conf. Information Theory, Statist. Decision Functions, Random Processes*, 479-494. (MR35-6270)

van Woerkom, Adrianus J. (1960). Program for a diagnostic model. *IRE Trans. Med. Electronics*, ME-7, 220.

van Woerkom, Adrianus J. & Brodman, Keeve (1961). Statistics for a diagnostic model. *Biometrics*, 17, 299-318.

von Mises, Richard (1945). On the classification of observation data into distinct groups. *Ann. Math. Statist.*, 16, 68-73. (MR6, p.235)

Votaw, David F., Jr. (1952). Methods of solving some personnel-classification problems. *Psychometrika*, 17, 255-266.

Wald, Abraham (1944). On a statistical problem arising in the classification of an individual into one of two groups. *Ann. Math. Statist.*, 15, 145-162. (MR6, p.9). [Reprinted in *Select. Papers Statist. Prob. Wald* (1955), 391-408]

Wald, Abraham (1952). Basic ideas of a general theory of statistical decision rules. *Proc. Internat. Cong. Math.*, 1950(1), 231-243. (MR13, p.480) [Reprinted in *Select. Papers Statist. Prob. Wald* (1955), 656-668]

Wald, Abraham & Wolfowitz, J. (1951). Characterization of the minimal complete class of decision functions when the number of distributions and decisions is finite. *Proc. Second Berkeley Symp. Math. Statist. Prob.*, 149-157. (MR13, p.667)

Walsh, John E. (1963). Simultaneous confidence intervals for differences of classification probabilities. *Biom. Z.*, 5, 231-234.

Wartmann, Rolf (1951). Die statistische Trennung sich in mehreren Merkmalen überlappender Individuengruppen (Diskriminanzanalyse). *Z. Angew. Math. Mech.*, 31, 256-257.

Watson, H.E. (1956). Agreement analysis. A note on Professor McQuitty's article. *British J. Statist. Psychol.*, 9, 17-20.

Weber, A.A. (1951). Efficacité de l'indice nasal comparée à l'analyse discriminante. *Acta Genet. Statist. Med.*, 2, 351-363.

Weber, Ernst (1957). Betrachtungen zur Diskriminanzanalyse. *Z. Pflanzen.*, 38, 1-36.

Wee, W.G. (1970). On feature selection in a class of distribution-free pattern classifiers. *IEEE Trans. Information Theory*, IT-16, 47-55.

Weiner, John M. & Dunn, Olive Jean (1966). Elimination of variates in linear discrimination problems. *Biometrics*, 22, 268-275.

Welch, B.L. (1939). Note on discriminant functions. *Biometrika*, 31, 218-220. (MR1, p.154)

Welch, Peter D. & Wimpress, Richard S. (1961). Two multivariate statistical computer programs and their application to the vowel recognition problem. *J. Acoust. Soc. Amer.*, 33, 426-434. (MR23-B3115)

Wesler, Oscar (1959). A classification problem involving multinomials. *Ann. Math. Statist.*, 30, 128-133. (MR21-4502)

Whitney, A.W. & Dwyer, S.J., III (1966). Performance and implementation of the k-nearest neighbor decision rule with incorrectly identified training samples. *Proc. Fourth Annual Allerton Conf. Circuit System Theory*, 96-105.

Williams, E.J. (1955). Significance tests for discriminant functions and linear functional relationships. *Biometrika*, 42, 360-381. (MR17, p.381)

Williams, E.J. (1961). Tests for discriminant functions. *J. Austral. Math. Soc.*, 2, 243-252. (MR25-1608)

Woinsky, Melvin N. & Kurz, Ludwik (1969). Sequential nonparametric two-way classification with a prescribed maximum asymptotic error. *Ann. Math. Statist.*, 40, 445-455. (MR38-6727)

Wolverton, Charles T. & Wagner, Terry J. (1969). Asymptotically optimal discriminant functions for pattern classification. *IEEE Trans. Information Theory*, IT-15, 258-265. (MR43-1329)

Woo, T.L. & Morant, G.M. (1932). A preliminary classification of Asiatic races based on cranial measurements. *Biometrika*, 24, 108-134.

Yarbrough, Charles (1971). Sequential discrimination with likelihood ratios. *Ann. Math. Statist.*, 42, 1339-1347.

Yardi, M.R. (1946). A statistical approach to the problem of chronology of Shakespeare's plays. *Sankhyā*, 7, 263-268.

Žežel, Ju. N. (1968). The efficiency of a linear discriminant function for arbitrary distributions. (In Russian) *Izv. Akad. Nauk SSSR Teh. Kibernet.*, 1968 (6), 124-128. (MR42-6995) [Translated (1969) in *Engrg. Cybernetics*, 1968(6), 107-111]

5. ADDENDA

5.1 *Additions to Section 4*

Anderson, Oskar (1926). Reprinted (1963) in *Ausgewählte Schr. (Anderson Reprints)*, 1, 39-100.

Herold, W. (1969). A proposal for statistical judgement of Penrose's discriminant analysis. (In German) *Biom. Z.*, 11, 123-135.

Landahl, Herbert D. (1939). A contribution to the mathematical biophysics of psychophysical discrimination. II, III. *Bull. Math. Biophys.*, 1, 159-176.

Landahl, Herbert D. (1941). Studies in the mathematical biophysics of discrimination and conditioning. II. Special case. Errors, trials, and number of possible responses. *Bull. Math. Biophys.*, 3, 71-77.

Luce, R. Duncan (1963). Discrimination. *Handb. Math. Psychol. (Luce, Bush & Galanter)*, 1, 191-243.

5.2 *Unverified References to Books*

Helstrom, Carl W. (1960). *Statistical Theory of Signal Detection.* Pergamon Press, New York.

Patrick, E. A. (1972). *Fundamentals of Pattern Recognition.* Prentice-Hall, Inc., Englewood Cliffs, N.J.

5.3 *Unverified References to Research Papers*

Bunke, Olaf (1964/67?). The choice of the separating point in discriminatory analysis. (In German) *Abhandl. Deutschen Akad. Wiss. Berlin Kl. Math. Phys. Tech.*, 4(?), 31-33.

Burnaby, T. P. (1966). Distribution-free quadratic discriminant functions in palaeontology. *Computer Applications in the Earth Sciences, Comp. Cont. Geol. Surv., Kansas*, 7, 70-76.

Harper, A. E. (1950). Discrimination of the types of schizophrenia by the Wechsler-Bellevue scale. *J. Consult. Psychol.*, 14, 290-296.

Harper, A. E. (1950). Discrimination between matched schizophrenia and normals by the Wechsler-Bellevue scale. *J. Consult. Psychol.*, 14, 351, 357.

Hopkins, Carl E. (1964). Discriminant analysis as an aid to medical diagnosis and taxonomy. *J. Indian Med. Prof.*, 5043.

Hoyle, W. G. (1969). On the number of categories for classification. *Information Storage and Retrieval*, 5, 1-6.

Hughes, R. E. (1954). The application of multivariate analysis to (a) problems of classification in ecology; (b) the study of inter-relationships of the plant and environment. *VIII Cong. Intern. Botanique, Paris*, Sections 7 et 8, 16-18.

Kanal, Laveen (1962). Evaluation of a class of pattern recognition networks. *Biological Prototypes and Synthetic Systems*, 261-269.

Kanal, Laveen & Nambiar, K. K. (1963). On the application of discriminant analysis to identification in aerial photography. *Proc. Seventh Nat. Conv. Military Electronics, Washington, D.C.*

Ledley, Robert S. & Lusted, L. B. (1962). Medical diagnosis and modern decision making. *Proc. Symp. Appl. Math., XIV. Mathematical Problems in the Biological Sciences*, 117-158.

Lewis, N. D. C. (1949). Criteria for early differential diagnosis of psychoneurosis and schizophrenia. *Amer. J. Psychotherapy*, 3, 18.

Martin, H. (1960). Probability of misclassification. I. Variables. (In German) *Qualitätskontrolle und Operations Research*, 5, 109-112.

McGuire, J. U. & Wirth, W. W. (1958). The discriminant functions in taxonomic research. *Proc. Tenth Internat. Cong. Ent.*, 387-393.

Murty, B. R. & Arunachalam, V. (1967). Computer programmes for some problems in biometrical genetics. I. Use of Mahalanobis D^2 in classificatory problems. *Indian J. Genet. Plant Breeding*, 27.

Reyment, R. A. (1966). Homogeneity of covariance matrices in relation to generalized distances and discriminant functions. *Computer Applications in the Earth Sciences, Comp. Cont. Geol. Surv., Kansas*, 7, 5-9.

Takekawa, T. & Watanabe, S. (1971). Experiments in discrimination and classification. *Pattern Recognition (The Journal of the Pattern Recognition Society)*, 1 (3).

Tomassone, R. (1963). Applications des fonctions discriminantes à des problèmes biométriques. *Ann. Ecole Nat. Eaux et Forêts et Statist. et Expériences*, 20, 583-617.